Robotics Research and Technology

ROBOT-AGE KNOWLEDGE CHANGEOVER

ROBOTICS RESEARCH AND TECHNOLOGY

Robot-Age Knowledge Changeover
Rinaldo C. Michelini
2009. ISBN 978-1-60692-905-6

Robotics Research and Technology

ROBOT-AGE KNOWLEDGE CHANGEOVER

RINALDO C. MICHELINI

Nova Science Publishers, Inc.
New York

Copyright © 2009 by Nova Science Publishers, Inc.

All rights reserved. No part of this book may be reproduced, stored in a retrieval system or transmitted in any form or by any means: electronic, electrostatic, magnetic, tape, mechanical photocopying, recording or otherwise without the written permission of the Publisher.

For permission to use material from this book please contact us:
Telephone 631-231-7269; Fax 631-231-8175
Web Site: http://www.novapublishers.com

NOTICE TO THE READER

The Publisher has taken reasonable care in the preparation of this book, but makes no expressed or implied warranty of any kind and assumes no responsibility for any errors or omissions. No liability is assumed for incidental or consequential damages in connection with or arising out of information contained in this book. The Publisher shall not be liable for any special, consequential, or exemplary damages resulting, in whole or in part, from the readers' use of, or reliance upon, this material.

Independent verification should be sought for any data, advice or recommendations contained in this book. In addition, no responsibility is assumed by the publisher for any injury and/or damage to persons or property arising from any methods, products, instructions, ideas or otherwise contained in this publication.

This publication is designed to provide accurate and authoritative information with regard to the subject matter covered herein. It is sold with the clear understanding that the Publisher is not engaged in rendering legal or any other professional services. If legal or any other expert assistance is required, the services of a competent person should be sought. FROM A DECLARATION OF PARTICIPANTS JOINTLY ADOPTED BY A COMMITTEE OF THE AMERICAN BAR ASSOCIATION AND A COMMITTEE OF PUBLISHERS.

LIBRARY OF CONGRESS CATALOGING-IN-PUBLICATION DATA

Michelini, Rinaldo C.
 Robot-age changeable knowledge / Rinaldo C. Michelini.
   p. cm. -- (Robotics research and technology)
 Includes index.
 ISBN 978-1-60692-905-6 (hardcover)
 1. Robotics. I. Title.
 TJ211.M49 2009
 629.8'92--dc22
                    2008055353

Published by Nova Science Publishers, Inc. ✢ New York

CONTENTS

Foreword		vii
Presentation		ix
Preface		xv
Chapter 1	The Industry Paradigms	**1**
Chapter 2	Industrialism Twilights	**51**
Chapter 3	The Knowledge Paradigms	**97**
Chapter 4	Robot in Manufacture Jobs	**147**
Chapter 5	Robot in Service Applications	**179**
Chapter 6	Remarks, Prospects and Conclusion	**231**
References		**273**
Index		**313**

FOREWORD

*Gerd Hirzinger**
German Aerospace Center; Institute of Robotics and Mechatronics

In the past, there has been a kind of very general disappointment about the fairly slow progress in robotics compared to human performance – despite many years of robotics research involving a large number of scientists and engineers. Industrial robots today, in nearly all applications, are still purely position-controlled devices, perhaps with some static sensing, but still far away from the human arm's performance with its amazingly low own-weight-against-load ratio and its online sensory feedback capabilities involving vision and tactile information, actuated by force-torque-controlled muscles. However, there are a number of observations in the area of industrial robots that prove that there have been a lot of advances with respect to certain aspects. In particular, industrial robots are much cheaper now (approximately by a factor of 4!) than 10 years ago, while the peripheral costs (e.g. precise part feeding) have remained nearly the same, thus indicating that this development highlights the features of a dead end, calling for more sensor-based adaptivity.

Heavy stiff classical industrial robots try to guarantee some kind of positional accuracy, but indeed what they can guarantee is only precision in terms of repeatability, no absolute accuracy. Manufacturing tolerances are the reason why computed and real position may differ by a few centimetres. Thus people have tried to calibrate robots by expensive external measuring devices – however when a robot is supposed to work on a car frame on a conveyor belt, and the car is not in its nominal position, the robot's calibration does not help at all. What really counts in most cases is the relative position and orientation between the robot's end-effector and the object to be handled. And this can be so easily guaranteed by sensory feedback, typically vision or other non-tactile sensors. But when a robot gets in contact with the environment, the next problem arises, because fast force control via a wrist force-torque-sensor trying to approach the human performance is extremely difficult. This is one of the main reasons why robotic assembly is still not realizable in many industrial applications. Thus what robotic researchers claim is that robots operating similar to humans should not only show up high-fidelity sensor fusion of vision and force, but should better be of the "soft robotics" type, i.e. to have torque feedback control in all joints. They are then capable of rendering their arm arbitrarily compliant with different types of stiffness or impedance. And the next generation might then have antagonistic drive concepts for arms and hands showing up inherent compliance similar to the human muscles. We see "robonauts" of this type in

---

[*] Oberpfaffenhofen, D-82230 Wessling; Tel. +49 8153 28-2401, Fax: +49 8153 28-1134; email: Gerd.Hirzinger@dlr.de

space for orbital servicing and planetary exploration in the next decade. And we see a huge field of applications in surgery, slender arms which generate confidence in doctors and patients and which are capable of cutting bones contactlessly by laser beams. And we see the tele-operated minimally invasive surgery arms, which have already caused a kind of revolution in urological surgery.

Not only Bill Gates believes that service robots, useful e.g. for elderly care, will show up a similar ascent in the next 20 years as PC's had in the last 20 years.

Asian countries discuss more frankly about the over-aging of our future societies than we Europeans do. A number of groups worldwide is now focussing their research on smaller, anthropomorphic mobile robot systems for elderly care, e.g. with two arms and hands, and a stereovision head, moving on wheels or legs. They must be capable of not only fusing vision and force in a conventional way, but they must show up cognitive and learning capabilities. And here the progress still is modest. Nevertheless, the roboticists dream that the age is coming soon when we will have a situation similar to that of the ancient Rome, but with artificial slaves instead of human ones.

In the struggle towards innovation, nonetheless, the robots achievements, out of technology-driven, happen to require strong economically-pushed motivation. They represent in that context the paradigm shift along the industrial progression, permitting to explore the on-process artificial intelligence, up to autonomic-mode operation, whenever the enabled improvement assures the return on investments. The said paradigm shift might open the way to the yet-to-be breakthrough, so to remove or mitigate the natural resources inexorable decreasing of the manufacture markets, by opening the novel tracks along the cognitive progression. At least, this is the desirable omen, to be devised, reading the book.

PRESENTATION

Laszlo Nemes[*]
Fellow of the Australian Academy of Technological Sciences and Engineering
CSIRO Chief Research Scientist (*retired*)

Professor Michelini's book has undertaken a huge intellectual venture to fill the gap in the eco-techno literature. He provides analyses and recommendations on how to maintain sustainable economic growth in the years ahead. There are many challenges, such as, fast moving technologies, ever changing social structures, unregulated global financial systems, all of which can cause uncertainties for the future. Yet, the book is full of ideas and suggestions for making a more stable and better world. The book does not underestimate the existing dangers, but clearly emphasises the seriousness of the problems and points out for urgent actions.

Societies accept the necessity of sustainable and ecological developments but they are very slow implementing them as they fear that their living standard will be negatively affected. There is no clear understanding that the quality of life could be maintained even improved if we use the available resources wisely, without changing the basic fact that industries are the dominant wealth creating factors in societies.

The author is well aware that societies are afraid of could be real threats. There is no point in believing that these are just impermanent risks only. We have to realise that our path have reach a cross road and the "industry age" is not offering the same life of easy anymore. We need a sea change in our way of thinking and in our approach to address the socio-economic problems that can be faced.

The author's engineering background together with his academic experience provides a unique skill base from which he offers ways to avoid the potential dangers. This remarkable book studies the world socio-economical history from the *industry* to the *robot* age.

The first three chapters deal with general concepts: they discuss the technical, economical and social environment we are living in. The next two chapters summarise the technical development of robots; where and how they can be applied. The last section offers alternative prospects and advice.

The general discussion is supported by facts:

- the "industry age" has been a very prosperous period in history as it has created an environment when large number of people have been able to accumulate wealth through their skills and talents using *artificial* processes. It is important to note that

---
[*] e.mail: laszlo@nemes.com.au

- the *whole process* depends on trust and when it is shattered, the system goes into recession affecting millions of people;
- the effectiveness of industrialism was based on *western world* defining the rules. These were heavily influenced by cultural, economic, political and social factors;
- these decisive conditions do not work any more. Human greed has used an unrealistic amount of the earth finite resources, and dumped back waste and pollution. It is necessary to define a new set of distinguished characteristics for modern societies;
- the "robot age" offers the new mix of attributes, for the "to de-materialise" axiom and allows the "to re-materialise" innovations under the watchful eyes of transparency and widely accepted control functions;
- these new attributes characterise the *knowledge* paradigm (*vs.* the *industry* one), showing the limits of the old *industrial* revolution. A *cognitive* breakthrough is widely expected and will create alternative stocks.

The following comments are the technical requisites and they are not comprehensive enough to fully describe the history and the operation rules of industrialism. Will this come to an end or turn into a modified paradigm then there will be a market for entirely new basic products. The complete analysis requires a more complete set of requirements including the underlying social and cultural factors.

The Chapter One examines the "industry" paradigms. It starts with the descriptions of their underlying rules and follows it through as they developed into a worldwide driving force for progress. The major points are as follows:

- there are rigid underlying rules in the structured *capitalism* but the most surprising omission is that the "natural" assets are widely neglected when manufacturing businesses are described. This serious shortcoming needs to be recognised, otherwise all related analysis will be incomplete and therefore misleading;
- changes are forced on societies when the existing production system will not result in growth of wealth for the nation. Alternative production ways had to be developed when the old *agricultural* society would be replaced with an *industrial* one. Improvements were made in the new system until sound growth of wealth achieved;
- there are fundamental cultural factors influencing the "industry" paradigms. They are so important that we can regard them as the decisive socio-political features of western life. This identifies the framework for the new scientific work-organisations.

The Chapter Two offers a bird eye view on how the shortcomings of the *industrial* revolution have affected our recent history. These have negatively influenced the present structure of the economy, thus seriously impacting our way of life. The reasons are organised in three sections:

- the work organisations are changing fast; production (and development of technology) is market driven off-setting the original "industry" *minimalism*. The aim is to exploit potentials of sophisticated robots, thus to grow the economy of *scope*;
- the scope of entrepreneurship has been widened to include *product-service* delivery. Vendors are forced to operate across the whole life cycle of products forcing companies to take *them back* at the end of productive life;
- the current ecologic accepts the fast exploitation of earth resources without any consideration of the environmental pollution.

The industrialism therefore has moved from the old "industry" paradigms, to the new "knowledge" paradigms, making it clear that wealth generation from transforming the earth resources into waste and pollution cannot go forever. It has also demonstrated that the "robot age" already provides added value by the "to de-materialise" axiom and prospect of the "to re-materialise" axiom. Such breakthrough is enabled by the *cognitive* innovation. The Chapter Three provides a "knowledge" framework which can be broken into the following components:

- the technical framework focuses on network issues with special functions and facilities for updating information from a world wide distributed system. It provides some productive *flexibility* setting for emerging organisations which may have potentials to utilise the new «knowledge» paradigms;
- the socio-economical framework is the one which is most neglected when formulating the strategies for organisational development although its importance cannot be emphasized enough. The capitalism has moved societies from city-states, to nation-states (even up to sub-continental size). Such developments have dramatically influenced the quality of life both locally and globally;
- the trade framework is for studying how the industrialism is able to turn the *global* or *no-global* to *post-global* trading. There is no question that there are reasons to worry about the many negative prospects which may have severe consequences on us.

The investigations in the first three chapters give such unique viewpoints, which have not been widely accepted as yet. The gap between the *industrial* and *developing* countries started widening with the industrial revolution and it is opening even wider due to recent paradigm shift. The "industry" paradigms, at least in their original form, have led to rather instable growth, being based on the *transformation effectiveness*. Extensive natural capital has been used for producing many consumer goods and soon disposing them as wastes and pollutions. Since the earth is a finite resource and can be regarded as a closed ecosystem, we cannot grow wealth this way for long.

The over-population, over-consumption and over-pollution are threatening our very way of life. The notion of *global village* is a slogan overused by both politicians and the media without providing any real path for sustainable economic solution. The book suggests alternatives, in which the "knowledge" paradigms might support further wealth creation. This is underpinned by the so called *value chains of intangibles* which may lead on to the *cognitive* breakthrough. In this case the *transformation efficiency* applies not only to material resources, but also on *artificial* life deployments.

The above mentioned analyses offer an alternative view of the "industry" pattern in contrast with the universal practice, which relies heavily on western culture and organisational models. This could explain why the industrialism has been successful within a limited timeframe in human history and in certain countries only. It might be useful nonetheless to explore alternative outlooks on how restructuring the world to come with collaborative *altruism*, in lieu of competitive challenge to make further growth possible.

Analysing the "robot age" using the methodologies presented in the earlier chapters, one can come to the conclusion that its technical foundation has already been well established.

The book continues with two chapters based on the engineering knowledge stemming from research in the Industrial Robot Design Research Group at the University of Genova, Italy. The group has discovered how robot technologies can help merging the information flow and enhancing the material flow in the value chain assuring further success of the manufacturing economy.

The group activities focused on *instrumental* robots, which are equipped with carefully specified sensors all developed in the group. The robots are task-driven and operate in domains. The Chapter Four describes some important aspects of manufacturing robotics and they are arranged in three sections:

- *intelligent* work organisations will be the foundations of the workforce. They will act collaboratively and share decision duties and development tools;
- *flexible manufacturing* environment provides many opportunities for new robotic functions and will invite new service facilities. In this setting robots will play key role in shop-floor logistics, inspection and repair;
- *integration of robot technologies* will be a major challenge and new design methodologies will be needed.

The industrial robotics is a very powerful discipline and guides engineering activities towards the integrated design paradigms. Therefore robots are used more and more in complicated processes such as supply chains. Their continuous operation is paramount as their reliability will determine greatly the overall performance of the system. The *flexibility* of the *intelligent* organisations opens up new ways of decision making and knowledge management. Well structured procedures and matching IT infrastructure allow collecting and sharing information. Naturally this cannot be completed in one go. Rather the system should be designed that the cooperative intellectual work should continuously accumulate knowledge in shared pool thus gradually taking over the departmentalised practice of engineering and management. Such infrastructure and design methods have developed the practice of *simultaneous* engineering which has been the first step towards the full *product-process-environment-enterprise* developments leading to the *extended* enterprise concept.

Three sections of the Chapter Five give a short overview of robots in the *service* industry:

- the operation methods can be selected on the basis of the tasks to be performed, on the control methods to be used and the on the environment where the robot will be installed;
- there is a short survey of various types of remote-operated robotic equipment;
- typical examples are given for autonomous task-driven robots.

These are only short summaries of the present day practices, but the given examples offer a glimpse on how the "robot age" might help merging tangible and intangible values. In many applications, as described in the book, the robot deployments are instrumental, together with the organisation reforms and the existing cultural factors, to support the "knowledge" paradigms. The changing "knowledge" paradigms will certainly push for a technically coherent "robot age", if (or when) the socio-political framework will support a consistent *post-global* economic system. Engineering has clearly outlined the huge possibilities for using sensory controlled robots in manufacture and service industries. The introduction of the autonomous machineries will certainly require coherent policies and will depend on the economic viabilities of such integrated systems.

The Chapter Six summarises the recommended future topics and concepts with the understanding that these are realistic scenarios. It is expected that the *global, no-global, post-global* views will be the future scenarios that human societies must choose from.

- the *hyper-market* arrangement is ruled by multinational corporations whose successes are closely linked with the political situation of the countries where the

enterprises are operating from. These companies will quite likely avoid full transparency in their dealings;
- the *cautious headway* road, is based on strict *ecological* concern, with frugality in mind. The focus is on autarchic self-sufficiency, to re-create protected zones, fully banning all new technologies, as potentially risky;
- the *post-global* approach will be looking at sustainable growth in the "robot age". It will be built on *cognitive* breakthrough to join the "to de-materialise" and "to re-materialise" axioms, with *altruism*-driven political world organisations.

The book does not provide certainties except that the industrialism will not play a major role in the future. The *cognitive* revolution is an emerging hypothesis. More attention will be paid on value chains with important intangible content and that positive outlook is driven by technology. The "robot age" will require new solutions in engineering. Sustainable growth has to be maintained, and the "knowledge" driven opportunities should be given more attention. To answer to the demands, we shall dramatically modify the current paradigms of the industrial revolution.

The anthropic principles, assumed in the book, lead to accept cultural divides while increased "technical" innovation is necessary for human survival. The *agricultural* revolution has been most likely multifarious phenomena accepting mixture of alternatives to achieve continuous improvements. The *industrial* revolution was somehow limited by the original minimalism and by the division of labour in the *scientific* work organisations. The innovative "robot age" vision provides coherent "knowledge" paradigms. It will remove the old guess work from designs and will guide robot applications towards complex alternatives, more suited to the eco-friendly requirements required by sustainable growth. Unfortunately technically sound prospects are not sufficient to promote the breakthrough. We have to start forming new socio-political structures.

The book carefully emphasises all along that it is necessary to consider the economic and political motivations when deciding on technical issues. Environmental protection regulations are increasing by the day and reaching record numbers. Their technical content usually has strong political and economic flavour, making it difficult to separate various solutions on their merits. The "industry" patterns of wealth creation in rich societies are mostly market-driven so the eco friendly growth clashes with the practice to turn raw materials into waste and pollution. As noticed, the achievement of the industrial revolution is the characteristic of the western world and should not be considered as universal option.

The general analyses are the core part of the book leaving the examples of applications for illustrations only. The inter-dependences of economics and technologies correspond to presumed cultural prominence throughout history. Concepts such as *capitalism* or *democracy* have strong relations to the stage of evolution of human societies. If we believe in some anthropic principles we need to find cultural correlations and proper appraisals justifying the changes of *capitalism* or the *democracy* trends.

This is a unique book leading us in the future. Based on deep philosophical principles, it contains technical details which are supported by social considerations and environment friendly criteria. It is a fascinating reading.

PREFACE

The mankind is facing a divide, in the struggle towards better life quality. The wellbeing is related to the spendable riches, the individuals dispose. The wealth of a community, in the devised analysis, is, roughly, dependent on four terms:

- the *natural* capital, factually typified by the quality and quantity of traded goods, in the citizens' possession;
- the *human* capital, coherently characterised by the workforce ready on the market, to obtain new products;
- the *financial* capital, conventionally exemplified by the accessible money and equivalent payment means;
- the *technical* capital, properly assessed by the knowledge and know-how supporting value added chains.

The four entries are not equivalent. The first two are *native* contributions; the next two are *artificial* additions. The *artificial* side in the riches build up is full of promises, permitting man's ruled expansion, out of the limits of the natural order pace. The advantage is not without inconsistencies.

The *financial* assets are contradictory factors. The 2008's hitches bring severe question marks on the financial mathematics, after the recent Nobel's prizes ('95 to Robert E. Lucas Jr., '97 to Myron Scholes & Robert C. Merton), with the smart formulas, assessing the stock options' trend. The new sort of financial instruments and the standard risk management models play sinister role in the recent crisis. In general, however, the markets for options and so-called derivatives did not insure against current lack of backing, and the one-sided risks or pre-specified trends are only illusory hedging. Indeed, the risk's embedding in the stock price happened to be *artificial* trick, with merely formal economic science contribution.

The *technical* assets are less conflicting factors, at least, once considered their cultural potential, in building innovative ways-out. Known reference divides are the *agricultural* revolution, when the old *opportunistic* economy of the nomadic populations became useless, and the *industrial* revolution, when the dependence on the land produces, at the biological pace, suffer adverse weather conditions. If the past issues are worth, these *artificial* contributions might turn helpful, once the alarming signals are mastered. The book moves around these points.

The current division between industrialised and developing countries/societies is based on relatively recent paradigms of mankind's growth which have occurred within the so-called industrial revolution. Actually, the «industry» paradigms, at least, in their original formulation, lead to rather unstable growth, being based on the transformation effectiveness, through which the *natural* capital is used, to be reshaped into prised consumables, and quickly disposed in waste and pollution.

The book explores alternatives, in which the «knowledge» paradigms support further wealth build-up, based on the value chains in intangibles, and, possibly, on the «cognitive» revolution, where the transformation efficiency, this time, applies not only on material inanimate resources, but, as well, on *artificial* life processes. The analyses move reconsidering the «industry» patterns in their contingencies, chiefly, related with the western world cultural background and organisation style. The experienced context, perhaps, explains why industrialism appeared in man's history with narrow timing and localization outcomes. It might be used, even so, to explore other outlooks on how reforming the world to come, with collaborative altruism, in lieu of competitive challenge, to make further growth possible.

Thus, the «industry» patterns of the wealth build-up of the affluent society are revised, recognising that the growth sustainability clashes against the practice to turn raw materials into waste and pollution. Indeed, the industrialism lights and shadows show series of reasons that the known patterns needs to modify, pushed by socio-economic and politico-legal drives, and enabled by technological aids.

The changeful «knowledge» paradigms permit to look at technically coherent «robot age» issues, even if with questionable outlooks, because of the impending eco-limits. The engineering know-how permits visions on the instrumental robotic aids, from the manufacture, to the service areas, as these are deemed to be worthy spurs, to self-consistent policies.

The eco-protection, quickly expanding into numberless regulations, is, here, faced, according to the *technical* capital visions, providing the coherent changeful «knowledge» paradigms, to get rid the existing industrialism's original limitation. The durable growth demands address engineers and managers, at several ranges: robot designers, with interest in the product lifestyle market, and economists, with mind in the intangible value added; or service engineering providers and reverse logistics attendants.

The topics give useful insight on eco-driven knowledge-based options, «robot age» socio-economic potential, innovation through «complexity», when the trivial reductionism is useless, etc.; in general, the outlined paradigms' shifts open the way to the yet-to-be «cognitive» breakthrough, to remove or mitigate the natural resources inexorable decreasing of the manufacture markets, opening new tracks. At least, this is the desirable omen, to be devised, reading the book.

Chapter 1

1. THE INDUSTRY PARADIGMS

The affluent society is recent findings, achieved with the industrial revolution, and based on the transformation efficiency of the manufacturing activity, which takes out the unique earth material stocks, ceaselessly downgraded into waste and pollution. The trend is explained with the massive exploitation of non-renewable resources that builds on the huge resort to *artificial* energy, highly drawing on raw materials and not leaving time for restoration and remediation processes.

The industrial revolution is the outcome of particular events, bringing the UK, and subsequently other European countries and the USA, to utilize the «industry» paradigms in the work organisation and manufacturing transformations. In short time, the innovative paradigms modified the process productivity, conferring wide competitive advantage to the *industrialised* countries, compared to the other ones, to impress full restructuring of the world trade and wealth.

Only recently, the growth sustainability, threaten by the «industry» paradigms, is questioned, being maximally based on non-renewable resources. In the not far future, such non-reversibility generates the double damages: lack of exploitable stocks; build-up of man-harmful environment. However, to reject the «industry» is not the alternative, because this should mean the elimination of the quality of life, acquired in recent times.

Moreover, the «industry» paradigms are not fixed. They have been pushed to modify, by social, economic, politic and technical reasons, and it might be useful to investigate, whether the ecologic motivations could suggest extra amendments, to bring forth durable growth. The fact is becoming impending demand, and many viewpoints exist.

In the following, a combined approach is devised, trying to describe: the picky circumstances of the «industry» capitalism:

- why such sort of wealth accumulation happened (Chap. 1);
- by which way it is now moving to the «knowledge» capitalism (Chap. 2);
- whether the evolution might promote eco-conservative issues (Chap. 3).

This first chapter deals with the distinctive «industry» setting peculiarities, with three sections.

The first sketches the wealth development, requiring «four» capital assets: two *native*, the human and natural assets; and two *artificial*, the technical and financial assets. This allows differentiating, along the wealth development, the fundamental elements, without any of which the growth is fictitious.

The second section gives a bird eye view on how the market staples changed, since when the transformation economy became the reference support, from the *agricultural*, or the *industrial* revolution, up to the possible *cognitive* revolution to come, based on new «knowledge» paradigms.

The last section provides the introductory overview on the reasons why the western cultural bias, by means of personal competition and reductionism, is the evident background for the technological applications and structured deployments of the industrial work organisation.

The three sections, together, provide the reference casing of the old «industry» paradigms, showing the requests: to keep under control all the four capital assets (the *natural*, further to the *financial*, *human* and *technical* capital); to devise ways out, towards new staples; and to give hints on some cultural peculiarities, showing that alternative patterns are, hopefully, existing.

The affluent society is recent findings, achieved with the industrial revolution, and based on the transformation efficiency of the manufacturing activity, which takes out the unique earth material stocks, ceaselessly downgraded into waste and pollution. The trend is explained with the massive exploitation of non-renewable resources that builds on the huge resort to *artificial* energy, highly drawing on raw materials and not leaving time for restoration and remediation processes.

The industrial revolution is the outcome of particular events, bringing the UK, and subsequently other European countries and the USA, to utilize the «industry» paradigms in the work organisation and manufacturing transformations. In short time, the innovative paradigms modified the process productivity, conferring wide competitive advantage to the *industrialised* countries, compared to the other ones, to impress full restructuring of the world trade and wealth.

Only recently, the growth sustainability, threaten by the «industry» paradigms, is questioned, being maximally based on non-renewable resources. In the not far future, such non-reversibility generates the double damages: lack of exploitable stocks; build-up of manharmful environment. However, to reject the «industry» is not the alternative, because this should mean the elimination of the quality of life, acquired in recent times.

Moreover, the «industry» paradigms are not fixed. They have been pushed to modify, by social, economic, politic and technical reasons, and it might be useful to investigate, whether the ecologic motivations could suggest extra amendments, to bring forth durable growth. The fact is becoming impending demand, and many viewpoints exist.

In the following, a combined approach is devised, trying to describe: the picky circumstances of the «industry» capitalism:

- why such sort of wealth accumulation happened (Chap. 1);
- by which way it is now moving to the «knowledge» capitalism (Chap. 2);
- whether the evolution might promote eco-conservative issues (Chap. 3).

This first chapter deals with the distinctive «industry» setting peculiarities, with three sections.

The first sketches the wealth development, requiring «four» capital assets: two *native*, the human and natural assets; and two *artificial*, the technical and financial assets. This allows differentiating, along the wealth development, the fundamental elements, without any of which the growth is fictitious.

The second section gives a bird eye view on how the market staples changed, since when the transformation economy became the reference support, from the *agricultural*, or the *industrial* revolution, up to the possible *cognitive* revolution to come, based on new «knowledge» paradigms.

The last section provides the introductory overview on the reasons why the western cultural bias, by means of personal competition and reductionism, is the evident background for the technological applications and structured deployments of the industrial work organisation.

The three sections, together, provide the reference casing of the old «industry» paradigms, showing the requests: to keep under control all the four capital assets (the *natural*, further to the *financial*, *human* and *technical* capital); to devise ways out, towards new staples; and to give hints on some cultural peculiarities, showing that alternative patterns are, hopefully, existing.

1.1. UNIQUENESS OF THE CAPITALISM GROWTH

The industrial revolution is event that deeply modified the human life quality. At first glance, it became possible to increase the current spendable riches, by the ostensibly *artificial* means, based on machine-driven productive processes and scientifically-assessed work-organisation. The trend successfulness is, still today, prised, and it is not wise to reject the enabled options, as these are the only way to manage and guarantee the achieved quality of life. We might look at the necessary prerequisites of the industrial way, to successfully expand the wealth, in view of understanding the growth spreading out. This list might sort out different entries, and the following four are noteworthy choice:

- appropriate technology and know-how, assuring the enabling manufacture knowledge to the transformation processes;
- ample financial assets, leading to the development and the ownership of business-consistent production means;
- adequate work-forces, permitting to rule the productive processes, through scale (or scope) manufacturing effectiveness;
- proper availability of tangibles, granting the supplying of artificial energy, raw materials and auxiliary means.

These four entries are not equal, even if limited surrogating replacements are viable with resort: to non-proprietary technologies, to productive break-up, to out-sourcing, to intermediate or reused stuffs, etc., all incorporated in the manufacture flow with shunts and buffers. The overall productive effectiveness is, for practical purpose, measured by money, with the effect to make use of the common metrics, motivated (in classic economics) by the supply and demand law, and the marginal value conjecture (provided that steady provisioning exists).

Likewise, the impressive successfulness of the industrialism has contributed to put in shadow the intrinsic instability of spendable riches, generated by drawing on natural storages of resources cumulated over several millenniums, and quickly dilapidated, through the manufacture exploitation, into litter and stain. The clear inconsistency comes out from neglecting the *fourth* entry, as if the earth resources could be thought without limits, or as if the forward (manufacture) transformation could be balanced by the equivalent backward recovery, yielding like amounts of material resources, within acceptable periods.

In fact, the industrialism is *artificially* providing, only as first issue, unlimited abundance; at subsequent times, the remaining populations will be given smaller and smaller heritages, and in-progress less friendly surroundings. The dramatic efficiency is winning option for the today's living beings, but it is, as well, critical indicator of how long the growth permanence will last. On these facts, one needs not to exorcise the changes turned out by the industrial revolution, rather to accept the efficiency and to balance the decay. Is it feasible ? We shall try to define the reference frame, and to look after possible further accomplishments.

Besides, the above recalled four entries, as pointed out, are neatly assessed in money, regularly to be considered «capital» assets. The industrialism, then, is the particular economic system, in which the transformation effectiveness is gained through work organisation features, namely, the scientific/intelligent employment of the *human* capital. Of course, alternative necessities exist: first, by requesting appropriate *technical* capital; further, by demanding adequate *financial* capital; then, above all, by procuring the necessary *natural* capital.

The interlacing of *industrialism* with *capitalism* has long history. Conflicting interpretations subsist, depending on the favoured definitions. The two concepts, actually, considerably differ, unless because the former has historical origin in the industrial revolution, well defined observable fact, having time and local reasons, and the latter is entangled phenomenon, tied with multifarious economic systems, only partially, confined to the Marxism biased reading. In this first section, some introductory remarks on the capitalism are recalled, however, keeping in mind the facts leading to the more recent intertwining with the industrial societies.

1.1.1. Capitalism and Growth Challenges

«Capital», hence «capitalism», comes from "caput" (head). Among other uses, it became the reference in the ancient city-states, at the time of the Italian (and in some cases, European) *communes*, to rule the local governments. In these periods, the feudal organisation disbanding, and the central authority (empire, church, etc.) vagueness pushed the, otherwise advanced, societies, to self-ruling political units, based on direct democracy, and to economical systems leading to "mercantilism". The citizens, co-operating in the business, were allowed to interfere according to their «capital share», most of the times, selecting the current decision with resort to direct advice of each partner. Thereafter, the capitalism corresponds to the self-steering mechanism, when the central government (or religious) rules have little impact, suggesting to establish bottom-up ruling, grounded on the consent of the involved partners, keeping their shared responsibility.

The «capitalism» has undergone many phases and establishments, depending on political and legal conditions, faced in each historic period and social settings. A preliminary outlook distinguishes:

- the ancient multiple-centric establishments of the city-states, and of the *mercantilism*; the period characterises by the direct commitment of the merchants for promoting the exchanges and the trades;
- the organised establishments of the nation-states, which assure back to the *industrialism*; the period meets with contrasts on *free-market* vs. *custom duties*, and the labour/finance *class struggle*;
- the spreading of the multi-national corporations, capable to operate in the *globalisation*; the period faces the limits in natural resources stocks, and the *ecologic regulations* demands.

The outlook is mostly axed on the western world socio-political trends, since the *industrialism*, with annexed lights and shades, is heavily Europe-centric. In the third chapter, the last phase is further split, outlining three possible scenarios for the years to come. The final chapter further stresses on scenarios at the *global village* range, because the mankind survival requires worldwide solutions.

The picture is sketched around the economic systems, and their parallel with the affluence fruition, made achievable by the emerging, in-progress spendable, «capital» assets. Indeed, the word «capitalism» shall, chiefly, be used out of the current conventions. Marx's arguments with «labour» in conflict, still, dominates most of the frequent models, up to the conjecture to deal with the financial capital alone, which sets out, as the society supplies the opposed work-force (labourers, pulled out the country by starvation). On the other side, Weber's view is equally biased, assuming that the «capitalism» spirit lies in some *ethic* effort, to improve the wellbeing. These mainly political or religious interpretations offer side-views with misleading effects, because not justifiably conditioned by merely financial interests. Better balanced concepts lead to identify with «capital», the all fortune into the people availability, while the spendable riches for the everyday life have to limit to the revenue that is obtained, without drawing on the piled up stocks.

In its early implications, the «capital» did never confuse with the just financial contributions. In the joint companies, each partner confers his specific giving, say:

- the domain means, expertise and mastery or «technical capital», permitting the activity to be carried over;
- the money funding supports or «financial capital», providing the trust, to back up the business;
- the labour and executive aids or «human capital», assuring the front-end operation work-force;
- the raw materials and provisions or «natural capital», giving the tangibles to handle and transform.

The resort to the four inputs is well assessed: the "mastery, backing, work and materials", according to old artisans formats; or: the "machine, money, labour and supplies", in the conventional industrial organisation; or, as well: the "knowledge, investment, employment and tangibles", in an updated description, represent the *necessary* prerequisites of every productive company, and ought to be considered jointly, whenever the entrepreneurial history is analysed. Distinct individuals can be at the origin of the specific assets, and each of them is entitled to profit from its personal contribution, within suitably assigned rules.

The mapping of «enterprise shares» into «partners' privileges», or vice versa, is suitably defined in terms of related *capital* rights, and the four assets role leads to identify four partnership allowances. The remark suggests to explicitly refer to *four* capital assets, and to measure the *industrial* effectiveness, without omitting any of them.

Now, the fourth capital asset is recent entry, at least, as (explicit) independent input. The indiscriminate taking out from the earth stocks and its inexorable spoil into waste and pollution create impoverishment and decay, since no equivalent accumulation is regenerated. In addition, the withdrawals mostly fall in areas with ownership at the opportunistic stage, and without local government's interest to clash against their own citizens immediate profits, attained at the expense of other populations. The situation may, in some way, repeat the old establishments, when the people were individually required to self-organise, face to the central political dearth. The substitution might look to multi-national corporations. The users of the "global village" natural capital are pushed to plan bottom-up rules, having the advice and approval of all involved parties, according to worldwide institutions, with further barriers on the *capital* rights delegation value, deferred to protect the future generations. The picture is not without apprehension, and the all book will turn around these topics, looking for possible ruling principles. The solutions are crucial, and need to be devised, to keep sound mankind life quality.

The «capitalism» of the sustainable growth challenge can not repeat old plots. New ones are deemed to exist, on condition that all four capital assets give return and the business expands with no squander of any of them, because the original stocks are reinstated at the same or higher levels. The discussion has, certainly, to take account of several other conditions. Moreover, the decision to consider four capital assets equivalent shall be clarified. The term «capital» is arguable, and the word «heritage» might be used, understanding by it the protection of the inherited goods, as the primary obligation to warrantee the successors' future. Anyhow, the word «capital» implies that the assets are evaluated in «money»; the bookkeeping addresses unified assessments, without confusion on the contribution origin; and the explicit account of the input/output tangibles flows is fraction of the balance, not to be omitted.

1.1.2. Native and Artificial Earth Resources

The equivalence of the *four* capital assets requires some deepening. Actually, only the «human» and the «natural» assets are native terrestrial resources, while the «technical» and «finance» contributions are *artificial* additions, established and increased by the men. When facing bookkeeping duties, one soon recognises the disparity in the «money» balances and trends. The native quantities, only, are evident, with records that show the demographic trends and the world discovered inanimate accumulations and breaded or farmed produces. In fact, the inert heaps come up from non-renewable sources, and the livestock and crops are considered renewable resources. Better insight needs minor adjustments, because of few up-grading due to synthetic substances, and the terrain decay induced by irrigation, fertilizers, single-crop farming, biocides, etc., bringing higher saltiness, bio-types smash, erosion, humus rarefying, aquifers contamination, varieties disappearance, and the likes.

The trend changes, when dealing with the artificial capital assets bookkeeping. There is apparently no limit at their piling up, so that the growth appears blessing opportunity for the future mankind affluence. The «finance» storing up is nice art, requiring, however, deliberately organised social and legal bylaws, to enable the «ownership» modes. The barter, original trading way, is replaced by the standard «money», if recognised authorities guarantee its (accepted) value. Besides, several communities produce surpluses, exceeding local needs, but require extra goods to improve their life stability. The long-distance trade needs artificial intermediation assessments, based on fixed acknowledgement. The «finance» capitalism, quite soon, follows, building up in excess stocks, condition attained, when the technical advances boost the productivity.

The *mercantilism* is known achievement, documented by the Italian maritime republics, assuring goods exchanges, with the backing in the funding autonomy. The small (Venice) or no (Genoa) territorial spreading out assured the bankers' role, with the political recognition of properly accepted rulers (Empire, Church, otherwise established authorities). This old capitalism was strongly linked to the (Mediterranean) operation area smallness, shortly kept up with Amsterdam and the Netherland, which opened the worldwide trade, with spreading of the colonial restricted areas.

The role of the «human capital» came out, with the nation-states, showing that *financial* without *political* power is weak, and men native contribution is further requirement. In Europe, the nation-states set up, where the city-states never got to high influence: England, France, Spain, etc., so that the institution of nation-wide rulers was easier. The territorial extension permits the recruitment of armies (out of mercenaries), and the creation of bigger markets (sheltered by border taxes). The nation internal boundaries abolition is due, not only to political, but as well to technical and economical reasons. Sets of technological innovations, in the textile, milling, mining, metallurgic, etc. sectors and in communication structures (roads, inland waterways, etc.) asked for big increases of the market domains, pushed by buyers, interested to compare alternative offers.

The most successful nation-state «capitalism» sets up in the UK, started with the Elisabeth kingdom and Thomas Gresham policy, assuring money stability, to grant the bankers' investments in their trading companies. The development of Great Britain nation-state (and end of the Amsterdam city-state) is progressively established on all four capital assets, and its final accomplishment, with the build-up of the worldwide Empire, is maximally the outcome of the «technical capital» excellence. The USA succession moves in the same lines, requiring bigger nation-state, to interact, with increased power, at the global level.

The role of the «natural capital» comes out now, when the wealth growth due to artificial contributions is facing the limits of the earth stocks. The change is the issue of extreme alterations in the transform economy, ending the affluent society with unlimited raw materials turned to waste and pollution. In the new conditions, the failure to account the withdrawn tangible resources and the downgraded or toxic damping brings to give a false picture of the growth, basically showing that the present populations profit of non-renewable riches, while little is left for the future generations.

The sketched bird eye view of the trend, from «city-states» to «nation-states», might suggest the dominance of the native capitals, up now, almost exclusively, embodied in the human capital. The industrial revolution comes in as dramatic discontinuity, with the artificial capitals as central driving enhancer. The natural capital is, up now, confined to play the implicit role, so that the financial capital could come to play the leading character, being immediately recognizable and spendable. With the natural capital becoming decisive, the growth sustainability forcedly has to give power to the technical capital, only this bringing hope to find *artificial* resources, out of the original earth accumulations.

At this point, the equivalence of the *four* capital assets might require cautions. Quite evidently, the metrics to assess the *artificial* either the *native* capitals have different coherence, when defined in each pertinent milieu, or when transferred to measure the un-diffentiated piling up/withdrawing of given riches. The unlimited growth of knowledge is clear advantage, but this does not automatically balance the equally unlimited withdrawal of earth original resources. The man quality of life is based on the availability of tangibles, at least, up to satisfying the material necessities. This requires a more subtle exploitation of the technical capital, say, a revolution to change the way that the mankind follows to get spendable riches, as compared to the current affluent society one.

The equivalence of the *four* capital assets requires some deepening. Actually, only the «human» and the «natural» assets are native terrestrial resources, while the «technical» and «finance» contributions are *artificial* additions, established and increased by the men. When facing bookkeeping duties, one soon recognises the disparity in the «money» balances and trends. The native quantities, only, are evident, with records that show the demographic trends and the world discovered inanimate accumulations and breaded or farmed produces. In fact, the inert heaps come up from non-renewable sources, and the livestock and crops are considered renewable resources. Better insight needs minor adjustments, because of few up-grading due to synthetic substances, and the terrain decay induced by irrigation, fertilizers, single-crop farming, biocides, etc., bringing higher saltiness, bio-types smash, erosion, humus rarefying, aquifers contamination, varieties disappearance, and the likes.

The trend changes, when dealing with the artificial capital assets bookkeeping. There is apparently no limit at their piling up, so that the growth appears blessing opportunity for the future mankind affluence. The «finance» storing up is nice art, requiring, however, deliberately organised social and legal bylaws, to enable the «ownership» modes. The barter, original trading way, is replaced by the standard «money», if recognised authorities guarantee its (accepted) value. Besides, several communities produce surpluses, exceeding local needs, but require extra goods to improve their life stability. The long-distance trade needs artificial intermediation assessments, based on fixed acknowledgement. The «finance» capitalism, quite soon, follows, building up in excess stocks, condition attained, when the technical advances boost the productivity.

The *mercantilism* is known achievement, documented by the Italian maritime republics, assuring goods exchanges, with the backing in the funding autonomy. The small (Venice) or no (Genoa) territorial spreading out assured the bankers' role, with the political recognition of properly accepted rulers (Empire, Church, otherwise established authorities). This old capitalism was strongly linked to the (Mediterranean) operation area smallness, shortly kept up with Amsterdam and the Netherland, which opened the worldwide trade, with spreading of the colonial restricted areas.

The role of the «human capital» came out, with the nation-states, showing that *financial* without *political* power is weak, and men native contribution is further requirement. In Europe, the nation-states set up, where the city-states never got to high influence: England, France, Spain, etc., so that the institution of nation-wide rulers was easier. The territorial extension permits the recruitment of armies (out of mercenaries), and the creation of bigger markets (sheltered by border taxes). The nation internal boundaries abolition is due, not only to political, but as well to technical and economical reasons. Sets of technological innovations, in the textile, milling, mining, metallurgic, etc. sectors and in communication structures (roads, inland waterways, etc.) asked for big increases of the market domains, pushed by buyers, interested to compare alternative offers.

The most successful nation-state «capitalism» sets up in the UK, started with the Elisabeth kingdom and Thomas Gresham policy, assuring money stability, to grant the bankers' investments in their trading companies. The development of Great Britain nation-state (and end of the Amsterdam city-state) is progressively established on all four capital assets, and its final accomplishment, with the build-up of the worldwide Empire, is maximally the outcome of the «technical capital» excellence. The USA succession moves in the same lines, requiring bigger nation-state, to interact, with increased power, at the global level.

The role of the «natural capital» comes out now, when the wealth growth due to artificial contributions is facing the limits of the earth stocks. The change is the issue of extreme alterations in the transform economy, ending the affluent society with unlimited raw materials turned to waste and pollution. In the new conditions, the failure to account the withdrawn tangible resources and the downgraded or toxic damping brings to give a false picture of the growth, basically showing that the present populations profit of non-renewable riches, while little is left for the future generations.

The sketched bird eye view of the trend, from «city-states» to «nation-states», might suggest the dominance of the native capitals, up now, almost exclusively, embodied in the human capital. The industrial revolution comes in as dramatic discontinuity, with the artificial capitals as central driving enhancer. The natural capital is, up now, confined to play the implicit role, so that the financial capital could come to play the leading character, being immediately recognizable and spendable. With the natural capital becoming decisive, the growth sustainability forcedly has to give power to the technical capital, only this bringing hope to find *artificial* resources, out of the original earth accumulations.

At this point, the equivalence of the *four* capital assets might require cautions. Quite evidently, the metrics to assess the *artificial* either the *native* capitals have different coherence, when defined in each pertinent milieu, or when transferred to measure the un-diffentiated piling up/withdrawing of given riches. The unlimited growth of knowledge is clear advantage, but this does not automatically balance the equally unlimited withdrawal of earth original resources. The man quality of life is based on the availability of tangibles, at least, up to satisfying the material necessities. This requires a more subtle exploitation of the technical capital, say, a revolution to change the way that the mankind follows to get spendable riches, as compared to the current affluent society one.

1.1.3. Knowledge Resources and Technologies

When looking at the history of mankind, the technical capital is understood as the enabling prerequisite of all civilisations advancement. However, the central place is, mostly, transferred to other drivers: coherently, perhaps, to the human capital, being this native origin of every artificial development; or, less clearly, to the financial capital, bestowing the pervasive role to include in it all the different abstract spurs towards prosperity and wealth. The technical capital weigh, surely, is not ignored. Since the oldest times, the people perceive the artworks and skill achievements, as features discerning the civilised, from the barbarian societies. The knowledge, thereafter, built up and handed down to posterity, defines as the intangible part of the heritage, having well established characteristics, to become independent capital asset, coded as community legacy.

The *knowledge* distinguishes from *intelligence*, even if both clearly link to the *homo sapiens*. The former relates to the collection and/or storage of information, facts, truths, principles, rules, etc., as for learning, investigation, application, etc.; it is, also, the body of truths and facts accumulated by mankind in the course of the time. It is an abstract entity, and can be coded for handling and transmission. The latter (from the Latin *inter legere*) shows the capacity to choose, understand, analyse and similar other mental acts, or it refers to the piece of information, data, news, etc. about a circumstance, event, person, etc. purposely selected, singled-out or up-dated. Hereafter, the *knowledge* properly classifies between intangibles, while *intelligence* needs to have associated energy for detection, transmission and reception. The two views are formal, as the knowledge, itself, is inexplicable with no material support, and totally useless if no observer exists. They are mentioned for better insight of the (four) capitals role:

- the «knowledge», *abstract* entity created and ruled by the men, having the faculty to be accumulated and transmitted as qualifying heritage;
- the «intelligence», *natural* entity shared by the *homo sapiens*; usefully, at times, duplicated in *artificial* intelligence (as with the *artificial* energy).

The interlacing of the artificial capitals is puzzling reality. The picture of the technological means and organisational methods, really, available to sustain the quality of life, are essential step, when defining the economical systems, used to assess the peoples income and wealth, This way, the goods and services price and origin, the wellbeing of individuals reason, the enterprises profit and efficiency, etc., might explain, each time, providing evidence to reasonable causal frames. The necessity rule is such that relevant changes in the technical capital bring in paradigms shifts, even when the intricacy hides the direct causality links with the accumulated knowledge. At least, the knowledge growth is prised truth.

In fact, the mankind faced periods, more or less long and deep, of crises (the middle-ages, depressions, etc.), during which known technologies disappear and confirmed productive organisations dissolve. The stagnation or regression reasons are multiple, linked, in most case, with turmoil: e.g., in the European middle age, the barbarian invasions removed the old know-how, only regained, once the civil societies reorganised. The monotonic knowledge build-up, thus, is necessary, but not sufficient, reason for the progress. Besides, univocal references, documenting the incremental transmission of the technologies, do not exist, and the continuous betterment of the civil societies might fail, because of intrinsic or induced reasons. The inherited knowledge, by itself, does not guarantee the wellbeing preservation. The eco-consistency, today, comes out as intrinsically critical condition, because the withdrawals from the natural capital attain non-negligible unbalances, and the «compensation» through *intangibles* is arbitrary choice.

Further remarks show that, first of all, the mixing of the cultural achievements and the technological advances is not without opponents, who like to distinguish the speculation of *humanities*, from the conjectures of *machinery*. The know-how practical issues are, in these views, inferior, as compared to the men talent. Then, as second observation, the perennial development is rejected, as nonsense, in our finite material world. The picture leads to the *thrifty society*, founded on resources sparing and consumption decrease, through the *shift down* way, say, the pace-wise drop of waste and pollution. The conflicting positions deserve attention.

The culture, in keeping with its primordial meaning, starts by abilities in the wild land farming and fierce animal taming. These are very practical techniques, leaving little room to imaginative thoughts. The two terms, the Latin «*ars*» and the Greek «*techné*», are equivalent, and both relate to the mastery to reach useful achievements. The abstraction of concepts, ideas, notions, theories and models is naturally linked to the way the men interact with the surroundings, to reach better life quality, from satisfying the urgent needs. Indeed, the knowledge-driven defy modes noticeably differ, and the world populations have been following separate deployments. Still, some generic features are shared, such as the following ones:

- the cultural background is fundamental feature of the civilised societies, drastically affecting the citizens' quality of life and wellbeing, by several means, including the education spreading;
- the scientific backdrop is general support of every noteworthy advances, providing purposely fitted opportunities, to promote valuable innovations or even to start paradigm breakthroughs;
- the technological setting is specific enabler of the current progress, by suitably supporting the know-how, with patents and licences, to obtain return from the technical capital.

The knowledge sovereignty, as independent capital asset, not to be confused with the human one (which is the originating source), either with the financial one (which is the trading means), presumes that suited metrics exist, and are exploited to fix acknowledged exchange price.

Besides, there is no doubt that the knowledge build-up has a cost. The culture has, also, value (e. g., in the selling/buying of masterpieces or handicrafts), or is charge (e. g., in the public school systems). The obtained remuneration, however, does not match up with standard metrics, rather with more complex estimation, in which the artist/actor mastery/qualification plays big role, or the fair-reward fees have proper tradition. Similarly, the science is appreciated as useful investment, having chiefly long distance return, making quite intricate to establish standards, when provisional assessments are needed. So, we infer that cost/return pricing, in general, comes out from mostly implicit assessments. The exception shown by the technology pieces (patents, know-how, etc.) gives little insight on the knowledge heritage shared and transmitted by the peoples; it does not supply meaningful idea on the «technical capital» market, at least, until the civilisation growth key drivers explicitly come out, into economical assessments.

The turn might happen with the knowledge capitalism. The guess emerges in direct link with the information and communication technologies, recognised as powerful way to move the value chain of the goods brought into man possession along better conservativeness outcomes. In the knowledge market, however, the supply and demand law does not set up self-sufficient metrics. The transfer of the knowledge pieces can be indefinitely repeated, as it operates on intangible units, which remain in the possession of the seller, unless formal agreements are stated, to specify the transfer, ownership, enjoyment, etc. rights. The regulation needs to be enacted by government and interstate accords, established by legal metrology protocols. Several side restraints, however, subsist. The knowledge transfer and exploitation is scarcely effective, if the recipients are unfit or unable to understand and/or to manage it. The other way, the incremental enrichment permits a trained operator to acquire the full modus operandi, only purchasing modest details. The knowledge market, then, is intricate affair, and the trading and measurement rules require careful attention (not solved by the supply-and-demand law routine use).

1.1.4. Financial Resources and Hegemony Rules

The financial capitalism has acknowledged deployments, with known history behind and assessed metrics to document the build-up of spendable riches, for the mankind benefit. Quite recently, the growth becomes critical, face to exogenous motivation in the incipient exhaustion of its driving backing, due to the depletion of the natural capital stocks. Besides, the previous phases show relevant features, which can be usefully brought up to better understand the cross-interplay among the other capital assets.

Originally, to simplify the barter transactions, the pecuniary means happened to be tangible unit pieces, easy to locate and to exchange. The early standard, the Latin «*pecus*» (goat), or unit herd sample, puts, of course, big problems, to grant steady duplication and submultiples drawing. The troubles are avoided, when the standards are given by coins (in noble metals), supported by governments, with recognised power and mandate. The value of the official «money» is decreed by kings/emperors (or the goddess «*Moneta*»), which grant widespread uniformity, by proper mints and vaulting. The paradigm shift is factually powerful, and the market profits by the small unit pieces, easily transferred and basically stable. The governmental authority and the civil servant organisation are enabling foundation of the monetary systems, which assure the abstract pricing continuity.

The monetary set-ups are, accordingly, direct falls-off of controlled social and political systems. Several side-effects, moreover, follow, moving from the ruling of the personal rights and duties, and ending with the protection of the individual ownership of land, houses and objects. The settlement of structured legal frames is chief premise of the nations' progress, even if the financial capitalism could remain implicit, when the backing civil institutes guarantee the harmonic lifestyle conditions to individuals and communities (with self-balanced local foodstuffs). This was not the case with the European peoples, and it became apparent, after the dissolution of the Roman empire, with the ruling of the feudal establishments. In that period, the capitalism explicitly moved the starting steps, with the city-states multiple-centric setting (and *mercantilism*), up to suitably agreed trade methods.

In that situation, the powerful order occurs from lineage, the civic mindedness comes from culture, and the wealth means to hoard pieces of gold. The business company requires the four inputs "mastery, backing, work and materials"; more in particular, the trading company is especially driven by two of them: the finance sponsorship, to grant achievement freedom, and the pioneer trade-agents, to reach vastness prospecting. The split establishment of the city-states «financial capital» is consequence of the weakness of the central governmental authorities and civil servant organisations, so that the backing took place with resort to local monetary storing up and to bank transactions, fulfilled having (only) personal commitment. The clear partition, among the political ruling, requesting high descent (feudal) conditions, and the traders' economical influence, based on bottom-up multiple-centric settings, is inherent *mercantilism* and city-state capitalism limitation.

The ensuing nation-state capitalism is clear overcoming, as soon as the single individuals' commitment could not anymore deal with steered-market conditions, requesting wide and structured political support. The turn is fostered by two legal provisions: the *free-market* order, within the national borders; the *custom taxes* enacting, against foreign import and protecting the local producers. As before, the trading patents are *privileges* of the entrepreneurs or companies, having the (king or emperor) governmental licence, to be requested (and paid), to possess granted the monopolistic position. The reference domains make, from now, the difference, entailing nation-states and asking for permanent colonies scattered in the world, with support of strong protecting licences.

The establishment of the nation-states occurred, in Europe, at different times and unlike modes. The countries coming out started periods of intense warfare, with relevant falls-off in the hegemonic roles. The Great Britain case is, however, the nation-state capitalism here recalled, representing the paradigm shift from the earlier city-state capitalism, up to the worldwide expansion of the British Empire. The spreading out did not occurred without obstacles, needing over two centuries. The last phase, in the average, gave way to the astonishing achievements, offered by the «industrial» revolution.

To fix a bench-mark, the starting point could be the year 1776: the declaration of the American colonies independence, imposing sectional trade opening, out of preserves, and the Adam Smith book "Wealth of the nations", asserting the leads of the productive specialisation, under parliamentary democracies, which prise the *free market*. The postulation of free exchange mechanisms, with *global* extension, ruled the *"long global assent"*, started about 1840 and stopped in 1914 by the first world-wide war. The two periods together roughly typify the industrial revolution, with the British Empire hegemony (in rivalry with other European countries).

Later, the nation-state capitalism, confined to the European scale, could not withstand the continental powers, of the cold-war period. This preceded the *"short global assent"* at the end of the 20th century, when the URSS dissolution left the USA as only world ruler. There are similarities in the two *"global assents"* lines on the international *free market*; say, the paradigm shift towards bigger political foundation, this time turning the nation-state capitalism, into the subcontinent-state one.

Actually, both *"global assents"* were announced by premonitions. The UK golden age begins from the end of the seven years war against France, in 1765, with the industrial revolution prosperity boom, typified, in textile shops, by the Arkwright hydraulic loom, the John Kay flying shuttle, the Crompton watermill. In the same period, the financial market moved from Amsterdam to London, and the combined technologic and economic supremacy founds its complement in the maritime domination. Later, the UK defeated the France of Napoleon, with the primacy of the London Exchange and the compulsory (China 1842, Japan 1854, etc.) market opening, to start the *"long global assent"*. The queen Victoria jubilee message was simultaneously delivered all over the Commonwealth: UK, Canada, Australia, India, etc.; the telegraph did the job of the today world-wide-web. At that time, the financial market was even more *global* than a century later, showing the London ability to rule the finances, worldwide.

Subsequently, the critics discovered several ambiguities, behind the assent. If a country specialises in machinery, and one in wines, both profit in trading their products, with the buyers benefit of low prices. However, the two parts are not equal opportunity, as the industrial player will indefinitely increase the production (and lower the prices, if the raw materials and the artificial energy are not paid their value), but the agricultural one remains at natural cycles. The specialisation, accordingly, moves the textile manufacture to the UK, leaving the cotton growth in India. The want of symmetry bears especially relevant and lasting outcomes, with the conventional industrial arrangements, requiring impressive financial capitals and localised sources of artificial energy (coal mines); the economy of scale is dominant strategy, and the new competitors need to face entry thresholds, magnified by the world centralised financial market.

The *"long global assent"*, of course, did not keep on quiet stability rules. The inner contradictions lead to the class struggle, with revolutionary issues, lacking organised reaction. The outer discriminations move to the commercial conflicts, with temporary compromises, up to the eventual deflagration of world-wide wars. The drawbacks are well recognised. In the same time, we cannot ignore the huge benefits. The electricity appears, with its networking features, also shared by the telegraph and telephone (the internet technologies continue the successful trend). By the pocket books, the cultural diffusion has global reach, and, by broadcasting, the news diffusion and social entertainment grant shared access. By transportation efficiency and remote-transmission abilities, the distance barrier drops. These are the «global village» musts, to make all peoples understanding their common sort (with the planet earth). Instead, the *"long global assent"* is followed by a period of fragmentation, before the *"short global assent"* phase, quickly undermined by the *no-global* opposition.

The market integration/fragmentation phases are explained by conventional economic theories, showing how to modify the marginal costs by customs duties, and how to protect the national offers through technical standards. In terms of merely economical reasons, the global assent presents advantages, leading, as an average, to wellbeing enhancement, and drawbacks, indiscriminately exploiting the natural capital. Only now, the economy/ecology dilemma starts showing the end of the affluent society patterns.

The explanations are quite entangled. The world countries stand out according the industrial level, and the emerging conflicts oppose to the wealthy populations. The *"long global assent"* closed with world wars between leading countries. The *"short global assent"* ends with actions that, in short, are defined «terrorism», say, out of the standard legality of the war. Besides, the opposition *global/no-global* does not give insight to the intrinsic instability of the industrial countries, face to the sustainable growth. The contiguity, between some *no-global* movements and «third world» organisations, better explains the urgency of the situation, without, however, offering any help towards solutions.

1.1.5. Human Resources and Social Administration

The «capitalism», in most views, trusts the *market* regulatory function, as it acts, bringing the economy, along factual paths, to equilibrium settings, with the most effective resource allocation. The (Adam Smith) *invisible hand* is underlying force that grants optimal economic welfare to the society, where each person acts in self-interest. A single performance index is sufficient to drive the trend, and the high industrial productivity grants the growth increase, with the whole community advantage. When the financial capital hegemony establishes, the *market* creates privileges, with richer minority's profit, and marginal majority's exploitation. The reached equilibrium cannot be stable and evolves to altered settings, driven by the majorities, as soon as they get the upper hand, in the parliamentary democracies, towards less biased wealth distributions.

The «capitalism» contradictions are apparent in the paradigm shifts, from the city-states, to the «global village», requiring bigger exploitation areas, to consent steady running. If these cannot set up, the *human* capital asks centrality, aiming at protecting the lower workers, as social class, modifying the leadership, to give the priority to mass-distributed interests, transferred as collectively shared rights, or, as individual title to receive quotas of the common benefits. In fact, the mankind history shows ceaseless fight between political entities, to conquer and maintain the power, and this means dialectic evolution, opposing establishment to progress, up to the steady running, when equal chance will reign in the *global village*.

The steady running is, maybe, utopia. The political economy challenge resides in explaining the «equity» either the «iniquity» of given revenue sharing out. The understanding of how far the inequalities are necessary, to make feasible building the fortunes of winning individuals, conflicts with the concept of fairness, which forbids to deprive lot of persons of the vital wellbeing, in view of *just* standards. Among the positions, the communism is well known, acquiring achievement in countries, where the dialectic *becoming* pushed to the revolution of proletarians, to establish the collective ownership and the stateism equilibrium.

Indeed, in the Soviet Union, the «workers» only were full-citizens, excluding the peasants and the armed forces; just tiny subsets of them were members of the communist party, and very few could enter in the bureaucratic nomenclature with governmental functions. This way, the new party order was frozen, replacing the finance order, so that a ruling minority again establishes with leadership duty. The top-down bureaucratic order (as, in the middle age, the feudal descent) happened to assure little effectiveness, and the lack of competition issued, on the average, low wellbeing. Surely, the financial means are dubious metrics to assess authority and power, and one can figure out alternative selective signs (guild membership, blood investiture, party nomenclature, etc.), reaching equivalent meaningfulness. Yet, the «human» capital is *native* attribute, as compared to *artificial* constructs; nonetheless, the drastic measure of suppressing the private ownership eliminates return on the «financial» capital and spur towards competition.

With the industrial revolution, the «natural» capital transform effectiveness is wealth build up method, ruled by the *free market* law: if the assets change, suited transient effects occur, before reaching new equilibriums. Karl Marx turned the focus on the *becoming*: the civil institutions, governments, enterprises, workers, etc., are in endless conflict; the progress comes from the class struggle, leaving sideways the illusory temporary respites. The industrial upper-class profits of the riches rise, and the low price goods are dodges, to destroy proletarian's resistance. As power is inevitable aspiration, the production means will inevitably expand, to establish monopolies, with the workers' class detriment, up when the revolution will expropriate the expropriators. With Lenin, the picture is moved on, stating that the imperial countries, keeping their wellbeing by colonies exploitation, can involve the lower-classes, exploiting their «capitalism» embezzlement.

The communism went through the 20th century, with countries (Russia, China, etc.), to experiment it, through personal property suppression, production means confiscation and state capitalism establishment, organised under the party council (soviet) direction. The central planning and control needs pervasive bureaucratic machinery, to please the primary needs (housing, victuals, garments, etc.) and to reach suited equity. It is difficult to manage the diversified requests satisfaction, due to priorities arbitrariness and privileges allocation. At least, this is the lesson offered by the history. Elsewhere, the *free market* capitalism mitigation brought to the welfarism and social-democracy alternative, with the main intents:

- to allocate the benefits in such a way, to assure widespread standard life quality and to forbid hoardings;
- to promote state interventions to avoid or to mitigate the effects of the temporary

depressions or recessions.

The welfarism tracked several paths, suggesting many issues, depending on the local socio-political set-ups. The approach can be derived by the idea (strong in Germany and France) that the State has direct responsibility to equalise the citizens' fortunes, transferring spendable riches to the paupers, with resort to the progressive taxation, payment of unemployment subsidies and creation of shared systems of retirement and disability subsidy. Consequently, the state intervention areas expand, requiring massive fiscal income. The citizens' private property is protected, preserving the related spur towards the enrichment (with falls-off in higher personal taxes), but enacting rules to safeguard the less fortunate people, and biasing the *free market* continuation.

The J.M. Keynes theoretic account brings to economic systems, with helpful policies against depression. The receipt '*virtuous cycle by government ruling*' was experimented before Keynes (e.g. the German expressways construction), but he has the merit of fixing the parliamentary democracies route, to exploit the planned economy measures with sectional range, keeping the market competitiveness out of the temporary incentives. From the government standpoint, the duty does not address the goods price or the profit division, rather to stimulate the production level and the consequent occupation. Benefits are achieved, if the investments are fostered, without waiting that the recession trend (drop of purchasing power due to unemployment, smaller wages, etc.) is turned up by private entrepreneurship.

The key of the success of the welfarism depends on the capability to join the efficiency of each person acting in self-interest, with the governments (or central banks) measures, to re-establish the full-employment, in case of stagnation. The resulting economic system is hybrid, because:

- the macro-economics follows planned development issues, to manage the employment (and prices);
- the micro-economics goes on with free market ruling between each-other-competing enterprises.

The policy is done assuming that macro-economics planning shall not affect the micro-economics actors, and that the government action modifies the socio-economical frame, not the enterprises interplay. The macro-economics, in other words, ought to be documented and objective. Quantitative models need to exist, with disaggregated data and timely statistical assessments. Suited authorities shall be appointed, with access to all fiscal information on the citizens' totality.

Such policy requires sophisticated means, and its effectiveness conflicts with the questionable assumption that the nation's macro-economics measures operate without sensible bias from the worldwide context, and with the individual privacy protection, further accepting the investments twofold limitation:

- citizens tolerable taxation, collecting *adequate* incentives to promote/steer the production and market;
- public indebtedness, to assure the short and long term sustainability to the treasury management.

The focus on the human capital leads to protect the majority of less fortunate workers, by respect to the minority of wealthy tycoons. The nation-state requires establishing authorities, whose duty is to identify the interests of the country by respect to the others, and to organise and rule the relations between the citizens, by choosing social priorities. Higher shelter means taxes and wages, fitting the lifestyle level. In the global market, this entails the competitiveness, requiring protectionism, by custom taxes and/or imperialism. Otherwise, the welfarism (or populism, in this case) increases the public debt, spending the income to come, with profit of today voters, and damage of future citizens.

The difficult balance of financial *vs.* human capital is debated. The concept of *fair* reward (or price, or incentive, etc.) or of *iniquitous* pecuniary interest (or cost, or charge, etc.) needs to have a metrics. The *free market* (in the society of equals) means demand-supply law, saying that the *equitable* payment is the one agreed in the seller/buyer bargain, with no conditioning bias. When the conjecture fails, the choice of *a priori* criteria needs to be defined, to assess the «just» standards. The money (e.g., gold bar reference) is conventional reference, with little consistency when measuring the people wealth, out of local and temporary relative figure. The difficulties of assessing the metrics for the technological development and for the man wages equity should not surprise. The resort to an *a posteriori* figure, such as the *free market exchange value*, has special flavour, being factual appraisal, to be exploited when dependable, out of cognitive or sociologic figures. At least, this is the welfarism lesson, successful, when the micro-economics domain is *sufficiently* unaltered by trans-national bias, and the macro-economics one does not outrun it.

1.1.6. Natural Resources and Ecology Constraints

The «natural capital» is last entry in the survey, even if it happened to be the earliest reference in old times, with the *opportunistic* economy (management of necessities, based on what is picked-up in nature), and up to the whole agrarian transformation one (management of necessities, with the intervention of human labour), as the average productivity could have been thought steady, but, critically sensitive to the current climate occurrences (the *nature*, worshipped as potentially clement or cruel) and highly affected by ordered and peaceful labour organisation, the country around. Only the industrial transformation economy is getting rid of the current climate changes, although, with the risky drawback to be dependent on raw materials provision and fully subdued to the waste and pollution falls-off.

The «natural capital», rather than negligible accident, is becoming the critical conditioning fact of the growth. The end of the affluent society is, now, identified by heterogeneous reasons, as if only passing facts affect the otherwise successful society. The following list collects example partial views:

- the transfer of the workforce from the manufacture to the service domains, lowering the productive segments;
- the drop in activity spirit, induced by high taxation and inflation, and by excessive regulations, lowering the entrepreneurial courage;
- the productivity decrease, due to less motivated workers, today, too much interested to rest and amusement;
- the merging of three causes: the profit shrinking with full employment, the USA protectionism (not convertible dollar), the governments focus on the overprotection of the citizens' welfare;
- the end of the current Schumpeter's wave (before the creation of the next entrepreneurship surf);
- the globalisation effects, yielding employment instability, with productive break-up, out-sourcing and wage disparity increase;
- the steep upsurge of the raw materials price (notably, the mineral oil);

- the effects of the «*ecologism*», with the *no-global* obstacles, hindering the economically sound enterprises.

These reasons, and similar other ones, are all partially true; however here, the analysis suggests a more drastic fact: the huge imbalances in the management of the «natural capital». The end of the merry season of the industrial revolution (and the nation-states capitalism) is, surely, the outcome of combined reasons, but the mixing of these might hide the main facts, namely:

- *over-population*: with rising needs, the fully *free-trade* conditions result to be unfair, to reduce poverty and to improve living standards; regulations, with potentials requiring local and global balances, have to be fixed, as the earth is a closed system;
- *over-consumption*: the market should be addressed towards *eco-efficiency*, to break the link between economic prosperity and tangibles consumption, by re-setting, recovery, rematerialising, etc. rules, which modify the trend, fixed by the affluent society habits;
- *over-pollution*: the environmental effects increase due to waste/emission accumulating, with falls-off in surroundings poisoning and bio-diversity loss, unless proper bylaws are enacted to forbid given behaviours and to support *eco-compatible* solutions.

These facts are directly tied to the way the wealth is created by the «industry» paradigms. With an outlook on the listed economical views weighed through the above three main facts, the explanation is devised on socio-political motives. The real situation is rather clear-cut, and subsequent analyses reconsider the all.

Up now, the capitalism success is discussed in political contexts, allowing the citizens to take active part in progressively larger areas, participating to wider free markets or to integrated economical systems. As corollary, the enterprises, world-wide interacting, lead to *global* competition, by multinational hostile strategies, endorsed to scattered local managers, productive facilities, finance options, sale policies, etc., to become global competitors (at least, for selected domains). This means, politically, to look after the trade *global* assent, maybe, up to far settings, say, to what is later described as the *hyper-market* scenario.

The approach is, for instance, reference tenet of the USA commercial policy, given by the 1989 *Washington Consensus Act* [1], where the world trade regulation is specified in series of measures, which the developing countries are expected to follow, say: extended denationalisation; widest taxing bases and smallest marginal rates; protection of private properties; public expenditures below revenue; market fixed interest and exchange ratios; subsidy removal; import decontrol; free flow of finance capitals and investment, production, trade, etc. assets. These precepts assured benefits to the USA economy, and the International Monetary Fund, IMF, experts did not foresee the severe negative effects produced into different contexts (Indonesia, Latin America, etc.).

The *global* approach has, clearly, many supporters. The "*short global assent*", during the last fifteen years of the 20th century under the USA hegemony, and the previous "*long global assent*", under the British empire, are similarly based on the trade opening: the recent, inspired by the *Washington Consensus Act*; the ancient, factually ruled by the structured empire. The information technologies are today big change, exploiting the economy of scope, with operators, which identify their interest in the company's one, overcoming national

---

[1] J. Williamson: "*What Washington means by policy reform*", Inst. Intl. Economics Rpt.: The Progress of Policy Reform in Latin America, Nov. 1989, Washington.

egoisms, through the finance prominence that makes plausible the market merging into a single *global village*, according to the later (paragraph 3.3.3 and section 6.3) devised *altruism* scenario. At the moment, the "*short global assent*" characterises by three facts:

- the USA financial and cultural hegemony, after the disintegration of the communist economical systems;
- the technology innovation, based on communication nets and technical options for the individual emersion;
- the ability to imagine the evidently high efficiency development model, identified as «new economy».

The three of them are, also, explanation of the *global* approach weakness. The "*short global assent*" props up worldwide market, with benefit in ending secluded status income of local actors, and in letting scale effect to multi-national players. The market self-regulatory ability, left to spontaneous processes, finally results in the establishment of worldwide competitors, in progress absorbing the weaker actors, towards oligopoly formation, capable to fulfil finance business, not exempt of shadows, bringing to booms, failures and mergers. With the new millennium, the "*short global assent*" is giving the way to not fully fixed economic systems, somehow trivially, defined as *plain post-global*. Almost in parallel with the *global* deployment, the opposing *no-global* political movements spread worldwide, with in common the protest against the recalled three features.

Actually, the ecologic concern is basic unifying *no-global* concept. The boost towards 'refusing', as a whole, whatever comes from the «industry» paradigms, becomes central axiom, with clashing outcomes, also, in what affect the technical capital. The *no-global* road-map means to be against innovation, looking back to the 'arcadia' order, accepted as safe (e.g., fighting genetically modified produce). The position leads to correlated issues, such as the dissent on patents and royalties (e.g., for drugs to be distributed to the *developing* countries), motivating the «no-logo» movements, to fight the classic «capitalism».

The *global* and *no-global* abstract readings might be too simplistic, to bring to coherent economical set-ups, achieving practicability, e.g.: the affluent society, as said, has no future; the ecological conservativeness has disciples, which think in paradox to join life-quality continuation, resource consumption and environment pollution removal, simply, with the industrial effectiveness rejection. Anyway, the demographic and financial factors add to ecologic and social ones, so that, today, the governs accountings cannot anymore ignore the overall balances over the four capital assets, with included tangibles flows (stocks depletion and waste increase), showing the out-of-balance off-sets that damage the future generations.

The *no-global* groups have the merit of the eco-impact assessment request, as constraint for the people to come, or inter-generation pact to safeguard the earth life. In short, if the *global* approach understand the global village within economic viewpoints, the *no-global* one protests the development dramatic new demands, not known, when the "*long global assent*" went to end, a century ago, because of socio-political reasons (leading to series of world-wide wars, fought for national "vital spaces"). Today, the global village is reality, and the *active post-global* way to face growth demands means to investigate and acknowledge the problems, in view to find out solutions, recovering the «natural capital» balance.

The emerging law frame, rather than to orient, has to *compel* the consumers (manufacturers and users) to more concerned behaviour, joining the ecologic, to the economic demands. In that context, the ICT means are, sometimes, referred to lead, more than to the knowledge-, to the 'watch'-society: the citizens' personal rights are postponed to the communities' rights, asking, to some extent, contiguity between the economical and the political freedom spheres. The protection of men survival in the earth surroundings is assumed to have priority, coming before any individual spheres. The choice of lawful or unlawful behaviours interferes in the consumers' habits, distributing the control on the modes and times for enjoying goods and services. The objects ownership is not absolute; their use is tied up by the third people and the environment protection.

The political and regulation acts are, by now, evolving, and the social consent plays the chief role, in their effectiveness. This develops on the awareness of the ecologic risk, requiring the different splitting up of the private spheres. The bylaw will progressively develop, applying set of rules, such as:

- the regulations need to have global coverage, without local exceptions or personal exemptions;
- the functions market shall expand, to satisfy the clients, avoiding strictly unnecessary transfer of material goods;
- the end-users shall be oriented to thriftiness, avoiding resort to disposables and extensively looking at recovery;
- the servicing infrastructures ought to be stimulated to widen the provisions of intangibles;
- the reverse logistics has to become standard practice, with reuse or recycle of the end-of-life items totality;
- the rules will be based on assertive interventions, to get the people used to eco-conservative choices;
- the prohibitions and vetoes would be minimised, to avoid the profit from the underground activities.

In short, *fair* trade acts, binding all players, shall address *altruistic* behaviour towards posterity, and *positive* habits to sustainability. The ecologic order offers new prospects, linking the purchasers' benefit with the suppliers' profit and outer parties' guard. The stakeholders and interest groups self-protection aims at choice consciousness and transparency, assuring fair spreading through joint and several liability competitions. The profits to sustainability come from the trend visibility, as the accounting of the natural capital contributions is not private affair between consumers, because of the inherent *global* falls-off.

The vision can be made to turn out dialectic progresses towards the balance of the ecologic/economic issues. The *post-global* approach explores sample patterns, to establish the saving priorities, e.g.:

- the role of the ownership rights on the resources, product and manufacture means, to promote and store intangible wealth (e.g., including intellectual property protection *vs.* the «no-logo» claims);
- the worldwide market management of goods/services ownership rights, to reach scale-economy effectiveness, through the *altruistic* legal groupings, holding the objective future generations interests;
- the bookkeeping habit to grant the all engaged capital assets neutral yield, «natural» supply inclusive, with companies and individuals commitment ruled, according to smart profit.

The *post-global* position, as fair result, gets over the *global* and the *no-global* standpoints. The *altruism* scenario, nevertheless, is deemed utopia, unless the new developments, enhanced the «knowledge» paradigms, discover alternative growth opportunities, with widespread benefit. If this is the case, the *solidarity* scenario is quite likely to become true, with no heroic commitment. Further notices, notably in the section 6.3, sketch also the related *hyper-democracy* political ruling.

1.2. LESSONS FROM THE MANKIND HISTORY

The industrialism is mankind recent (and short) conquest, so that the link with the four (technical, financial, human, natural) capital assets and, there through, to the «capitalism» way to interact with the wealthy society, is partial explanation. The industrial revolution biasing effect, in point of fact, is typical success of the so-called western world, and, thus, it presents with locally and timely different falls-off, not fully clarifying the capital assets interplay, when one explores the spendable riches creation.

This chapter section, then, aims at reinterpreting the market staples, which, in the past, made possible the wealth accumulation, and which, presumably, are to be addressed, if we try to save the life-quality level achieved by the today affluent society. The analysis is, of course, affected by the industrialism and the connected potential, still looks towards more general prospects, always, allowing the place to the socio-political contexts of the technical and economical trends.

The lessons from the mankind history show that considerable steps ahead in the wealth build-up are the issue of two «revolutions»: the agricultural, at the very beginning of the civilisation; the industrial, here especially considered. The need of *revolutions* comes out, when the current life-quality dramatically crashes into insurmountable obstacles, and breakthrough paradigm shifts are crucial, to create alternative ways out. The agricultural revolution was compulsory achievement to get free from the opportunistic economy, where the people was looking forward exclusively to *naturally* produced foods and stuffs. The industrial revolution was vital option to establish growth methods, which exploit transformations, on the whole, *artificially* mastered. This started the merry time of the affluent society, which clashes against the earth raw materials shortage and the polluting effects of the men activity. The environment is progressively modified, and the generations to come will receive a downgraded heritage as for the «natural capital».

The agricultural and industrial «revolutions» are, both, based on deep changes in the *natural* order: the farming and breeding drastically alter the domains of the living beings; the manufacturing effectiveness radically modifies the time cycles of the environmental processes. This leads to discern the two «revolutions»: the former draws up on renewable sources; the latter profits of non renewable stocks. The situation is even more intricate, if we look at the man-driven changes from an evolutionism viewpoint. Today, wide consent exist on the conjecture that «life», «intelligence», etc., are *natural* phenomena, to be explained by the conjunction of ordinary occurrences, giving off tailored issues, having probabilistic *a posteriori* evidence. The man, itself, is the result of the *natural* evolution, and caution needs to apply, not to alter the on-going order stability. By themselves, the revolutions, *artificially* promoted by the man, contradict the ordinary evolutionism, except if no drifts follows (but a «revolution», not affecting the earth order, has negligible end benefits).

Now, the end of the affluent society, promoted by the industrialism we know, is recurrent fear of today economical analyses. The mentioned three main facts are there to remind that «revolutions», with negligible end benefits, will not open any hope to transmit our wealthy lifestyle to the generations to come. The section tries to look at the mankind history, to understand, whether alternatives might be devised, or, forcedly, the population, consumption and pollution need to stop and, the life-quality, in the shortest time, shall forcedly decrease.

1.2.1. Agricultural, Industrial and Cognitive Staples

The mankind aspiration towards "utopia" is inborn effort to look at enduring wellbeing, without the ceaseless struggle to survive. The objective is reached, if sufficient wealth is made available, so that every person has the chance to obtain food, goods and lodging above the bare minimum to stay alive, allowing to plan out the satisfaction of extra whims. The early step corresponds to the *agricultural revolution*, when the world inhabitants fought to get rid from the *opportunistic* economy, by the farming and breeding activity, to seclude sets of riches, for the benefits of given communities and individuals.

This archaic revolution unlocks the benefits of *transformation* economy, say, the added opportunities *artificially* created by the men activity, by which the basic necessities are satisfied out of what is spontaneous nature generation. The option is the result of specific «technical capital» assets, timely included in the human heritage, and transmitted to the subsequent generations. It is, also, the origin of the segmentation of the earth land, to protect the tilled territory from wild animals and from other people. The agrarian option, quite obviously, bring to distinguish the world populations into:

- sedentary villages, fast aiming at conservative social establishments;
- nomadic tribes, mostly requiring authoritarian hierarchic institutions.

To large extent, geographic and climatic reasons motivate the division, with noticeable outcomes in the Chinese empire and Mongolian hordes, otherwise in the Roman empire and Barbarian waves. The partition, however, distinguishes on several aspects, especially when the more established communities aims at the harmonic integration of the individuals into self-sufficient rural societies, where the foreigner should be assimilated, or when separate populations are assembled, having subsidiary abilities and products to be exchanged for better overall profit. The related establishments, accordingly, differ in many aspects. In the former case, the focus is on the territorial management: from the uniform fairness in the political decisions, to the effective regulation of the shared civil, hydraulic, etc. infrastructures. In the latter case, the trade is big concern: from individuals' rights and definition of ruling laws, to the attention on the transport and communication efficiency.

The last hints are just schematic picture of conservative social establishments. Further insight is obtained from what is now known as networking theory. In fact:

- when local hubs, strongly connected to the neighbouring nodes (clusters), are further liked by long-distance each-other (weak) bridges, the resulting set-ups assure intermediation by *representative* ruling;
- when the locally strongly connected nodes' communities are linked each-other by some long-distance (or weak) bridges, the resulting set-ups add options to (spread out) cohesive *monolithic* societies.

The latter net has locally well-structured patterns, with the scattered random bridges, to provide the over-all connectivity. The former net is based on suitably chosen links among hubs, which grant the local areas service of the autonomous clusters.

The long-distance (weak) links assure the connectivity, allowing character to the establishment. The ordered behaviour is given by the bonds assuring the social commitment, provided by the embedded privileges of the integrated structures. The distributed (uniformly tied) arrays are presided over by the representatives of the central government. The local clusters defer the weak connectivity to the hubs, enabling the multiple-centric setting. Both the establishments generate the "small-world" effect, which permits to connect one node to another, through low number of steps (the "six passes to target" paradigm), employing the weak links, to travel the long distances (out of the characterising community order).

The spread out array patterns chiefly identify many natural phenomena, such as: the brain connections, people relational nets, epidemic outbreaks, etc., with the threshold condition, below which the linkage does not propagate. The cluster/hub patterns are common in many human settings: telephone, electricity, internet, etc. distribution, foodstuffs/consumables logistics, airports placing, or many derived mappings: world wide web links, wealth allocation between citizens, and so on.

The monolithic set-ups seem to come out from a top-down process, devised to manage stability, with resort to longer paths or makeshift bridges. These assure connectivity as added option, while the local aggregation among equals makes easy to surrogate the function of the neighbouring nodes. The hub and cluster set-up is, mostly, given by a bottom-up process, getting high functioning by minimal redundancy. The two arrangements differ under external attacks:

- the random hits at once downgrade the distributed array performance, but the collapse is easily avoided by creating a new weak link; on the contrary, they slightly affect the representative nets, unless a hub is touched;
- the systematic assault reverses the situation: the net collapses if the hub function cannot be replaced by simple measures, easily enabled, on the contrary, whether the spread out govern rules the connectivity.

Leanness and reliability are distinctive features. On these facts, the societies, coming out from the agricultural revolution, differently organise, if prominence is on the widespread connectivity of self-sufficient communities, or if (self-ruled) cities, districts or countries, need vital goods from abroad. In both situations, the pre-set establishments are compelled to face the external assails of less wealthy populations, attracted from the cumulated riches. Here again, the differences are evident. The monolithic societies can incorporate the assailants, keeping stable the establishment (Mongol dynasties in China). The multiple-centric settings undergo break-up changes, chiefly leading to sectional or layered aggregations, depending on national or territorial spurs (Europe countries, with Celtic, Latin, German, etc. descents and languages).

The economic systems of the agricultural revolution characterise the mankind wealth build up, over the (documented history) four thousands years, and before, providing clear sign of the man-driven *artificial* transformations advantages either drawbacks affecting the surroundings. The remark implies that the agrarian ages are differently valued by the *monolithic* societies, having uniform return and self-reliant produce, or by the multi-centred establishments, with ruling hubs aiming at active compensation and trading business. The latter lay-out, much more than the former, is urged to find out new lifestyles, able to lower the critical dependence of the wellbeing on the current land herds and crops.

Quite obviously, hence, the industrial age happened to be firstly explored and enabled by the European countries, with, again, especial drive by the nation-state capitalism, which could resort to adequate levels of the technical, financial and human capitals, in default of natural capital. The industrial revolution exemplary picture refers to the British empire progression and spreading out, achieving the paradigmatic issues in the "*long global assent*", and the related paradigms of the *free-market*, to assure the wide enough trade-area for the growing quantities of manufactured goods.

The (unlimited) exploitation of the earth raw materials is the revolutionary up-grading. The staples are in the tangibles manufactured goods, sold at low price, because of the industrial transform efficiency, and promptly made available as mass products. The man, for the first time, is not tied up to unpredictable natural changes and to harmful weather variations. He can programme the desired output, and obtain it with little margin of error, basically with no trouble coming from the outer physical cycles. Winning fact is the availability, in large quantity and little cost, of *artificial energy*, joined to reinterpretation of the man labour, according to the minimalism of the *scientific* work reductionism.

The industrial age is confined to persist, at the most, three hundred years (as compared with the agrarian age many thousands). The schema of picking up the earth stocks, and to transform all of them into waste and pollution, cannot remain suitable for long, as we live in finite surroundings and the restoration/reclamation processes take quite longer times, weighed against the industrial efficiency. The ecologism ideas clearly identify the industrialism carry out as non durable growth prospect, requiring drastic upturns, not to bring the humanity to definite end.

The ecologism, certainly, moves from properly assessed facts. It, nonetheless, does not suggest effective alternatives to the industrialism. At the present state-of-the-art, we cannot say if safe alternatives exist, assuring protracted growth for the times to come. From the physics laws, quite the opposite, the entropy principle gives for sure that the universe has an end. Then, the challenge is to preserve the human habitat at the life-consistent levels, for as long as possible, lowering the downgrading trends and looking after *artificial* recovery transformations.

The devised upsurge is searched in the outcomes of the *artificial intelligence*, permitted by the cognitive breakthrough to come, primarily, based on replacing tangible goods and factories by intangible functions and facilities. This change in the market staples is, most clearly, not sufficient. In fact, the industrial economy did not eliminate the farm produce, and the foodstuffs are permanent demands. Similarly, the ICT deliveries will not abolish the manufactured products, and the materials supports are perpetual request. This shows that we cannot get rid of the entropy decay, and that waste and pollution are endless threat.

The cognitive breakthrough, most likely, shall evolve, looking at regenerative processes, repeating, to some extent, the ideas of the agricultural revolution, since its beginning engaged in the efficiency upgrading of the natural phenomena, by means of *artificial* farming and breeding. The new challenge might resort to the bio-mimicry, namely on exploring *artificial* life phenomena, to efficiently create the build-up of novel resources, equivalent to the earth accumulations of the raw materials, and to re-establish the environmental fitting-out, with safe remediation of the contaminated areas. The cognitive breakthrough (hypothesised) scenario is still, largely, unknown in terms of prediction and feasibility, and the fear against totally new *artificial* developments is, besides, promoting conservative reactions, motivated through the «precaution principle».

1.2.2. Land Produce and Renewable Stuffs Provision

The old agrarian revolution allowed the utmost changes in the men progress: it certainly happened in long periods and with important worldwide falls-off. The overall issues are, nonetheless, quite impressive: with savage wolfs transformed into tame dogs, or wild lands, into productive farms. The stirring spirit and talent come from the proactive attitude of individuals, which trust their capabilities to perform clever choices and to improve the event course. The canalising of rivers, the preparation of lands, the selection of seeds, the building of homes and fences, etc. all these actions drastically modified the original environment, and the result was in the progressive build-up of the man-centred habitat.

In other words, the civilisation exploit «technical» capital to interfere with the «natural» equilibriums of the earth without man, by means of *artificial* inventions, which modify previously existing trends, putting the earth surroundings on the way to assure larger amounts of resources, with resort to the cleverer exploitation of living beings, properly tilled and domesticated. At the same time, all harmful vegetables and dangerous animals are removed and, possibly, eradicated, to make the territory friendlier.

The man perturbations result winning. In contrast, the precaution principle, if limited to innovation refusal, would mean to reject ploughing and sowing, to not increase the humus erosion, and to trust birds or wind actions only, with the hope into floods, to add fruitful slime. The *artificial* means, certainly, are a mix of low-impact improvements (irrigation, terracing, etc.), but systematically aiming at establishing the permanently anthropic order. In short, the benefits are, at times, weighed against the damages, only quoted by the pessimists, further demonising efficiency, which speeds-up the self-decline (e.g., by land salinity increasing and seams drying up).

The ecologism, more precisely, happens to be against the modern (intensive or extensive) agriculture, which employs fertilizers, biocides, irrigation, single-crop farming, etc. to enhance productivity and reliability. Such practice leads to several snags: biotypes breakout, erosion, humus rarefying, aquifers pollution, indigenous varieties disappearance, saltiness increasing, self-selection of weeds not affected by insecticides or pesticides, etc., with long terms destructive falls-off.

Besides, in the agrarian areas of the industrialised countries, the homologated, chemically-dependent farming, with standard seeds, is addressed as biasing facts. These are symptoms of the oligopolies (in chemistry or bio-sciences) profit, with damage, now, to growers and planters and, surely to, future generations. Besides, the countrymen contribute minimal fraction to the gross national revenue, with 10-15 times bigger figures in the providers' domain and growing potentialities due to the diversified interests which might quickly build up.

The alternative ecologic approach aims at the drastic impact reduction through the biologic agriculture, with *safe* produces, prised by elite buyers. As prospected outcome, the choice, basically, entails:

- actual advantages, when the eco-system damages are avoided, by suitably forbidding high impact practices, and the produce quality is guarantee by safer farming processes;
- prospect limitations, as the main issues privilege conservativeness, without answer to the demanding requests of the earth populations and expanding quality of life expectations.

To sum up, the ecologism leaves little hope to growth. With lights and shades, the agrarian tradition distinguishes from the industrial one, because it is based on renewable resources, and, (at least) on the short terms cycling, the grown produces can be iterated with suitable trust on the process smooth continuity (without abrupt rises or drops). The agrarian wealth build-up, with more or less success, remained reference of the world populations for their maintenance and deployment, out of the mercantilism limited side-help, at the time of the city-states. The change into nation-states was differently lived by the world people, leading to the industrial revolution economic trends.

The change characterises, in the average, the European countries, compared to the other populations of the world. In Europe, as well, we can distinguish several starting points, with the pioneering setting off of the United Kingdom. The most typical agrarian capitalism, on the contrary, happened to be considered in France by a group of theorists, who named themselves "economists" and are now better known as "physiocrats". The French nation-state enjoyed flourishing agriculture, which granted self-sufficiency to comparatively high quality of life, only slightly affected by the trade of foreign goods.

The physiocrats, active in the second half of the XVIII century, suggested the theoretical system, where, for the first time, the «natural» capital is supposed to play the central role. Among them, F. Quesnay (important medicine doctor) is the most prominent, while A.R.J. Turgot, the most influential, coming from a family of merchants, became prestigious civil servant and minister (France's Secretary of the Treasury, for a while).

The physiocracy hypothesises that immanent orders rule the society, governed by the «nature», which is the origin of all provisions. The agriculture is the non-replaceable supplier of wealth. The trade and industry, merely, accomplish, in the end, distribution and transformation duties, over already existing goods (the mines cultivation belongs to the natural order). The management and regulation of the community resources, especially with a view on its effectiveness, directly comes out from the natural orders and no legal/political bid can infringe them.

The economy, from the Greek *oicos nomia*, literally indicates the homeland ruling. The thriving agriculture is the only way to increase the average wealth of the nation, as the people engaged in the other activities do not contribute to the wellbeing. Incidentally, taxes on the supply chains modify the goods cost, biasing the remuneration of the farmers by fictitious withdrawals. Indeed, all taxes rest on the agriculture, and should be, coherently, perceived at the origin. The landlords and homesteaders are the leading classes, since:

- the responsibility of the country social and political development totally relapses on the land produces;
- the farming, directly remunerates the productive classes (of the growers, breeders, etc.), and, indirectly the non-productive people (of the traders, craftsmen, employees, professional individuals, civil servants, etc.);
- only after that the primary (agriculture), the secondary (transformation, distribution, etc.) and the tertiary (tax and administration, etc.) costs are covered, the net revenue (*produit net*) is obtained, and, accordingly, the positive or negative change in the nation wealth is evaluated;
- the charges compression is winning issue to guarantee the net revenue to landlords and homesteaders, otherwise the profit transfer moves to non-productive people, and the community, as an all, becomes poorer.

The focus on harvested produce and livestock unreservedly means favouring closed-loop transformations, in view to look after balanced capital assets. The all picture is, certainly, outdated, when the attention is turned to the economics, later enabled by the industrial revolution. On the socio-political side, too, the approach is bizarre, as it places the «natural» law before any positive rule, so that kings and governs shall only enact apt wordings, to better explain the requirements, binding citizens and nation, to not infringe the *natural* order.

The ordered bylaw concerns the individuals, with warrantee of property rights and trade freedom according to the aphorism «*laissez faire, laissez passer*», which opposes to the stateism and also to government licences, because the «rule of the nature», or «*physiocracy*», comes first. Each economic actor (including secondary and tertiary jobs) serves the others, and takes profit by co-operating. The nation-wide market (purchases and sales) is self-sufficient stage, with stand-alone actors, fully specified by the complementary flows of goods and payments. It is essential to establish the national accounting system, with the coherence and completeness of the family budgets.

The embedded self-regulations make possible to reach the accountability of the relational frames, which bind the human communities. The resulting "*tableau économique*" shows how commodities (raw-materials, foodstuffs, artefacts, etc.) flow from the farmers (and landlords), through merchants and craftsmen (or other unproductive people). The value chain takes back to the farmers the money, along multiple paths (under the governmental overseeing). The modern translation of the "*tableau*" leads to the cross-sectional assessment or the input-output analysis, which made the Prof. W. Leontief to receive the 1973 Nobel prize. This way, the flows of the revenue are specified, providing the exchange structure, which affect costs, prices, rates, wages, taxes, etc., and give confirmation of the cross-links that bind every economic system.

The lesson, taught by the agrarian capitalism, is summarised by two concepts:

- the priority is on the renewable stuffs provision (having the «farming» as the only productive activity);
- the closed-loop capital assets balance comes from the implicit assumption of the conditioning «natural» rules.

Both are partial views. Further staples contribute to the wealth accumulation; the «artificial» patterns can help supplying spendable riches. The first is removed by the industrial revolution, directing the productive staple on the manufacturing activity; the second is assumed to be widened by the cognitive breakthrough, with focus, back, on the «natural» capital, but now as inspiration for new *artificial* life processes. The complements, accordingly, have maximally been the result of the *artificial* energy, as compared with the workers capabilities, and are expected to be promoted by the *artificial* intelligence, in the revolution to come, through the bio-mimicry potentials, or similar innovative opportunities. The acceptation of multiple-staples alternatives, timely enabled by man-driven «revolutions», is the anthropic interpretation of the human positive growth. Based on history issues, farming and breeding are good enough ideas to establish safe staples; by itself, the «industry» effectiveness has, surely, to be recognised harmless. The idea to look after future (man-driven) revolutions shall not be *a priori* refused, being the only way to obtain improvements.

1.2.3. Earth Stocks and Transformation Challenges

The «industry» economy, while fairly new, has already undergone paradigm shifts, the most typical being the proletarian class struggle against the production means owners, and, today, the «new economy» patterns of the knowledge society, with the struggle to qualify and possess the enabling intangible background. For a better understanding, we might look a bit inside the tried out trends and accepted patterns. The word «industry» has, by itself, three meanings:

- the aggregate of the manufacture and technology production companies in a field of activity (for instance, the automotive industry);
- the enabling entity assuring structured set-up to the production means, by work organisation (aiming at, e.g., scientific job allotment);
- the abstract man's attribute, corresponding to diligence, zeal, application, etc., (in opposition with mastery, skill, expertise, etc.).

The industrial revolution consists in the organisational efficiency, permitting to enrol generic workers, so that the factory could have been seen as the beehive or anthill, with job-allotted operators, whose productivity depends on application and zeal, rather than on skill and expertise. According to that picture, the classic industrial capitalism obtains his success because of the finance investments into production means, while the labour is important only in terms of its availability. Then, the revolution leads to separate the possession of machinery, plants, fixtures and technologies, as money-supported outcomes, wholly ruled by the few owners, according to their *capitalistic* rights, with inherent remuneration.

It should be noted that the word <industry>, from the Latin «*industria*», was, in old times, connected to the third definition only. For Cato or Leonard of Vinci, it could not have been straightforward associating, to the well established human ability, the link towards the tangible aggregation of productive facilities, either the factual competence, providing structured set-up to the manufacture processes. The link is, today, obvious; the altered industrialism, centred on the men-like cognitive abilities, rather than on the materials processing effectiveness, is change requiring «knowledge» backing.

The paradigm shift is, moreover, consistent with what experimented by the recent capitalism evolution. In the earliest industrialism (of the anthill or beehive snaps), the opposition of the destitute work-forces, against the production means owners, came out, to show the elitist establishment of the new issues, when the disparity of the profit depends on the up-rising steepness. The previous mix-up of merchants and governments applies to the new leaders. The industrial capitalists request protection of the manufacture processes and technologies, by licences and patents, and guardianship of the goods, by trade-marks and logos registration. The result is the somehow odd blend between «free-market», to open broader areas to mass-produced goods, and «imperialism», to encourage and maintain the "*global assent*" under efficient leadership.

The «industry», along the modern history, deeply modifies, and quite recently, the information and communication technologies addressed the «new economy» lines. From the high-intensity financial investment structures, the change entails high-intensity qualified-knowledge organisations, where the human capital team is principal requisite. The struggle «production means *vs.* labour» vanishes, and also the stability of old-fashion enterprises, due to critical changes, principally, on two points:

- the knowledge possession and qualification of each employee;
- the complex net-organisation of the competition framework.

The «new economy» challenge builds on diverse capital assets. The *technical* one comes first, though, with intricate links with the *human* work. The *financial* one has smaller impact, due to lower investments, moreover, lessening the *natural* contribution penalty, through highly intangible value chains. The bond among the knowledge society and information infrastructure, then, suggests:

- as for the artificial side, the centrality moves to the technical capital, since the financial assets to start the business lower, while the know-how shapes up the entrepreneurial activity, both, to begin and to move on;
- as for the native side, the focus is on the human capital, with flavour on IT aids, and falls-off in supply chain visibility and in eco-impact assessment for the natural capital withdrawals (after recovery deduction).

The products, put on the market, now, have the prevailing value in intangibles, and, when one deals with the manufacture transformations, two issues emerge:

- the extensions in every deliveries towards value added services;
- the resort to multiple-duty organisations, to cover the all products lifecycle (call-back and recovery included).

The manufacture phase, critical for earlier companies effectiveness, becomes the limited operation, by which the design is accomplished, to enable the products supplying, with functions and performance, properly adapted to satisfy the buyers needs, assuring lifelong conformance-to-specification maintenance abilities, and embedded bent for take-back and recovery. The business moves along enhanced immaterial tracks. The tangible parts, the hardware, are support of the intangible ones, the software, which confer the value by the attached services: to the direct profit of the users and to the indirect benefit of third people (managing the eco-protection). Even if the business organisation rules do not largely differ from the old ones, and the value chain turns on intangibles, mainly, when the sold product-service delivery leaves apart the actual transfer of goods, to perform the provision of the (equivalent) functions, needed for the clients' satisfaction.

The multiple-duty organisation is chance of the knowledge entrepreneurship. The point represents sharp scientific divide, by respect to the reductionism, which allows to analyse the complexity into self-explaining elementary constituents and to fully understand how the whole works. The evidence shows that to assess «how individuals behave» does not explain the joint behaviour, unless the interaction is, as well, known. With the dominant cross-linking, the complex systems are better specified by synthetic features, while the analysis becomes almost useless.

The exploration of the net concerns leads to establish new facts, understanding the joint effects of strong and weak links, and of the interplay of them, out of the fixed (deterministic or stochastic) relational schemes. The heuristic modes and soft computing procedures provide insight to the groundbreaking network science, with related falls-off towards the knowledge entrepreneurship. The two keys are, perhaps, not fully perceived or underestimated, when the «industry» concept has to be moved from the *transformation efficiency*, to the men' *application zeal*.

The choice of competition factors belong to domains of eligible «knowledge», associated to the human and social capitals, and transferred in goods, which might infinitely duplicate at (almost) no additional cost, on condition of the background, widely shared by the individuals. This is, by itself, an epochal change, as nations, like China and India, can bring their citizens at highest levels, creating industrial potentials, exceeding the USA or EU, due to the number of qualified experts, not, any more, on the cumulated riches already available.

The dependence of the «new economy» business on the earth stocks cannot be avoided, but the power distribution among the world countries will modify, to the benefit of concerned communities, with the best technological deployments. The scenario can, possible, bring to trans-national organisations, shaped according to the world wide web, where the corporate frame, at least for the market staples, will represent widespread multiple-duty facilities, not biased by local links. The picture leads to the global-village, without sectional castling. The related scenario shows the wealth generation by the knowledge entrepreneurship, with change in trade effectiveness (*hyper-market*) and the value chain structure, showing that:

- the «new economy» follows the industrialism logic, assigning, however, fundamental role to the technical capital, with the human one as enabling support, more than the financial, and to the natural one, as entity under full and continuous surveillance;
- the infrastructures acquire especial value added, being the instrumental background, which guarantees shared, or, at least, accessible support to intangibles, leading to new ways of wealth build-up, by means of net-concerns and complexity commitment.

The transformation economy, after the old *agricultural* and recent *industrial* revolutions, remains standard reference, with, nevertheless, premonition signs that important changes might come out, say:

- the productive process aims at "investor/entrepreneur *vs.* worker/operator" modified relationship, with resort to work organisations, setting in shadow the "financial *vs.* human" capital struggle;
- the «new economy» puts in the foreground «new» contrasts, such as: the "knowledge/technology *vs.* personal expertise", say, the distinction of the technical capital by respect to inventors and users;
- the new work organisation patterns, through the economy of scope (in lieu of scale), with resort to «complexity», instead of reductionism;
- the acceptation of the industrial staples gives little chance to sustainability, unless that the eco-conservativeness alarm becomes evident urgency, due to the supply chain visibility.

The industrialism, out of the premonitions, operates in every area, old or new. The emphasis on the «new economy» means providing examples, where changes give insight on the chance allowing the human capital to the trade of knowledge and functions. For the first time, it does not explicitly require, at least as inherent condition, the «natural capital» wear and tear essentiality. The emerging staples might become intangible provisions, having, little, need of material supports for the productive and trading operations. The «to de-materialise» track is useful aid, to turn the earth stocks transforms, into more conservative practices, as the wealth build-up, mostly, resorts to knowledge-driven value chains.

1.2.4. Bio-World and Artificial Life Exploration

The industrialism, along the original paradigms to transform raw materials in waste and pollution, cannot assure sustainable growth. The latest opportunities, opened by the «new economy», according to the just recalled «to de-materialise» axiom, possibly lead to more conservative staples, but cannot remove the intrinsic downgrading, necessarily at hand, each time (inanimate) material transformations are accomplished. The paradigm shift, from the anthill or beehive community, to the knowledge and technology society, merely, permits to attune the value chain with higher intangible addition, to the all necessary corporeal provisions.

On these facts, the affluent civilisation has little respite, and the thrifty culture comes in, to deal with over-population, over-consumption and over-pollution. The way-out looks after recovery, reuse and recycle practices, to lower the withdrawal from the earth stocks and to limit wasting and polluting outputs. The thriftiness is, here, prised as virtuous conservatism with scientific rigour and moral strictness of avoiding all non standard traditions and refusing non-fully acknowledged issues, to not infringe the precaution principle.

Whether the knowledge society «to de-materialise» axiom is understood not sufficient, the «to re-materialise» axiom shall be devised, to try building back the original natural capital, for amounts equal (of larger) of the used earth stocks. The idea is to look at circulatory economy, where the *artificial* processes continuously move back to the original assets. This means establishing the stable bookkeeping of the involved resources, to measure the trend steadiness, either the decay slope or the growth slant. The scientific rigour requires establishing:

- materials balance, now: "nothing creates and nothing destroys, everything evolves", and the bookkeeping shall classify the items, each time, at hand;
- energy balance, the earth is an almost closed system, still limited exchange exists, and the bookkeeping has specially to assess the sun potentials;

- entropy balance, the physics teaches us that our world, necessarily, moves towards increasing entropy, and the bookkeeping shall register the slope.

The picture might seem tricky, and the (just discussed) industrialism changes provide further insight:

- the old shop-floor effectiveness takes benefit from the work-force zeal, if, however, the provision of huge *artificial* energy is granted;
- the «new economy» advance exploits to the best the operators know-how, if, however, the adequate *artificial* intelligence is available.

The value chain build-up, accordingly, comes out by switching from *artificial* energy, to *artificial* intelligence. Both are instrumental enablers, assuring passive support to the industrial transformation efficiency. Now, the *artificial* energy has primarily been obtained from fossil seams, and, typically, resort to non-renewable source domains. The *artificial* intelligence, on the contrary, basically, operates in nearly intangible domains, because the hardware-software pair permits the «new economy» deployments.

We can make one bound back at the man history, noting that the *agricultural* revolution affects to the animate world, discovering more efficient transformation processes; the *industrial* revolution involves the inanimate earth resources, aiming at enhanced efficiency by ordered resort to *artificial* energy. The «new economy» further addresses the ordered support of *artificial* intelligence, adding the benefits of the knowledge entrepreneurship, along the «to de-materialise» way. The, quite obvious, idea is to equally try the «to re-materialise» way, to take into account the specific cognition properties of the living beings, through bio-mimicry, fostering the resources reinstatement, by the life processes emulation.

The idea means coming back to the *agricultural* revolution methods, acting on the animate world, and, this time, to discover efficient *artificial* life processes, in view of re-establishing the natural earth stocks by comparable raw materials, and of reclaiming the polluted surroundings at the living being safe levels. The way is new, but can be explored by the knowledge entrepreneurship skill, with detailed visibility on the pace-wise achieved results. The precaution principle is deemed to be serious impediment, since no shared consensus already exists on the scientific explanation of the *natural* phenomena such as the «life» (and «intelligence»).

The empirical data show that the «life», at its lowest level, generates ordered bio-topes, combining materials and energy into structured way, and progressively incorporates further materials and energy to survive. The *ordering* ability, brought forth by the «generating code» of the self-reproductive potential, is the *natural* means for the procurement of foodstuffs, which assures the alimentary chain, to nourish the further life expansion. On the earth, the chain (possibly) counteracts the entropy decay, achieving the energy balance, by exploiting outer inputs (say, the sun radiations, for the chlorophyll synthesis). The bio-world shall be explored as potentially conservative solution.

Then, the systematic resort to the life «ordering» property offers options for the growth sustainability, through the industrial exploitation of bio-mimicry, thus, enabling the organisational effectiveness, into the physically consistent space. The living entities require energy to survive and grow: the generating code exists with its (living) support; and the «self-reproductive potential» vanishes with the death. The separation of codes and supports is feasible, and the bio-mimicry can lead to processes with the *artificial* bio-world productivity, perhaps, out of the terrestrial entropy increase, thus, little affecting the earth reserves, whether energy/material provisioning comes from the outer space.

Should the cognitive revolution develop, the sustainable growth is equivalent to forbid the raw materials withdrawals that exceed quotas, physiologically equal (out of the reverse logistics recovery) to the bio-mimicry stimulated generation, in view to reach the neutral yield from the natural capital. The approach is feared, as the bio-mimicry way, to usefully contribute to the growth, presumes intensive and extensive deployments, deeply modifying the natural rate, beyond cautious trends, which protects from, voluntarily or not, side-effects.

The «to re-materialise» by the bio-mimicry way, on such facts, promotes the new revolution (after the *agricultural* and the *industrial* we know), based on the biology-inspired genetic engineering, originating the revolutionary paths, which exploit the *artificial* living entities self-generating abilities. When looking at the civilisation's *artificial* sides, one is deemed to devise trivial truths. The *artificial* intelligence of the knowledge society is sound reference, after the *artificial* energy of the old manufacture market; it switches on (mostly) intangibles, after centuries of wealth built on drawing on material stocks, with no restoring. The *artificial* life of the genetic engineering might recover full safety, if duly incorporated into the *cognitive* revolution, say, into «bio-mimicry» transforms, wholly transparent, and suitably ruled by the «robot age» *artificial* intelligence.

The people's reaction in front of *artificial* energy is, generally, without fear or strong opposition, even if the entropy decay is assessed law. The conflict against *artificial* intelligence is quite limited, making the «robot age» accepted outcome. The panic face to the *artificial* life has to be explained. The «intelligence» and the «life» are, both, impressive anomalies, in the *natural* evolution. The trend towards «intelligence» is highly incredible, when we note that the *nature* went along, e.g., hundreds parallel ways to electromagnetic waves sensors (eyes), and all stopped at image recognition, with no attempt to create and up-keep relational data-bases. The stumpy «intelligence», as the ability to single out and recognise given pieces of information (sensorial data) into a context is, in fact, shared by many animals, allowing 'educated' behaviours in front of repeated stimuli: several «intelligence» levels exist, and the living being classify at many «life» orders. The *artificial* life does not entail man prerogatives, simply the *self-reproductive* ability; equally, the *artificial* intelligence is admitted non-human privilege; and, factually, typify as:

- ability to build, correct and up-date images (*inner model*) of the world;
- ability to devise evolution forecasts, by *simulation* with the inner model;
- ability to select and order the information, according to axiomatic *laws*;
- ability to decide *learning patterns*, consistent with inner model and laws;
- ability to acknowledge the learning progress, by conscious *introspection*.

We might add a few remarks. The factual proof of human-like «intelligence» is, today, assessed by the Alan Turing test, checking the *<knowledge structuring ability, in view to answer questions on the fundamental physics laws, theorems of mathematics, or other cultural branch, requiring that the sets of propositions are 'creative', under human standards>*. The cognitive faculty is roughly equivalent to «intelligence», understanding by that the capability of recognising the sensorial data and classifying them to extract regularities, cross-relations, inconsistencies, etc., into ordered bases. Thus, the human «intelligence» is singular ability, with the original side-effect of building-up «knowledge». Today, we have no evidence to be more «intelligent» of known history (say, of Aristotle) people, but we surely possess much wider scientific culture and technological know-how.

The divide between the *industrial* and the *cognitive* (or other naming, which might be chosen) revolutions is foreseen to be ruled by the *artificial* intelligence, and the «new economy» explains the smooth transition towards companies, which will exploit *artificial* biological processes (by bio-mimicry, etc.), involved in the «to re-materialise» paradigms for alternative resources piling-up. Different views, of course, are possible. Here the continuity, made easy by the «to de-materialise» paradigms, is especially privileged.

The *artificial* intelligence role is, at the outset, prised for its instrumental aid in warranting the processes transparency. The *artificial* complements of every *natural* fact characterise, because they are promoted and governed by the man. The *artificial* energy is used not due to similarities or intensities (tornado or earth-quakes have strikingly odd character), but because of their regulated availability. The *artificial* intelligence, similarly, does not pretend to repeat human reasoning, but, by digital mock-up and processing, provides helpful influential backing. The *artificial* life processes, accordingly, shall be finalised to the «to re-materialise» scopes, according to the sustainable growth requirements.

The *cognitive* revolution looks at innovation, as necessary option for materials recovery, out of energy provision and intelligence exploitation, without the three of which, no future will develop. The bio-informatics technologies carry out life phenomena duplication, at the different levels of bio-mimicry, up to emulate vital function and brain steering. The «to re-materialise» and «to de-materialise» paths join, by duplicated genetic codes, enhancing the resource backlog effectiveness, more or less, as it takes place in the abstract domains of the knowledge build up, but now turned in the tangible fields of bio-mimicry carriers. The knowledge is, for sure, devoid of risk; it is, moreover, granting tangible processes transparency, instrumental means to make the "better safe than sorrow" into the spur to operate, when the "do nothing" simply means to fall into ruin.

1.2.5. The Market Law in Parliamentary Democracies

The industrialism comes out from quite typical technological and economical motivations, but not less significant are the political and legal reasons. The issues are related to the notable structure of capitalism, where the "*free-market*" and the "*economical global assent*" are relevant options. More precisely, these concepts historically refer to the exceptional deployment of the British Empire, and to the subsequent and related developments promoted by other European countries and, in general, by the so called western world, to in addition include the recent "*short global assent*" ruled by the USA. Thereafter, the industrialism can be viewed as specific wealth progress, rooted into properly defined national backgrounds, while the mere growth of industrial factories in other countries, notably where the full or partial stateism dominates, happened to create restrained outcomes, markedly, on the longer periods. The passed history shows the industrialism characters in:

- the reliability and quality of the country's political framework;
- the entrepreneurship competition level and worldwide orientation;
- the nation's successfulness to keep steady macro-economics course.

Theses facts have shared consent, even when other orders are given. The first, yet, makes key gist, as, without it, the micro- and macro-economics interplay adds little advantages. The citizens' rights need clear and large progress and protection frames, so that each individual has the certainty that his activity and achievements will not be object of the authorities' summary restriction or arbitrary control. The law sureness ought to cover the different civil claims, not to leave privileges that might wobble, depending on unexpected legal action.

The private ownership rights need special mention. Only if reliably protected, the competition between citizens grants each one the individually built-up fortune, and, thereafter, the benefits to the society, brought on acting for his profit. If the sphere is fuzzy or equivocal, especially where the public morals leans towards the stateism or collective ownership, the industrialism, inescapably, transform into bureaucratic accomplishments, with poor checks on returns, having actual value.

In the western world, the private ownership is facet of the public morals, with the tutelage of the parliamentary democracy frameworks. For sure, the protection is different, with graduations depending on the less fortunate solidarity measures and the governmental stipulations. Hong Kong and Singapore, for instance, have the "Common law" protection of the British heritage, under otherwise modified political frames. The continental European countries aim at larger social transfers through progressive taxation, as compared with the UK and USA habits.

The private ownership safeguard extent is important prerequisite; its stable ruling is even more relevant. The citizens shall have fair and abiding counterparts, in authorities and contentious litigation. The lawsuits, for sure, profit of goodwill and care of public opponents and ruling governance, and the court cases equality, reliability and efficiency are fundamental requirement, in every sphere concerning the citizens competing in the fair market. At fault, biased trends begin, affecting the growth safe achievements, because the relevant part of the human effort has to be spent to avoid mistaken pitfalls, not in productive work.

The competition *vs.* solidarity figures are big impasse in modern-days politics, as the *free market* is viewed as the conflict way of the wealthy minorities to up-grade their position, with damage of the less fortunate majorities. The aggressive side of the 'capitalist', then, has to be sanctioned, to the 'proletarians' safeguard. Nobody denies the industrialism effectiveness, but, even before recognising the sustainable growth limits, the measures towards suitable regulations are, today, standard practice. Their falls-off in the micro-economics domains affect the single citizen's freedom, depending on different juridical restrictions.

The national loyalty is the strength of democracy, enjoying shared consent of govern and opposition forces. The democracy is weak, if the national idea is frail or inadequate, because, without the national loyalty, the opponents are unending threat. The European countries derive that necessity from the legal constructs, say, the positive terms of the Roman law or Napoleon's code, or the immanent rules of the Common law. The sovereignty is enduringly subject to the voters' check and confirmation.

The safe growth is innately tied to the pair: ruler's responsibility and citizen's rights. The Latin idiom: *"quis custodiet illos custodes"*, faces the permanent need to control the controllers, the Govern Secretaries and Court Judges. The authority merits trust and respect, depending on the awareness and responsiveness, kept in its every day behaviours. Now, the liability does not create declaring or enacting its existence, rather only if the citizens can actively control and ratify. This entails watching over the authorities, to not accept drifts from alert civic mindedness.

The civilians' society enjoys political freedom, not because that right exists, rather because the citizen's rights are real, with the cross-linked duties, binding leaders with individuals, according to the common jurisdiction. This is why one shall distrust universal principles, to be accepted without discussion and control, as whether superior truths have mandatory consequences, not to be validated. The law and the justice are not due to governments or judges, but to the citizens shared will and national loyalty. The fact is big drawback, when aiming at supra-national authorities, for managing the *global village* fortune, and the related concepts, such as *hyper-democracy*, or the likes, require the careful duties balancing, to preserve the right control on the central authorities.

The supply and demand market law is, likewise, factual issue of the citizen's free will, once the ownership rights are acknowledged. Several contrary opinions bring, today, in question the individuals' rights, assuming the priority on highly imperative community's rights. The trend is towards transferring the authority to some independent judicial board and/or supra-national committee, freed from any democratic authority. In fact, the trouble is how to protect the future generations' rights, when today governments respond to current polls, where urgent demands continuously arise. The «first in, first served» rule lawfulness is questionable; as it expropriates the future people from enjoying the available «natural capital», since the all is fully exploited.

The trend is clearly evident, when looking at the many international agencies and organisations (ONU, WTO, ILO, etc.), and, in Europe, to the UE, ruled by European Commission, without direct connection with the controlling voters. The idea of worldwide governments is accepted aspiration of a peaceful future, after the continuous wars having affected the humanity. The UE, likely, interprets the said aim of the European citizens, which start figuring out the shared loyalty at the broader continental level. The situation is, however, at poor stage, with many national rulers not accepting to be moved out, and the revival of patriotic spirits at small or large peripheral ranges.

The market law in the EU context, then, is facing fairly intricate perceptions. The dependence on the national governments meets rightful democratic control; the common economical area is ruled through mandatory agreements, enacted by the supra-national committee; the outer trade regulation refers to loosely steered precepts, further weakened by the unlike taxing prospects. On that state of affairs, the fundamental guidelines can basically provide rough information on the actual wealth progress figures of the industrialised countries, and the past history elapsed characteristics are far to be helpful, particularly, when the environmental policy is tackled, in view of the growth sustainability, without the today voters control. The point needs ticklish concern, possibly, to be answered looking at mankind loyalty, extended to cover the future generations safeguard.

1.2.6. Governmental Regulation of Citizens Welfare

The formerly recalled political framework trends, out of the list of reasons or, better, the three main facts recalled closing the first section of the chapter, show that the growth through the traditional industrialism cannot repeat any more. The alternatives are, perhaps, chances, or dreams, to come, included the just sketched *cognitive* revolution, under the yet-to-be *hyper-democracy*. The lessons from the history, moreover, need to deal with the macro-economics regulation courses, to put in clear the further points that require revising the freedom of the individuals, face to the ownership of «*natural* capital» pieces.

Up now, the class struggle has required to modify the old industrial capitalism *free market* concept, for protecting the low-classes quality of life, taxing the upper (and middle) classes. Alternatively, the communism happened to be the main XX century challenge, with state industrialism, managed by the party "nomenclature". The common citizens' rights expropriation, turned into want of competition, due to the individual merits. The single person, maybe, enjoys shared comfort; the all country suffers, in the average, modest profits, under otherwise favourable trends. Under adverse affairs, the Nobel prize Amartya Sen observed that the dramatic famine affected the totalitarianisms, while no democracies faced heavy shortages, since enough information was available as warning caution.

The parliamentary democracy path, to the lower classes protection, followed the mixed-mode approach to enact macro-economics regulation, supporting at the micro-economics range the free-market and private ownership. The western world solution is consistent with the nation-state capitalism, as above recalled, plainly switching, from the "*long*", to the "*short global assent*", to wider national frames, when the world economical and political contexts required bigger backing. The nation's successfulness to keep steady macro-economics course has been taken, up now, as prised indicator of the industrialism successfulness.

The macro-economics planning shall not affect the micro-economics actors, in such a way to up-grade the nation social frame, not the enterprises interplay. The governmental steering, in other words, needs to be objective, having documented fairness, also, for the generations to come. This last condition, however, is open requirement, difficult to obtain the consent of current voters, because immediately conflicting with short terms advantages.

The planning to keep steady macro-economics policy becomes uneasy, if the world trade organisation requests the fair interplay among the involved partners. It is impossible for the EU countries, bounded by the common market treaty. More precisely, the EU macro-economics policy is singled out, leaving out the citizens' taxation and the public indebtedness to be *subsidiary* object of national rules, as if the biasing effects could be neglected.

The setting of supra-national agencies and rulers, however, is more and more necessary, to refrain from conflicts and wars. In the same time, the *global* cross-linking of bigger world areas makes evident that the function of the single nation-states at the European level is politically and economically out-dated. The obvious limits of the in progress EU *subsidiarity* policy bring to several non-senses, when the industrialism is compared with the previously mentioned three requisites, face to the shadowy EU political frame, to the biased sectional competition, and to the non-locally fitted macro-economics figures.

Besides, the question is just partially related with the supra-national inaptness, and dramatically dependent on the nation-size unsuitability to cope with the new challenges of the *global, no-global, post-global* trend. The ecologic *vs.* economic dilemma shows that the humanity is forced to survive in the shared *global village*, and segmentation of privileged areas shall, in any case, take into account that the earth resources are finite, and need to be shared. We have to deal with the *natural* capital bookkeeping, because the stocks are common wealth, not to be personally spoiled, without repaying for the enjoyed benefits.

The segregation of (small or large) natural capital pieces, as total property of a single person, community or nation, means suppressing that the other individuals, communities or nations (present and future) could profit from them. This entails dramatic changes on the ownership rights, with the twofold limit:

- the individual possession of tangible products or buildings, now identified with the (timely or lifelong) allocation of natural capital pieces, subject to specified sets of «on-duty» and «end-of-life» conditions;
- the personal responsibility of the use or misuse of the allocated resources, with the detailed bookkeeping of the tangible flow, in view of providing visibility to the generated consumption and pollution.

The old nation-state size is totally inconsistent, opposed to worldwide wants. The trans-national agreements are preliminary step, with, however, the request of the supra-national enforced control. The first face of the mentioned limit, perhaps, is managed by tuning the ownership rights, making in progress explicit the further sets of conditions. The second face is much more severe. The conflict makes up, when the «privacy» limits to «sensible data», related to the personal aspects. The duty entails the control on every material goods, entering for any title or reason in private enjoyment or possession, for whatever affects the eco-management.

The data collection, inevitably, infringes the personal sphere, recognising and recording every individual habit, having impact on the surroundings. To neglect the monitoring would mean to omit assessing the consumption/polluting ratios; vice versa, to accomplish it brings in several sensible information that need to be kept confidential and not transmitted as disaggregated data, potentially, referred to the individuals.

The regulation of the «privacy», today, is object of contrasting positions, with the approaches varying from country to country, and fast changing views (also for other reasons, such as the fight against the terrorism), and rapid technological innovation. The socio-economic and politico-legal frames, moreover, move for wider intrusion: the «privacy» cannot be invoked on the financial revenues, to not pay the taxes; similarly, could not be opposed for the ecologic impacts, to not refund consumption and pollution. The safeguard of the privacy, thus, does not mean prohibition of data acquisition, processing and storing, rather it deals with managing and scattering the information, assuring reserved transparency, once visibility on the synthetic results is provided, by the involved certification bodies, to the concerned government authorities.

The knowledge society, for sure, is offering the widened infrastructures of the «new economy» business. In terms of privacy, while not simple, we have now to separately address:

- the civil (political, religious, etc.) individual rights,
- the social (economic, ecologic, etc.) personal rights.

Moreover, without going into details, one shall acknowledge that the most of the social rights are already object of wide regulations, and the knowledge society has the potential of managing every aspect, with the proper «privacy» protection.

The government regulation of the public welfare started from the mere macro-economics balancing is, in progress, expanding to the sustainable growth ruling, incorporating the ecologic accomplishments at the *global village* range. The *free market* concept becomes more and more a myth, for the series of biases at every, from local to global, from private to public, levels, being easy to list the series of conditions that have to be fulfilled for the trading of any product. The scenarios to come are, further, discussed in the following, up to devising the *solidarity* setting, in the *global village*, where the chief benefits are made covering the not yet born citizens of the *hyper-democracy* world administration to come. The change might appear as sharp discontinuance in the capitalism of the market-driven societies; it is, nonetheless, the sign of rational behaviour, looking for lawful settlements, to safeguard third peoples (other world citizens and future generations). In fact, the *altruism* simply means to protect the mankind life, out of the appropriation of the first arriving robbers, no mind if the stocks are destroyed.

1.3. MODES IN THE KNOWLEDGE CHALLENGE

The «industry», in the views up now given, reaches the full effectiveness by the rather peculiar capitalism, where some socio-political sceneries (*free-market, economic global assent*, etc.) could join with special scientific and technological knowledge, to promote what is now named the *industrial* revolution. The overall phenomenon is short parenthesis in the history of the mankind, and many reasons lead to the conjecture that a breakthrough is necessary, to keep the growth going on. The past trends analysis is deemed crucial to understand, both, the grounds of the *industrial* revolution and of the overall advances due to *artificial* means.

Now, the industrialism we know is recognised to be faulty in terms of entropy, energy and materials balance. With the entropy, if we believe in the physics laws, there is little to do. With the energy, we already trust in outer *natural* sources (the chlorophyll synthesis, etc.) and hope in *artificial* occasions. With the materials, the reverse logistics provides very little advantages, and we shall look after totally new processes, e.g., the *artificial* life enabled by bio-mimicry. This motivates the parallelism between the *artificial* energy of the *industrial* revolution and the new *artificial* intelligence trends of the *breakthrough* to come.

We may infer that the culture-driven opportunities are the only possible way-out to devise growth sustainability. The study has, then, to look at the knowledge set-ups, capable to trigger off and to feed the desirable *breakthrough*. No certainty exists about prospects on the physical world, unless that we have proper trust on the models we are using to describe it, to accept equal confidence on what we are extrapolating from these models about possible future evolutions.

The last statement sounds obvious (for sure) to many persons. Apparently, this was not the case three or four centuries ago, making the *industrial* revolution, the typical western world prerogative. There is no reason to believe in the superiority of one or the other world population; on the contrary the flair exists that internal contradictions of the western societies may hinder, today, the progress to factual achievements (as it has been the case for, otherwise, advanced societies when the *industrial* revolution started).

In many spots, the principle of precaution turns to worst-case sceneries, each time depicting dark portraits of any changes, assuming "better safe than sorrow". Certainly, any action is potentially dangerous; the old warning against walking outside, because of the thunder risk, would lead to never come down a stair to not tumble: the safety margins are not zero. The (fission) nuclear plants, the chemical fertilizers and pesticides, the electro-magnetic fields, etc., have possible not fully controlled falls-off, as it happens with the genetic manipulations or 'greenhouse' effects. The psychology "when in doubt, do nothing" is high in the today western world mind, showing full reversal, as compared with the pioneering spirit of our ancestors. The conversion is, for sure, amazing.

The decision keeping habit, if risky conditions appear, is related to the cultural ranks and institutional frames of each given community or nation. The lack of the adventure spirit marks the political class aiming at the consent of the uninformed majorities, in front of potentially negative outcomes. With the knowledge society, this anti-scientific restrain is puzzling: it is surely precarious, when dealing with future populations' heritage, if the today people wipe out the all.

In the near future, the prospected *cognitive* revolution will be the prerogative of minds oriented on complexity management, rather than on reductionism ability. The focus is turned to the «intelligence»: man's *innate* gift, and *artificial* attribute, to be exploited by computer engineering. The (man) intelligence is the impressive evolutionism glitch that allows recognising and classifying the information pieces as *abstract* entities: the *data* (coded information) and the *knowledge* (structured information). What can we infer about the *artificial* intelligence developed by the individuals and communities of the *breakthrough* to come ?

The question is deferred to later discussion. In the following of the paragraph the attention is turned on the path of the *industrial* (compared with *agricultural*) revolution, as the technology-driven answer to the people development wealth, according to the western world style. The analysis tries to sketch the «cultural» framework, to offer socio-intellectual reasons of the revolution carried on within specific geographical boundaries. The picture helps to realize why the «industry» paradigms are timely concentrated phenomenon, and, in any case, the sustainable growth idea (or utopia ?) is issue of defined (and constricted) cultural background, and of connected socio-political settings.

1.3.1. European Tendencies to Reductionism

The birth of the technical knowledge is somehow oddness, being in progress generated into a restricted world area, the Europe, due to rather peculiar cultural conditions. Several interpretations can be devised, starting from the individualism, leading to the independence of actions and commitment, so that each human is the author of his own success or failure. On these premises, the man is forced to find series of principles, such as the assessment of identity, the causal attribution, the logical reasoning, etc., in view of finding out simplified necessary and sufficient conditions to oppose as fixed truths, not to be modified, thus permitting to reach the personal achievement.

The science, therefore, develops along the «reductionism» paradigms, namely, the assertion that models exist, fit to describe the universe where we live, assuring enduring relational frames. This is the same as to believe into a world, reduced to lists of (steady) properties, because the «complexity» can always be explained by «elementary» constituents. The «analysis» is powerful method, allowing to reach the roots of the each situation. Of course such scientific spirit is not enough by itself; it has to be connected with the inclination towards the «action», so that the technology build-up occurs by trial-and-error or activity-and-correction attempts.

The approach is reasonable enough, and the mankind history shows that the series of «physical laws» are efficient way to work-out the technical knowledge. The European peculiarity, according to these lines, is striking issue. The shared falls-off are not without haughtiness: the collective mindedness is properly set up, by dating the world history from the Christianity common origin. Characterising differences are a second thought, nicely expressed by the divisions into Catholic, Orthodox and Protestant churches. Then, the rational to the technological flair and successfulness permits further motivation of the common cultural background, not to be separated by the «action» paradigms.

With the standard *a posteriori* reasoning, assessing the outcomes and looking for the sources, one moves to a series of driving elements, such as the following:

- the belief in causal orders, culminating in the transcendent reliance on a single God, whose necessary and sufficient assessment is obtained by the ontological proof;
- the learning of syntactically built rhetoric schemes and communication languages, where sentences orderly distinguish names, adjectives, verbs and complements;
- the training requiring ways of life extensively based on the confrontation with competitors, so that the resort to laws is worthwhile opportunity and vital safeguard;
- the culture to the syllogism, forced by the individualism habits with socio-psychological requirements to the formal logic, as driver of communities and nations ruling.

The religious, linguistic, economical and political stimuli push to chief biasing heritages, which encourage the way towards mastering the personal fortune, even out of the recognised shared original connections. This means that the European approach to the people wealth build-up demands the concurrent existence of joint spurs at the individual/social, guidance/collaboration and cultural/planning levels, each time distinguishing the personal responsiveness.

The mixing of the recalled driving elements along the three operation levels is conveniently expanded to encompass, one by one, in the foreground, the citizens' lifelong efforts to achieve personal fulfilment and objective benefit, from, in the background, the eternal quandaries of the mankind civilisation to give a rationale to the current demands. The foreground challenges concern each people, as front end responsible of his destiny, e.g., addressing:

- the communication habits, say, the distinguishing features of the typically used languages;
- the competition standards, say, the commonly prised ways to create wealth and establishment.

The background provisos fix the funding dilemmas, whose acknowledgment distinguishes the wise men in the European, either in the Far-east Asia countries. Example permanent impasses are:

- the «being» opposed to the «becoming», chiefly, when the interpretation models are explored and exploited;
- the «dialectic» opposed to the «dialecticism», especially, when the doubt becomes non-avoidable worry;
- the «competition» opposed to the «fairness», particularly, when the socio-political factors are basic concern.

Introductory outlooks are summarised, simply to state that the western world deployment up to the industrial revolution cannot be seen as achievement with no alternative; on the contrary, it addresses quite an hazardous earth resources use, so that the reconsideration of the different impasses can only be noteworthy exercise, notably, if «complexity» has to be dealt with.

The opposition between only two cultural paradigms, as conflicting concepts, is, definitely, simplifying procedure, since the world is multicultural entity, and all other protagonists play their role, providing characterising peculiarities, which distinguish the contributions of the Moslem, Indian, South American civilizations and so on. The schematic presentation, yet, offers illuminating hints, because the alternatives are relatively sharp, and the conclusions are evident. Of course, the all is typical *reductionism* issue, and criticisms are acknowledged as wise warning.

1.3.2. Causal Plots and Communication Styles

The abilities in construction of causal reasoning patterns are characteristics of the Europeans, from the ancient Greece world, where the continuing philosophical schools, paradigmatically embodied into the Aristotle's teaching, provide the (still unrivalled) standard models. The causal analysis exploits steady rules:

- to define the problem, listing facts/ideas to be reported as 'self-sufficient' specification of all useful details;
- to declare the problem-solving theories/procedures that apply, according to the selected hypotheses;
- to describe the methods/algorithms used in the inferences, weighing up the related correctness;
- to acknowledge the results, showing their consistency with the facts/ideas, because of the assumptions;
- to discuss the outcomes, rejecting all achievable opposing deductions (as deviating digressions);
- to add comments about the exploited references and on the similar lines of proofing/reasoning.

These rules are gradually taught along the curricula of the European students, so that, for instance, it is obvious to learn the history connecting the sets of facts, to their originating causes, as if no alternative might ever occurs, being out of the logic cause-to-effects paradigm.

The approach is, certainly, worthy, when the «action» spirit is privileged, and the backward model shows the input/output portrait, to assess pieces of evidence, from already happened sequences. This might appear subjective or childish, if the «fairness» spirit is preferred, established on objectives of «harmony», where each person is seen as community's member, with no means to acquire independency. The surroundings is conditioning frame, so that the singling out of limited lists of facts/ideas cannot give insight on the actually in-progress processes. Opposite to the Aristotle's teaching, the Tao's one stresses on the «becoming» as the current attribute of the world. Thereafter, it will certainly be foolish to refer to snapshots, imagining to infer general properties, independent from the context. The chaining of instant portraits never explains the history.

Likewise, the Plato's «ideas» permit to conceive the abstract «being» from the empirical «phenomena», since, at the back, two assumptions are made:

- the universe we live in is «static», so that series of characterising features exist and, if captured, supply the *right* world description;
- the underlying layers can be separated, detaching the abstract attributes (ideas), from the local (accidental) neighbouring contexts.

Jointly, they bring to the well known sophists' paradoxes, of the arrow never reaching the target, making impossible the motion (thus, the «becoming»). They lead, as well, at the importance to classify the objects, to extract their fundamental properties. In the Taoism, the cataloguing is confusing; the ancient philosopher Zhuangzi points out that: "to classify and to restrict the knowledge prevent from achieving true wisdom". The separation of parts from the whole cannot make any sense, because the segmented entities emerge from how they fit in the temporary context: the frame is omitted, the details provide confusion.

The alternative approaches bring, either, to look after the world of objects and characterising attributes, either, to conceive the world of milieus and conditioning contexts. The second picture requires to never disregard the intrinsic *complexity*, so that the standard perception methods follow from the set of «liaison» patterns, notably, using the concord or association or alliance processes, to guarantee the community harmony. The dependence from the environment lowers the interest to establish permanent attributes, assigned to individuals and objects, as innate ideas. Besides, the focus on the complexity ties induces to never drop the frames and to enable all measures granting the best choral achievement, because all attainments belongs to the community, within which, only, the part subsists.

The two approaches are, surely, mutually exclusives, and the wise man should never follow one or the other, but shall choose in-between paths, according to the current situations. Nevertheless, in the average, the «action» spirit is deemed to characterise the European (following the Aristotle's teaching); the other side, the «harmony» mind is typical of the Chinese (subsequent to the Tao's teaching). The remark has several possible explanations, maybe, starting from the differences in the language that the people use to communicate.

The Indo-European languages are syntactically structured, using the subject-verb-object-complement lay-out, where stress is on parts' state characterisation, with series of adjectives. The Mandarin language coding orients the stress on the «becoming», favouring the attention on the frame prominence, to emphasise the relations that rule the world. The statement brings to several other observations:

- the Far-east Asia languages are basically context-driven: a phoneme has several meaning and is worthless out of the given linguistic location; the English terms are easily listed in the dictionary, with assigned definitions, since the interpretation, moved out of the sentence, is current habit;

- the Indo-European languages have the standard option to wording abstract entities, adding suited suffixes to real objects attributes (e.g., whiteness); the Chinese philosophers never theorised Plato's abstraction procedures, and no similar option exists in the language;
- the nominal sentences are usual in English, not in Chinese; the sentence: "squirrels eat nuts" distinguishes from "my squirrel is eating a nut" has no Chinese equivalent, and only the context gives evidence if, or not, general assertions are stated;
- the Chinese is a tone language, with (unchanged) monosyllabic ideograms (even if the pronunciation might vary), assembled to specify new concepts or functions (e.g., verbal sayings in lieu of verbs: *it is the Sue's eating*, to mean: *Sue eats*), by milieu-driven specification;
- the English is subject prominent language, and even impersonal sentences require the subject (*it is raining*); the Chinese is topic prominent, and the sentences, typically, start from the framework, and give the object in the defined context (*in the Italian Alps, the skiing is nice*);
- in Europe, the entity decides the action; in the Far-east Asia, the action is enabled in concert with others or is the outcome of impersonal accidents. The European speech is agentive (*he made the item to fall*); the Chinese, non-agentive (*the occurrence happened to make the item drop off*).

There is no reason to say that the language originates the habits, either that the brain fiddles with the communication styles; anyway, the children, since they start speaking, are fostered along one or the other approach.

The education and learning, in all cases, play fundamental roles in shaping the citizens' mind. The way that the Aristotle's either Confucius's ideas are preferred, becomes lifelong unconscious cultural and behavioural reference. Of course, large differences might appear at the level of the individuals, and no uniformity exists among the Chinese, Korean or Japanese natives. The same is recognised between the different European mentalities and inhabitants; the severe subject prominence in English does not exist in Italian, where the pronouns are typically omitted, and the strict syntactical order is dropped to stress on some object or complement. The modulations reflect, as well, in the current habits.

1.3.3. People Success and Business Settings

The centrality of the individual, either of the communitarian, responsiveness, is a further characterising figure. For instance, the majority of the US citizens are ready to subscribe a set of propositions, such as the following:

- each person has distinguishing marks, and is proud to be singled out from the crowd, because of these;
- each one deems to control his behaviour, and feels pleased when he leads the team and affects all others' conduct;
- the success is personal objective, and the multitude membership is feared as hazardous hindrance;
- the individual trust is main requisite, and the private records are motive for pride in the society;
- the interpersonal relations on equal terms are sought; otherwise, top-down hierarchy is appealing wish;

- the same duties/rules ought to apply to everyone, and no privilege shall be allowed out of the merit.

These, and similar other intentions, show that the personal autonomy is basic expectation. The individual believes to operate on the surroundings with «active» conduct, having the hope to positively affect his future, because of his own merit. The schematic statements bring to think up alternative human organisations:

- on one side, the «enterprises», say, institutions aimed at making easier the achievement of pre-set projects: the competition is main capacity, and the success is necessary stipulation;
- the other side, the «communities», say, people gathering because of shared identities and civic belonging: the friendly behaviour is principal glue, and the harmony fundamental prerequisite.

The USA Corporation looks at effective work-organisation, as basic step for optimal results. The Japanese company requires the workers concern, as featuring means of the business success. These recent years have shown that the western world effectiveness is not the only entrepreneurship rule; the participation govern might become winning alternative, when, in not far away future, the organisation complexity could not, any more, be tackled through the reductionism procedures. In recent years, the Toyota's intelligent task planning already profitably opposes to the Ford's scientific job allotment, and the issue is worldwide acknowledged.

The emphasis on the relation intricacy brings to say that the operators' feeling, state of mind, reactions, etc. cannot be disregarded, and is able, on the contrary, raise to the top place in future entrepreneurial settings. The statement is, here, mentioned in advance (as compared with the subsequent ideas), to suggest that lots of thoughts, obvious in our western world, might deserve revisions, so that the considerations that follow only represent temporary appraisals. Still, we ought to acknowledge that the people independence habit, shown by the above list of propositions, has granted, not only, the economical growth of the industrialised countries, but has been, as well, the winning spur of the technological advance, up now, acquired. Surely, the cause-and-effect reasoning has to be reconsidered, looking at the eco-impacts, recognising that the *today* plenty of spendable riches, with the *tomorrow* pollution and waste, is foolish (deceptive, deplorable, etc.) achievement. In other words, the truth based on snap-shots is, at least, partial.

In the western world mind, the reductionism principle is essential assumption, because of the innate belief to be able to model the material transformations into algorithmic procedures. In fact, the trust on the causal inference is well posed into the «static» universe, where the regularities that come out from the experimental testing, are transformed into the «laws of the physics». The process is supported by the «falsification rule», logically taken from the «non contradiction» principle, because a statement cannot be, at the same time, "true" and "false". The belief is equivalent to the «identity» principle, which assumes the situation coherence: "A is A, independently from the context".

The laws of the physics, in recent views, do not necessarily ascend to absolute frames, or belong to transcendent domains; nevertheless, the «falsification rule», on the recalled supports, is believed sufficient to grant their factual worthiness for all practical applications (notably, of engineers' concern). The link between the technical knowledge deployments and the people autonomic attitude towards the «action» spirit is lucid issue of backward reasoning, founded on pictures we are accustomed to use, when describing the western world successfulness during the last centuries.

On these facts, one should be interested in more helpful deepening, in view to better devise the growth sustainability and future wished scenarios. On that aims, the analysis, moved from the communication training means (education reliance), turned on the behavioural competition habits (people autonomy), ought to recall the mankind evolution, before going ahead to single out the characterising cultural and political features, which emerge as the most important aspects of the human social organisations.

The survey, according to these notes, very synthetically, brings to suggest that the mankind history characterises by turning points or «breakthroughs», when the existing life conditions cannot anymore supply enough, safe and reliable supplies for subsistence. The outlook brings to well-known divides:

- the end of the opportunistic economy (management of necessities, based on what is picked-up in the wild nature) started the «agrarian age», where the land produces, crops and livestock come of farming and breeding, due to wild vegetable cross and fierce animal taming;
- the swap to the transformation economy (management of necessities, with involvement of human labour) leads to the «industry age», with systematic exploitation of the earth stocks, where the staples are manufactured goods built from raw materials, turned in pollution and waste;
- the end of the affluent economy (wealth, no mind of non-renewable riches squander) will give rise to the «robot age», where the «artificial life» joins to «artificial intelligence» to widen the tangibles by intangible enablers, up to the *cognitive* breakthrough (bio-mimicry route).

The account shows that the «industry age» is short compared to the «agrarian age» length, and the «robot age» is epoch to come, with no certainty on how and how long it will develop. The divide between the «industry age» and the «robot age» is foretold by several warning signals, and this chapter section is looking at the main cultural options, obtained recognising possible ways out from the turned up impasses, previously listed to characterise the European growth decay. It shall be the step forward, replacing the «industry», by the «knowledge» paradigms, if we believe in the mankind future. Do we have to look for new cultural patterns ?

Now, the reductionism is typical of the «industry age», while the «robot age» is much more related with complexity preservation pictures. There is no reason to believe, today, that the switch is from the Eastern world style, to the Far East Asia habits. Anyway, the cultural advantages, whether functional, need to be judged, so that extra deepening is useful, to better devise effective sceneries. Besides, further to the cultural background, the breakthrough shall deal with the mindedness bent of the involved peoples, face to the mankind survival challenge. To such goal, the fairness spirit might be better issue than the competition style, aiming at societies' settings, where the political discussions could incorporate the future generations' benefits, as personal interest and *solidarity* spur of the today voters.

1.3.4. The Universe of States or of Transforms

The scientific approach, with explaining reasons systematically derived from general principles, is Greek concept. The unveiling of the speculative backstage was motive of high gratification; the Greek word "schole" means *happy hours*, say, the time, free from work, to be spent developing the knowledge according to the personal whims. In the Confucius' doctrine, the happiness, far from putting individual quirks before social obligations, consists in getting pleasure from the harmonic participation to the community life, in the allocated social network. The Chinese citizen is conscious to belong to his family, village and nation; then, the protective community automatically requires exact and strict obligations, while giving all ethical instructions of civil behaviour. Within the hierarchical social group, the discussion is useless, being the roles accepted, to achieve the ordered continuance of the collective life. This should not be confused with conformism; Confucius, states the prominent citizen is the one distinguishing himself by the achievement of harmonic fairness (*solidarity* spur).

The Chinese wise man was never involved into knowledge implying operation import. The masterworks, in all, much more relevant than the ones of the ancient Greco-Roman world, covered many domains from irrigation, to chinaware, from (magnetic) compass, to well drilling, from quantitative map-making, to biological immunisation processes, and so on, always aiming at case-driven solutions, with no interest in theoretical implications. The Greek philosophers, in a different way, are looking at the originating explanations of the natural phenomena; thereafter, when the empirical evidence could appear to be misleading, there is no fear to try conceiving abstract models (the Plato's ideas), provided that these better satisfied the hypothesised theory.

The «identity» principle is explored, to allow separating any *object* from the *background*, to single out «unit» elements, easier to investigate, having removed the disturbing effects. Along these lines, Democritus hypothesised the "atomic" theory, to reduce every *object* into elemental *atoms*. We mention the method, not the devised frame; similarly, the Aristotle's physics offers worthy schemes, with erroneous binding laws. Indeed, the hypothesised frameworks today are *childish*, and the Tao's distrust of the simplifying reductionism results wise. The context-driven alternative brings to the holistic views, and, whether the sought relational complexity could uniformly be portrayed by distributed effects, the "field" theory is foretold, assigning the proper intensity distribution, to each *point* of the space. However, in the modern science the two views are complementary, and no clear evidence exists that the holistic guesses are *better* approach.

The today physics accepts the "atomic" and the "field" views; quite obviously, we deals with 'photon' either the 'field' description of the electromagnetic waves (and light). Are still *childish* these models ? one either the other, or both, as they lead to equivalent pictures ? The answer is not straightforward, and is affected by how we consider the physical laws truth or trustfulness. Anyway, out of consistent responses, it is deemed imperative to look after the frameworks required by the scientific knowledge build-up, notably, to lead to the technologies. To that goal, we have to refer, along with the physical laws legitimacy, to the disputes on the *invariants*, say, the (dimensional) coefficients binding the physical quantities, that happen to be characteristic constants, should we live in the «static» universe (and the devised «laws» are permanently true). Of course, the disputes are useless, if we can only perceive continuously changing frames, where «laws» are temporary acronyms, depending on the observer/snapshot pair, without steady consistency.

Among the different laws of the physics, the entropy decay is believed to have strong confirmation (according to the «falsification rule»). Certainly, it is possible to think up experiments where the reversibility exists: in the kinetics theory of the ideal gasses, the "Maxwell imp" acts at the microscopic level, in view to override the macroscopic set-up, moving the system back to the original state. More subtle reasoning leads the macroscopic thermodynamics to define the «state» functions, namely, physical quantities that do not depend on the actually covered paths, and are fully defined by the reached local «state». They are clever constructs, to yield context-free changes, so that the «transformation» universe is, always, reduced to a technically correspondent «state» universe, where, positively, the entropy decay is correctly accounted by *equivalent* assessments.

The capability of separating permanent constructs, suppressing the links with the conditioning context, requires to take out the non essential perturbations and to invent the original patterns (like the Plato's ideas). These are risky procedures, if accomplished without continuous (and reliable) experimental checks, in order to reject the false models and to adapt the in progress assumptions. The other way, the acceptation of the «complexity» might paralyse, or, at least, lower the talent to figure out possible reductions and underlying taxonomies. Nonetheless, the wise man, in the Tao's view, is the one who acknowledges the person, the society and the environment all around, assuring the highest harmony, in agreement with the inherent «becoming», which makes unwise trusting on instant snapshots, divided by the (changing) cross-coupling.

We might hypothesise that the universe of context dependent transformations has inherently high complexity, so that it will be childish to imagine quantitative models, unless if further 'reduction' properties are recognised. The western style approach to the technical knowledge follows such hypothesis, due to the typical individualism mentality of people with «active» conduct concerning the personal affairs and the surroundings matters. In the scientific areas, this is recognised in the many causal frames, e.g., the symmetric Hamilton's formalism, permitting the backward time solutions, with no worry on the energy conservation, or even the Fourier's transform analyses on sampled transient signals, with tiny concern about the spectral (infinite) repetition. Shall we drop these formal results as *childish* ?

In the manufacture areas, the «active» ways bring in even higher appeal. The effectiveness, through the split-duty allotment paradigms where recognised by Adam Smith (in needle's manufacture), fully theorised by Frederic W. Taylor (*scientific* labour organisation) and properly enabled by Henry Ford (flow-shop economy of scale). The line of action is believed to be the industrial revolution winning innovation, bringing forth the exceptional performance of the western way to wealth growth. With some ingenuity, the innovation can be referred as the atomistic segmentation of the human work, as compared with the earlier holistic production management, carried over by the clever craftsman through the mastery manufacturing. Every item is reduced into elements; instead of expert artisans, low wage workers perform the exactly segmented task, so that nothing is left to perturbing changes. In recent times, the Toyota's company replaces the Ford's set-ups by job-shop economy of scope and *intelligent* work organisation, where the operators' commitment is exploited by means of the total quality schemata, to bring back the holistic order in the production transformations. The issue is today worldwide accepted, and permits to imagine the «robot age» season.

The «being» vs. «becoming» dilemma teaches, among other things, that the capability to extract stable interpretation patterns requires biasing afterthoughts. A bird-eye view picture brings to the (linear) sequencing of «states», with the inborn hope of the progress. Indeed, the civilisation moves forward, with the ceaseless accumulation of ideas and knowledge, so that, for us, it might be fairly instinctive assuming to be the recipient of incessantly increasing heritages. Quite the reverse, when every thing is unremittingly changing, without making evident evolutionary trends, the complexity becomes fundamental attribute, preventing the resort to the simplifying hypotheses. The steadiness is merely apparent; the current situations emerge as context conditioned: when the number driving factors is very large, the «becoming» is the *permanent* feature, while the «being» is useless snapshot.

The man perceives other kinds of "stability": day and night alternation, yearly return of seasons, etc., so that it might be quite instinctive to assume the cyclic character of every phenomenon, without late additions concerning the (actually arbitrary) progress. In our prospected interpretation, the "stability" of the growth is characterising feature of the western inclination towards the reductionism (and the *scientific* abstraction of laws), and the "stability" of the alternation is basic attribute of the far-eastern preference of complexity preservation (and *intelligent* resort to immanent human faculties). The *developing* countries concept is related to the heritage increasing trends, not to the «becoming» persistence; shall we infer that such guess, too, is foolish ? and the mankind lives changing times that might repeat, without, however, leading to possible improvements ? The questions show the critical nature of the «action» spirit, and the western conceited course towards anthropic breakthroughs, to invent new spendable riches.

1.3.5. The Opponent Logic or Mediation Art

Moving ahead with describing the peculiarities of the European spirit, the role played by the "logics" is directly evident. In China, the Mozi's school well stated all the fundamental principles. The logic was, thus, fully, known, but did not enter in factual achievements, for suitably deployed inferences. In the ancient Greece, the logics was immediately prised, as the powerful means in public debates, being the way to impose the individual importance and supremacy, by direct *democratic* arguments. The logic, indeed, provides constructs to infer judgements based on formal evidence, not necessarily true. For instance, we have proved evidence that the following sequence is untrue:

- the natural produces guarantee crops, having dietetic properties to enhance the good physical shape;
- the opium (tobacco, etc.) is natural product, properly farmed and selected with careful harvesting processes;
- hence, the smoking of opium (tobacco, etc.) will do good, improving the men's healthiness.

The logic aims at context-free reasoning, looking after absolute ideas, valid by themselves, so that the inference is not affected by the timely picked-up samples. The Tao's mind avoids abstraction and does not worry compromises. Moreover, the conditioning context, to face the ceaseless «becoming», requires to deals with changing *situations*, for which the abstraction of fixed concepts is only subjective (arbitrary). The approach is summarised by the following propositions:

- the *permanent change*: to acknowledge the given situation means ignoring the underlying transformation laws, as the reality is permanently fluid; the snapshot is just a view;
- the *contradiction rule*: the in-born fluidity means that contrasts, paradoxes, non-senses, etc., incessantly emerge: the evil and good, the old and new, the week and strong, etc., are shared attributes, because the opposites are complementary, being integral property together, never to be disjoined;
- the *holistic reliance*: due to change and contradiction, the fact/idea cannot be segregated from the context; all details depend on multitudes of driving sources, which cannot be omitted in the knowledge building.

Together, the propositions leads to the «dialecticism» (as compared with the «dialectics»), supporting the faith that contrasts are only apparent; proper insight brings to believe in "*if* A *is correct, neither* non-A *is wrong*", since the simplistic "*if* A *is correct,* non-A *is wrong*" is nothing more than possible event.

The wisdom to accept negation and opposition clangs against the reductionism to create context-free abstractions, in view to get rid from uncertainty. The way to deal with measurement errors is fairly helpful. The experimental data scattering brings to the statistical treatment and to Gauss's distribution acknowledgment, on condition to be subject to independent and unbiased disturbances. The repeated trials test cannot be used with time-varying phenomena, to build the distributions, and the uncertainty reduction cannot be obtained by merely *a posteriori* statistical treatments. The Kalman's filter is known solution, based on *a priori* probabilistic assumptions: the hidden process is thought linear Gauss-Markov model, so that the distributions are fully defined by the covariance parameters; the state updating avails of the transition matrices, with optimal estimates computed by iteratively solving the mean-squared problem. The all is self-sufficient algorithmic construct.

The clever outcomes are verified and expressed with the real time assessment of the spacecrafts trajectories, having the cameras, moved to the correct location, ready to film the astronauts landing and getting out. On that success, R. Kalman obtained his season of renown at Stanford University, before to end the carrier at the Zurich University. The all process uncertainty is fully reduced to algorithmic frames, with suited causal addiction. The alleged model is properly made to deal with the non-linear field effects by time-varying coefficients, and even with the insufficient statistics on the initial conditions by adaptive convergence feed-back, or with the biasing coupling drifts by timely re-instantiating the filtering weighs. The computation burden is, maybe, the biggest drawback, in front of the context-driven *correction* arbitrariness, since the results validity is strongly deferred to the backward recognition.

The *a priori* probabilistic assumptions arbitrariness might suggest relaxing the uncertainty modelling, with resort to the *fuzzy* sets concept. Now, the *membership functions* do not require statistical validation; they, simply, show that sharp values are out of reach (because of measurement errors, or other reasons), and permit to drop some inconsistencies of the probabilistic models (e.g. unlimited distribution density functions, with vanishing asymptotic probability). The *fuzzy* logic control systems are quite noteworthy achievement, expanding, out of the *deep knowledge* frameworks, the capabilities to deal with context-driven situations, where the A and non-A outcomes co-exist. For computer programming, the situations use the object-coding, where the objects assemble: *attributes*, *methods* and *belief*, so that the nominal instantiations join to the current contexts, with the related uncertainty estimates. The *fuzzy* logic control is, certainly, used everywhere, nonetheless with the greatest spreading in Japan, also for current applications, devoid of significant sophistication.

The *shallow knowledge* frameworks are arrival-issues, once the context-driven situations are dealt with resort to *expert systems*, and knowledge-based methods. The processes are favourably shaped up by the *declarative* knowledge (resources attributes, reference hypotheses, etc.) and the *procedural* knowledge (governing rules, performance criteria, etc.); the context is coded according to the:

- «*if* <antecedent> *then* <consequent>» assumptions.

The evolution is supplied by the «inference engine»; the uncertainty estimate follows from the set of plausibility assessments. The situation updating is gained, looking through the coded:

- «*if* < .. .. > *then* < .. .. >» rules.

The *shallow knowledge* is, anyway, embedded into causal frames, ruled by the «action» spirit with polydromic falls-out, since the *true/false* opposition widens, adding the *undecided* outcomes, possibly, at several range of *belief*, established by the chosen hypotheses and measured through heuristic criteria. The achievements might be considered, as well, as factual issues of the mediation art, promoted by the «dialecticism» standpoints, because the context-driven situations replace the (absolute) sharp states sequences. The change, not apparent with the probabilistic modelling, is better acknowledged by the *fuzzy* outcomes membership functions, and is fully transparent in the knowledge-based *expert systems*.

The «dialectics» vs. «dialecticism» dilemma leads, in short, to distinguish two habits:

- to think smart and to behave competitive, in a world where each individual is the positive builder of his own destiny;
- to think holistic and to behave wisely, in a world where each member has to respond for the whole community fate.

The personal commitment, in the former case, profits, aiming at autonomous enterprises, with clear-cut responsiveness that allows rewarding the activity of the best performing operators; in the latter, the perception of liaisons, binding people and surroundings, requires solidarity, assuring protection. The causal inference, with backward attribution of original sources and responsibilities, quite plainly brings about the ideas of abstract models, having, most conveniently, algorithmic structure, to assure explicit quantitative mapping, to driving inputs and off-setting disturbances. The holistic approach, judiciously, looks after the complexity, and acknowledges (apparent) inconsistencies, to not indulge in sharp simplifications; the quantitative models, however, show quite higher complication, and are recent technical acquisition of the computer engineering advances.

The «industry» paradigms well suit the *reductionism* transformation economy, where sets of *simplistic* assumptions (unlimited earth sources, self-balancing free-market, etc.) permits to establish the affluent society, along steady growth trends. When the dependence on the environment becomes incumbent threat, the holistic revision is necessary, perhaps, looking at «dialecticism» views, or «knowledge» paradigms, should the *altruism* policy assure *mediation* prospects, wisely brought about, through the «robot age» complexity.

1.3.6. Competing Freedom or Conciliated Fairness

The logic chaining has been recognised to be standard expression, displaying the individualism in the political or scientific domains of the western way of life, which, basically, is the enabling spirit of the industrial revolution. In the physical domains, the universe of states, with causal attributions, resorting to sufficient and necessary conditions, permits the clear-cut definition of «laws», which confer the deterministic evidence on how to construe the world. The approach, by explicit or implicit postulations, assumes the man's centrality, still today, when we know that the earth is totally negligible entity in the cosmos, the life birth, high improbable occurrence, and the intelligence, quite awkward discontinuity. The claim to build universal laws by elemental principles abstraction is amazing, and the affirmative corroborations are all the more astonishing. Still, the personal independence from the surroundings is fundamental guess, to prise the logical reasoning or individual actions, finalised to technological knowledge or entrepreneurial commitment, due to faith in activity-steered anthropic hypotheses.

All along with the previous reflections, the populations' peculiarities could be linked to environmental and social reasons. The western style reasoning brings to linear chaining, proposing causal series, such as the following ones, worded in opposition to the Chinese cultural framework, assumed as recognized alternative to the European world.

a) *The socio-economic frameworks* in the *natural surroundings*. The China subcontinent characterises by fertile lands, self-sufficient agriculture, navigable rivers, etc. capable to assure the resources centralised management, under unified ruling; this pushes the communities to interconnect through harmonic relations, accepting the vertical frames under the village elders, province governors and nation emperors. The (ancient) Greece and, as a rule, the European countries are narrow lands, interfaced to the sea, with meagre farming, pushing the trading with afar villages, to reach balanced nourishment; the trading with strangers requires strong attention on the local and personal identity, and, in the same time, widest interest about the foreign people activity, products and habits.

b) *The political/legal practices* vs. *physics/metaphysics conjectures*. When the official infrastructures are managed, the Chinese citizens continuously report to relatives and authorities, with utmost importance on bonds and obligations; the habit to refer to the social universe as spread driver yields to look at the world as diffuse field with context ties. The Greek citizen is forced to focus on objects and to deal with other persons, autonomously deciding about his own flock, and on what to buy or sell in full freedom; the recognised autonomy routine assumes to be master of his own fortune, and, to look at the environment as something that has to be understood and dominated, aims easier, to reach into steady and stable frames, enduringly ruled by absolute governing laws.

c) *The scientific/technical knowledge & wisdom/learning schemes*. If the surroundings is permanent overall conditioning context, each one is compelled to consider the whole sceneries, without omitting odds and ends; and the learning processes are axed on the complexity, without extracting single entities. The other side, if the individual gets used to autonomic views, he tries to discriminate the objects, figuring out behavioural rules and models by the abstraction procedures of logic inference and causal association. The alternative approaches consistently lead to human set-ups, either aiming at harmonic fairness and diffused mediation, either, prising individual freedom and conflicting competition.

The holistic description might cover the same statements, yet, by differently shaping the picture: not a linear chain, but a concentric diagram: the outer shell of the natural surroundings, enfolds inside the economic, legal, political and social layers, followed by the physics/metaphysics suppositions and scientific/technical constructs, and ending with the epistemic structures and learning processes central kernel. The *solidarity* is innate issue of such interlinked compactness.

The fact to differently arrange similar (or, even, the same) statements means that the acknowledged differences might direct towards series of alternative views and approaches, e.g.:

- in the explanatory constructions, the Asian style prioritises the relational connections, the European way prefers classes and taxonomies;
- in the universe conception, the intricacy ties suggest to consider fields and affecting intensities; the reductionism aims at objects with absolute native properties, suitably assessed through permanent law;
- in the implicit assumptions about change either stability, the former choice requires holistic closure patterns, handling complexity; the latter allows to single out issues, by repeated trials and statistics;
- when the outer world is considered, the «becoming» feeling emphasises the fear of the stranger behaviours that need to be assimilated, while the «being» mood trusts the foreign controllability and competition measures;
- in the unfolding models, the complexity recommends the resort to sets and frameworks; the abstraction allows to extract causal chaining and physical laws, to be actively exploited as problem-solving means;
- keeping interest on the reasoning means, the fuzzy logics is consistent with the mediation of contrasts; the true/false logic is nice un-biasing tool;

- in the public/civil concerns, the median mindedness settles the quarrelling partners; the competition spirit brings to convince the opponents, with use of the «government through discussion»;
- in the social/political frames, the community dominance promotes aiming at durable/shared heritages; the individual pre-eminence prefers looking at instant/direct achievements;
- in the everyday citizens' behaviour, the ethic is privileging the fairness and solidarity towards (present and) future people, either is aiming at legality and responsiveness, in parliamentary democratic infrastructures;
- in the perception modes, the attention on the context makes easier to look after and keep the surroundings, while the focus on the objects could bring at neglecting the decay and falls-off penalties.

The given alternative is not one better than the other. On the contrary, there is way to expect that highly enhanced achievement are reached, in general, whether both standpoints are taken in proper notice.

When, in particular, the industry age alone is investigated, the European style alternative seems to be the winning one, at least, according to the wealth build up, which characterises the affluent society. The straightforward, perhaps simplistic, evaluation brings to prise the competing freedom effectiveness, and to despise the conciliated fairness inefficiency. The appraisal is, nevertheless, to revise, when the growth sustainability cannot anymore be confined to the earth raw materials exploitation, according to the reductionism patterns of the industrial efficiency. Alternative ways, built on the complexity continuation, might become winning prospects, and need to be considered. The «competition» vs. «fairness» dilemma ought to be revised. The effectiveness in the spendable riches creation, through the competitive *free market* of the affluent world, is worth in a society of equals, where each individual has the equivalent defence and opposition opportunities, established by the unbiased interplay of (representative) democratic governments.

The unbounded exploitation of the natural stocks means to take out from the future generations' potential sources, without allowing protection and resistance. Actually, the parliamentary democracies respond to the in progress electorate, and the voters' current interests are not directly concerned by the blurred «fairness» demands, having unclear attribution to citizens to come. In the next *global village*, should the present analysis be correct, the *altruism* framework might be the *post-global* way-out, under some *hyper-democracy* administration, to be invented, in which the *wise* protection of the future mankind is obligation not to be bypassed by the existing voters' immediate selfishness.

To conclude the hints, we might point out that the parliamentary democracies have proved to be the most effective enabler of the industrialism. Today, they might happen to be, maybe, structurally inadequate to involved in protecting the (not voting) future generations. However, no evidence exists that other political organisations offer better guarantees. The setting of supra-national authorities is considered with especial attention, so that the growth sustainability could come into the imperative survival requirements, not to be object of opportunistic deals. Further hints are sketched in the following, notably in the 6.3 section, after some more insight on why the old (western world style) industrialism is deemed unfit to assure durable growth. The guesses look at the *solidarity* prospects as the *rational* issues, which best protect the mankind progress, over properly wide spans, out of the instant profit of the today voters.

Chapter 2

2. INDUSTRIALISM TWILIGHTS

The establishment and expansion of the industrialism allows dramatic changes in the human quality of life, bringing into the market great quantities of low cost goods, right to satisfy the needs of larger and larger amounts of individuals. The trend is followed by several countries, from the initial moves in the UK, to other European nations, USA, Japan and some Commonwealth countries. The standard make-up, however, keeps the fundamental «industry» paradigms.

Large changes are quite recent, where computer engineering plays the most relevant role, permitting to explore the «knowledge» paradigms, at first as useful enhancer, and, in the following, as the way assuring to be less dependent on the *reductionism*, and, when useful, to address the «complexity» in its intricacy. The chapter considers the noteworthy aspects:

- the work organisation changes, to exploit the economy of scope, instead of the economy of scale, with in progress addition of *artificial* intelligence, to operate on process, further to *artificial* energy, in view to take profit from the *scope* economy of (in lieu of *scale* economy);
- the market weigh of *intangible* value added, given by *extended* enterprises, with the lifestyle views of *service engineering* and *reverse logistics*, made easy by net-concerns, having *ecology* attention, supplied by the *knowledge* value added *product-service* provision flow;
- the socio-political spurs, due to the tangled *global*, *no-global*, *post-global* path, which makes the industrialism alteration necessary, depending on the forthcoming «natural» capital enforced regulations, openly emerging from the *ecologic* demands for the mankind survival.

These aspects bear high relevance, and cannot be easily separated into classes, because they co-operate together in establishing the industrialism trends, showing, in the same time, the resulting patterns fragility, if the sustainable growth scopes have to be reached. In the chapter the «industry» paradigms continuity is allowed, even in face to its clear frailty. In the next third chapter, the *durable* development prospect is reconsidered, highlighting the changeful «knowledge» paradigms, as a noteworthy way out. It should be clear that, today, the industrialism is the current economical structure, and no alternative exists, to grant the life-quality we enjoy. The trend frailty means, consequently, that new paradigms are necessary, not that winning prospects exist (nor already found). Nonetheless, the present study spirit is in the direction to trust the «technical» capital innovation.

2.1. TRENDS IN MANUFACTURE TECHNOLOGIES

The industrialism, along the relatively short period from the original start-up, progresses through a series of characterising steps, depending on the requests, the market is putting each time forward. The price is industrial product bet, keeping it within the means of many buyers. As compared with earlier craft-made deliveries, the improvement assures steady standard quality, with the drastic compression of the manufacture costs, assured by the internal organisation and the technological efficiency.

Within the approach privileged in the exposition of the «industry age» aspects, we might distinguish four ranges:

- manufacture responsibility on the product steady standard quality;
- manufacture focus on the process for optimal cost/quality ratio;
- manufacture focus on the lifecycle assessed quality and eco-impact;
- manufacture responsibility on end-of-life waste and pollution.

At the first range, the achievement is reached, by taking apart the design from the production, so that the product properties are established off-process. The all operation means having resort to the principle of abstraction, figuring out models, suitably equivalent to the forecast objects. It is worth emphasising the procedure, because, at the beginning, the lack of assessed technologies and factual know-how made very difficult, or, perhaps, a bet, the claim to reach standard quality.

The off-process product design is successful, if it covers all the details for the manufacture processes. This range is, thus, firmly related to the previous one. The production means become the characteristic feature of the industrialism, being the prerequisite of the transformation effectiveness and, most of time, requiring huge financial investments, to make the business profitable and winning. Here, we meet the entrepreneurial bet: the investments are repaid only if the products are sold for the planned volume, horizon and price.

The affluent society means over-production abilities and market-share linked to customised quality supply. The change, quite soon, has to meet the sustainable growth precepts, expanding the delivery requirements to assure eco-consistency. The entrepreneurial bet, now, is no more confined within the enterprise, and the product qualification requires the lifestyle design, with assessed quality standards at several on-duty *points-of-service*, out of the *point-of-sale*. The range shows that the companies' competition expands, and new technologies make their entry, with the related modelling and simulation features, to deal with the further abstraction domains to be incorporated by the business.

The product lifestyle eco-consistency is successful, when the all supply chain guarantees that the balance, between the raw-materials provisioning and the actual waste-and-pollution issuing, is controlled, and remains under suitable thresholds. The fourth range is, then, directly coming out from the third one. The producers' liability is, undeniably, clear at the product design level, because all the on-duty characteristics and the recovery (reuse, recycle) functions ought to be selected and assessed since that phase. Moreover, the EU regulation makes obligatory the free take-back for the end-of-life mass-produced durables (e.g., household appliances, cars), under the manufacturers' extended responsibility.

The outlined points are, perchance, obvious, but the existing analyses neglect the sketched evolution. Indeed, from inside, the industrialism cannot motivate the switching, from *internalities* (the product-process management), to *externalities* (the environment-enterprise adjustment). Therefore, the considerations that follow keep their motivation, because of (enterprise) external spurs, which make crucial the prospected variations. These spurs are technology-supported (the robot age) and economy-pulled (the bylaw requests); as an all, however, the driving causes come from the socio-political imperatives of the ecology protection. That is why the study, forcedly, widens the attention, out of plainly engineering views.

2.1.1. Evolution of the Work Organisation

The transform efficiency, enabled by the industrial revolution, comes to be the winning feature towards the costs compression, but, also, the critical starting point of the *rebound effect*, through which the innovative product chases away the old one, incessantly enhancing pollution and waste. The fundamental explanation of the efficiency, recognised by Adam Smith before even its full deployment, is the "work division" approach.

The concept develops, a bit surprisingly, out of the conciliated fairness of the community-centred populations, where each one accepts the role best suited to the harmonic becoming. Then, opposite to the personal competition surroundings, the work division, perceived as severe violence, is obtained pushing the reductionism at utmost level, as if wholly scientific truth could be devised to cover the optimal manufacture organisation. The opposition is, maybe, obliged, visibly showing the separation between the leading «capitalists», and the «workforce» diligence, only (and basic) required virtue.

Then, the «industry» age characterises by respect of, both, «pre-industry» and «post-industry» ages, by series aspects of the earlier handicraft, and the following robot-sensible settings. The list of noteworthy aspects offers blurring boundaries, and the example choices, suggested by the collected hints, provide, no more than, preliminary understanding.

During the craftsman age, the work develops in the personal shop-laboratory or craft-studio, the market staples are non-repetitive masterpieces, and the labour return comes from the economy of skill. The entrepreneurial concern is set up at the *point-of-order*, being risky to invest efforts and supplies before, with reliance on the *design-while-manufacturing* technical know-how. The work organisation is based on the master leadership, having decision structure in the chief commitment and motivation style in the individual creativity. The background knowledge has fostering virtue in the personal talent, and the qualifying mastery is the foremost competitiveness index.

At the «industry» age, the business runs in the large corporations shop-floors, with staples in the mass-delivery, and the company return built on the economy of scale. The responsibility span is fixed at the *point-of-sale*, when the buyer enjoys the choice between several offers. The supply technical specifications are always done *off-process*, presetting the most appropriate production means and methods. The work organisation, as said, follows the *scientific* job-allotment paradigm, and the decision structure is the hierarchical assignment, after *duty-division* analysis. The background knowledge resorts to the paired team addition, to grant leanness and competitiveness for optimised manufacturing effectiveness.

The list of notable aspects is consistent with the picture conventionally done, dealing with the industrialism, notably, the Tailor's reductionism and the Ford's flow-shop establishment. The «industry» age comes out by the watching over the eager application and zeal of un-skilled workforces, straight away obtainable and quickly trained, to be converted in effective enablers of the "industrial revolution" production means and methods. It is, perhaps, worthy pointing out that, in the all, the offered job has depersonalised effects, with short time falls-off in the dramatic opposition and class struggle, and longer time issues in the obvious removal of the men, replaced by equivalent machines or robots. With the replacement, the «robot age» does not, routinely, starts, but all premonitions are there, to foster the change into new work arrangements. With the reductionism, the job has a depersonalising effect, supported by the *scientific* work organisation abstract patterns, implicitly outlining the robot technical characteristics.

The «post-industry age» beginning is matter of study. Most likely, no sharp divide exists, rather transition steps, showing that the growth sustainability of the affluent society is fictitious or deceptive; still no alternative can be explored and exploited to grant better outcomes. For suggestion purposes, the «new economy» is sign of adjustments in the companies' business, and the «robot age» is about to show the work organisation break. The survey of noteworthy aspects, essentially, includes the following example details (today only hypothetical).

The business now, several times, runs into distributed (multi-national) shops, suitably, *networked* to be turned into *extended* enterprises. The market staples are product-service flows, with the company return by economy of scope. The supply responsibility extends at the *points-of-service* and at the *end-of-life* recovery, with technical specifications covering lifestyle assessments and provisions, as the value chain externalities play critical role. The work organisation sets in *intelligent* task-ruling, founded on distributed competency decision structure and on collaborative reward motivation style. The knowledge features exploit co-operative frameworks for problem solving scopes, and the business competition is won, by guaranteeing the product-service delivery lifecycle appropriateness.

The work organisation step forward follows twin tracks:

- the one of the outlined aspects, in progressively shaping and widening the business awareness from single masterpiece, to mass-products quality and to (eco-consistent) lifestyle supply chains;
- the one requested, when translating the business responsiveness into suited sets of entrepreneurial paradigms, to devise the offer and market share that fix the company *internalities*, which set the supply chain *externalities*.

The industrialism trend along the latter track is, in its turn, specified through a series of characterising technical requirements, starting by the whole presetting of the product design and the off-process planning out of the shop-flow, resorting to the joint product-process design of the simultaneous engineering, and reaching the product-process-environment design for ecology protection and the total product-process-environment-enterprise design for the optimal resetting of the productive facilities, all along the product-service delivering.

The listed hypothetical aspects are just technology-based options. Their actual carrying out is pretty different question, and will only be the outcome of the rather different spurs, generated by the imperative eco-regulations.

2.1.2. Efficiency and the Economy of Scale

The «industry» age is paradigmatically associated to the *work-division* and the *economy-of-scale* pair. Together, the efficiency is achieved by wide-range resort to reductionism, notably, to the *reduction* of the workforce into efficient operation fixtures. The concepts behind are said to be the *active* attitude face to the problem solving demands: no condition or position is so entangled that its answer is out of reach; once the skein end is taken out of the context, the all unravels, one at time.

In the technical capital build-up, several options open, from the reference to series of standards, to the trust in the system analysis. The standardisation is basic necessity of the *scientific* job-allotment, doing away with the uncertainty of fuzzy shaped parts and fixtures, and allowing to define exact work-cycles, as assemblies of unit operations. This reflects in the human counterpart, identifying constituent behaviours, and putting into practice the safe and sound ergonomics.

The system analyst belongs to higher sophistication problem solving spheres, where the *reduction* applies out of static situations or items, and cause-dependent or time-varying phenomena have to be investigated. Here, the superposition of the effects is the desirable issue, to find out linear describing models, according to the well prised omen: <happiness is linearity>.

Anyway, for optimal effectiveness, the work division, quite obviously, brings to the mass-production, and the 'economy of scale'. Then, the very *reductionism* suggests moving to the work-mechanisation and fixed-automation. Actually, this means recognising that:

- every (complex) task can be described by series of unit actions;
- one or more mechanical contrivances exist, to fulfil the unit actions;
- the man-replacement depends on the cost analysis break-up.

Up the mid XX century, the economy of scale progresses through huge mass-productions, authorising similarly colossal investments in fixed-automation. The flow-shop set-up requires compelling cautions, to avoid unscheduled events, and to guarantee the full steadiness, in order to obtain the planned returns, with the expected throughputs, due-dates and overall productivity figures. Of course, it is not sufficient to manufacture, it is necessary to sell the all products.

The buyers' satisfaction is chief condition, and the product's assessed quality is necessary requirement. The flow-shop is critically dependent on the efficiency, exactness and reliability of the intermediate unit fixtures and methods, so that the process control and up-keeping become vital accomplishment, whose cost is fully repaid through appropriate outcomes, such as the 'zero-defect' manufacturing, if the whole production is at the scheduled quality standards, avoiding the necessity of final checks, and related scraps, inferior class stuffs and re-manufacturing.

The fixed-automation production set-up, thus, punctually incorporates two parallel flows: materials and information, cross-coupled through the control tests. The computer options, promptly, happens to provide special purpose instruments, such as the following:

- CAD, computer-aided design, to help choosing and verifying the product properties;
- CAM, computer-aided manufacture, to help fixing the flow-shop resources and work-cycles;
- CAP, computer-aided planning, to help assessing the production agendas and targets;
- CAT, computer-aided testing, to help performing the on-process checks and up-keeping.

This way, the old inclusion of man operators is progressively replaced by the automatic devices and procedures, both, on the material and on the information flow. The fixed automation is the domain of multiple-function machining heads, accomplishing in parallel series of unit operations. It is, in addition, the domain of front-end sensors for the automatic quality-chart setting. The «industry» age looks as if the original meaning of *industry* has no more to be applied to the on-process operators, and a different interpretation becomes possible, showing that the work-flow reductionism is the preliminary step of industrial robot design.

2.1.3. Flexibility and the Economy of Scope

The flow-shop set-up and the fixed-automation successfulness stop working, when a series of facts, external to the prior manufacture practices, happens:

- the market saturation: it is not enough to make and offer quality-to-cost optimal products, to reach the enterprise-planned marketing targets;
- the society prosperity: the average buyer has large available cash to make possible the choice of personalised goods, instead of mass-products;
- the workforce education: the wider instruction lowers the workers' number for low-wage repetitive tasks and increases the high qualified offer;
- the technology changes: the computer engineering permits the information systems of the *intelligent* work organisation and flexible automation.

The four facts bring to the economy of scope, as the old industrialism forced change. It is, perhaps, amazing that the modification is affluent society outcome, bringing up less conservative behaviours, enhanced consumption and dumping, fostered by «new economy» abilities: the operators' expertise and technological advance. The «robot age», itself, is promoted by facts that are obvious evolution (two passive reasons compel to modify; two active spurs support to innovate), to grant the industrialism continuation.

The economy of scope is coherent attitude, to exploit intelligent manufacture and engineering. The resulting work-organisation has knowledge intensive set-ups (to foster distributed commitment and co-operative reward), having enabling help in the distributed intelligence, to right away evaluate and optimise:

- the *technical* suitability of the product-and-process construction files;
- the *operational* feasibility of the (lifestyle) overall supply chain;
- the *productive* consistency of the involved facilities and functions;
- the *economical* return of the devised finance accounting policies.

The economy of scope exploits the information that the active computer aids, CAD, CAM, CAP, CAT, etc., permit to obtain, for looking at the current market requests, adapting the offers, for widespread buyers' satisfaction. The new trend is based on the remark that the company's profit comes from outside, namely, from really sold products. From inside, the accounting budget can only look at lowering the costs, reducing the investment and enhancing the resources engagement. The economy of scope is a mix of requests, binding the product and the process:

- *to diversify the offer quality*: the scope is to vary the product mix, reaching higher market-share, with wider purchasers' satisfaction;
- *to concentrate functions*: the scope is the resort to group technology, etc., to process all items with general purposes machining centres;
- *to grant operation effectiveness*: the scope is to fix the production agendas, to up-grade the throughput and due-time delivery;
- *to suppress redundancies*: the scope is to manage the shop-floor logistics, exploiting strategic/tactical/execution flexibility.

The scopes reverse on the production facilities/methods, the critical demands coming from the modified market conditions. The technique is said *simultaneous engineering*, as it is obtained by the joint product-process design, in view to have the most effective pair, in terms of actual sales. The economy of scope is powerful way to aim at the eco-safeguard. It can include impact analysis, planning lifestyle efficiency checks, and the likes. The characterising features are:

- co-operative knowledge management: e.g., client-driven quality, product-service delivery, lifestyle design approach;
- piece-wise continuous improvement: e.g., shared responsiveness break-up, service extension, backward cycle exploration;
- on-duty diagnostics and inclusive servicing: e.g., pro-active maintenance, recovery flexibility, quality-driven ambient intelligence;
- lean engineering check-up assessment: e.g., removal of set-apart resources, just-in-time planning, team-work empowerment.

The total economy of scope, covering the direct and reverse logistics, builds the enterprise profitability through to the quantitative forecast and optimisation of identified supply chain characteristics: delivery quality, price, due date, lifecycle performance, maintenance and renewal plans, disposal and recovery targets, etc., by the knowledge-processing and decision-support ability and related information infrastructure, so that lifecycle information transparency becomes prerequisite for the trade and use of material goods, according to the customers' satisfaction and the environment protection rules.

For the enterprise success, the dependence on the market strategic positioning is standard admission, and the supply chain concept is modified into *value* chain, to join parts and materials delivery, with the related intangibles flow (*value* web), supporting the main contractor, with vital complements. Today, the expansion of design tasks and connected data-bases is compelled by the many enacted product lifecycle regulation constraints, and is made possible by existing computer aids. The changes, initially developed for the economical pressures, face to the market saturation and the purchasers' opulence, become worthy help to provide answers for the ecological urgencies, requested by consumption and dumping limitations and compelled behavioural thriftiness.

The productive means, from the fixed-automation flow-shop, modify into the flexible-automation job-shop. Within each separate plant, the shop-floor logistics of parts, tools, fixtures, etc., becomes critical requisite, to facilitate time-varying schedules, optimised on the tactical horizons and updated at the execution level to face unexpected occurrences (without the backing redundancy of the earlier fixed automation). The pro-active maintenance plays similar leanness role, possibly, by the resort to risky operation conditions under monitored warning and/or to inferior (but *overseen*) performance, before restoration and renewal.

The huge investments in all-inclusive flow-shop set-ups, oriented on the final-product delivery, are replaced by the productive break-up and out-sourcing. The economy of scale restricts, whether possible, to parts and sub-assemblies, left in charge to domain specialists. Indeed, the complexity of the end-delivery is, most of the times, increasing, for the addition of lifestyle knacks and checks. The focus on the core business suggests the co-operation among several dedicated facilities, so that each one assures the most effective proprietary technology and skill.

The productive break-up and out-sourcing are, after that, explored, to combine facilities and functions provided by lower wage operators, each time the piece or duty quality would be consistent with the end assembly. The economy of scope is, anyway, the winning way to look at preserving «complexity», and at looking after «knowledge» paradigms in the manufacture business. The options are devised as technology-feasible opportunities; their actual working depends on circumstances going beyond the merely engineering value.

2.1.4. Service Engineering and Reliability

The Service Engineering is emerging technique, which grants that «services» are designed and developed in a similar manner as products, and deals with:

- improving the functions enabled by delivered provisions;
- developing and realising the supply supporting frames;
- providing and executing the services, as corporate duty.

The service engineering, SE, profits by product lifecycle management, PLM, fixed at the design phase. The technique includes three tasks: service conception, service planning, and service execution. The PLM tools, habitually, focus on the *product* side; on the contrary, the SE tools emphasise the *service* aspects. To deal with *product-service* delivery, suitable PLM-SE tools are needed.

From a factual point of view, the SE will increasingly expand, together with the widening of the manufacturers' responsibility. Up to quite recent years, the *point-of-sale* is the divide, after which the buyers can little argue, if the supplied items do not correspond to the desired quality, due to ineffective functions. The features want, increasingly, comes from the conformance-to-use assessments, of the environment protection acts, with restriction or prohibition, to take benefit of deliveries.

This will result in changing clients' approval, from the *point-of-sale*, to *point-of-service*. As a consequence, the companies, to face world-wide competition, are compelled to modify their strategies, looking at lifecycle supply chains. The new style obliges the manufacturer to the buyers, for conformance-to-use at the point-of-service. The traded goods characterise because:

- the market-driven *quality* aims at customers' satisfaction by a mix of life-long operation features;
- the sustainable *quality* becomes monitored attributes, under transparently acknowledged schemes.

The economy of scope, as already said, turns, from economical pressure, into ecological urgency. The EU promotes the service supporting frame, discerning the alternative measures:

- to foster eco-conservative behaviours, by promoting the drawing-up of *voluntary agreements*;
- to force eco-sustainability achievements (for given products), by enacting *compulsory targets*.

Anyhow, the supply of conformance assessment certificates becomes relevant business, in the value chain, with the joint liability of purveyors and users for the eco-system protection, according to increasingly enacted rules. For the ecological requirements, the service engineering, besides, fulfils two goals:

- to cover the *servicing* side of the supplied products, namely, the (tangible and intangible) delivery to a client, granting the enjoyment of the specified *functions*, by lifecycle indentures, achieving return on investment, through resort to *scope* economy;
- to accomplish the environmental impact *monitoring* and *control*, reporting the results

to overseeing authorities, being entrusted proper access on the supply chain, with explicit charge to assess and to record the full products lifecycle/disposal footprint.

The service engineering, hence, splits into provision and overseeing tasks, so that a three-party scheme establishes. For the eco-servicing, the framework lumps together purveyors and users, both subject to the monitoring of independent (and duly accredited) supervisors.

From the user's viewpoint, the SE is supply value added, when the reliability of the purchased functions increases, as the conformance-to-use is fully assured. The maintenance service, thus, is relevant opportunity, having two goals: to grant the fair enjoyment and performance; to certify the on-duty eco-impact suitability. The trends in maintenance duties, neatly, characterises by:

- off-process set-ups by restoration rules: when, at *breakdown*, resources are turned-off from duty-state; or *pre-emptive* mode, at fixed times or given life consumption, provided by (estimated) reliability data;
- on-process set-ups by predictive rules: *reactive*, if resources are monitored to detect the onset of failures symptoms; or *proactive*, to keep the 'normal conditions' by organising on-going functional prerequisites.

Breakdown maintenance limits to cost repairing/replacing equipment, when failures occur. The strategy does not deal with zero break situations and needs to face the linked falls-off: occurrence time and activity stop. The basic scenario is: the maintenance manager receives malfunction reports or up-keeping demands; he allocates personnel, defines time for maintenance and sends field engineer. The decision is affected by several factors, say: job priority and constraints, engineers' ability and availability, costs and time uncertainty. The suited process knowledge allows avoiding downtimes, by pre-emptive mode, based on *a priori* failure rate estimation and operation periods pre-setting within mean-time-between-failure, MTBF, bounds. The stops are assessed with proper safety margins, and *preventive* maintenance applies, while resources redundancy grants operation continuity.

Better strategies are in use, which detect the anomalous running symptoms and enable reactive or proactive maintenance operations, before actual failures develop. For monitoring tasks, they require sensors and links to knowledge data-bases, aiming at on-process diagnoses, basically, related to:

- ill-running situation/trend monitoring, to prevent breakdown by detection of early symptoms;
- continuous monitoring and deviation control, to up-keep conformance-to-specification status.

With intelligent aids, the monitoring maintenance is feasible, by detecting the diagnostic signatures tied to operation thresholds, not only by finding symptoms related to established misfits. The up-keeping actions, after that, aim at *pro-active* mode, as situation/trend monitoring provides the knowledge to preserve fit-with-norm conditions. The onset of anomalies is avoided with reliable leads, because trimming or restoring actions are performed as case arises, possibly, during idle or hidden times, by resolving at the root the failure causes.

The availability of suited sensors moves *preventive* maintenance to subsidiary or confirmation functions, in view of exploiting the inherent systematic/scheduled arrangements for off-process duties. The *predictive* rules are standard reference, with the multi-layer setting: sensor data, signal processing, signature detection, condition/trend recognition, health assessment, prognostic, decision-support and restitution layers. The condition maintenance helps to operate before the collapse occurs; to reduce the number of failures; to do restoration only when needed; to reduce up-keeping and un-availability costs; to increase the equipment/product life, and so on, provided that the right servicing is enabled.

The proper PLM-SE tools combine the *predictive* and *preventive* modes, as early design requirement; and the *predictive* rules sophistication resort to the *proactive* regeneration, so that the zero-defect schedules set-up on the strategic horizons, while the *reactive* restoration establishes, only, when at the execution horizons, the anomalies trend cannot be removed. The philosophy is consistent with the product-service delivery, and the split-level backing: remote diagnostics, and front-end detection and up-keeping.

The off-process knowledge intensive PLM joins to the agent-based SE; this exploits system hypotheses for feed-forward plans, and uses on-process data for closing proactive measures, to preserve safe running situations, or for enabling reactive repair at the (planned either unexpected) breaks. The provided service develops on three steps: to detect and identify abnormal situations, defined as signature or threshold; to acknowledge the faulty situation, level of degradation, type of failure, etc.; to troubleshoot, detailing the restoring policy. The setting of effective diagnostic knowledge is time/money consuming and demands lots of efforts. With the product-service delivery, the remote operation becomes efficient standard, with smart agent-based technique, incorporated at the design stage.

2.1.5. Reverse Logistics and Environment

The *reverse logistics* is the process of planning, fulfilling and controlling the reuse/recycle of worn-out products, in order to conserve resources and protect the environment. The *backward* chain profitability requires boosting facts, such as:

- fully operating closed loop supply chains, grounded on standards for the parts/materials recovering;
- tendency to remanufacture, based on original design for reuse, instead of native disposal destination;
- general collection and disassembly organisations, with effective facilities for worn-out goods;
- improved sorting, retrieval and reintegration processes, to extract assessed quality resources;
- account of supply chain material and energy provisions, to repay the off-balance caused decay;
- recognition of every emerging impacts, to forfeit for the remediation and restoring accomplishments;
- exhaustive information frames, for monitoring and vaulting all the product lifecycle certified data;
- objective tax collection for all tangibles withdrawals, to establish fair-trade competition.

These are a mix of legal and technical prescriptions, with sound links to older habits. The reverse logistics, indeed, comes out from the known ways of obtaining secondary provisions and raw-materials, from exhausted scraps and end-of-life objects. The today innovation looks at the full recovery (reuse, recycle) of every material items previously put in the market. Then, the new reverse logistics, RL, becomes part of every supply chain, and the PLM tools, established at the design phase, ought to incorporate the related RL section, giving rise to newly generated PLM-RL tools.

The switch from the *opportunistic* reverse logistics (only if return follows), to the new systematic backwards flow, is possible on three conditions:

- mandatory rules are passed, stating that a product can be put in the market, if the manufacturer/dealer is responsible of the end-of-life take-back;
- compulsory recovery (reuse, recycle) targets are fixed, and their fulfilment is enforced, under the control of supervising bodies/authorities;
- suitable economic instruments are promoted, covering the backward flow costs, each time the opportunistic processing might fail.

The EU ecological policy is moving with account of these conditions, at least, when the focus is on typical durables, such as the automotive and the household appliances. In these areas, the current manufacturers' responsibility includes the products' recovery, according to detailed targets, after the (free) call-back of the end-of-life items. The reverse logistics is basic facet of the offers' lifestyle design, and the related cost is necessarily to be covered by the business value chain. The all connotations are, perhaps, still to be valued.

In fact, at present, the EU environmental policy does not show fully coherent goals. The attention restricted to noteworthy durables is, possibly, informative, to help training the citizens, or moralistic, to start by assessed affluence symbols, but leaves into shadow that recovery entails all comprehensive measures. The area of disposables is likewise liable of waste and pollution; the use-and-dump practice is even more relevant example of indiscriminate squandering. The hierarchy in the consumables reverse logistics, if any, has to address to ecologic or the economic issues, either to the reasoned balance of the two, but cannot address sample cases, picked up for, perhaps, informative or moralistic reasons only.

The EU, made up by quite populated countries, seems to be mainly concerned by dumping and landfills. The enacted directives aims: at the prevention of waste (*first priority*); at improving the backwards flow performance, especially, for the end-of-life goods treatments (*second priority*); and (*as well*) at the reuse, recycle and other forms of recovery of the exhausted stuffs. The undertaken analyses have contrasting issues. The enacted legal frames do not offer coherent regulations; too much emphasis is on details, prising facets in landfills safeguard, or covering sub-classes of products, without the complete attack to sustainability.

The lack of consistency risks undermining the credibility of series of rules, if the eco-advantages weighed against the drawbacks do not clearly appear, and the procedures opacity hides that the onerousness is equally distributed among all the citizens and across every EU partners. Besides, the affluent society deployment is conquest non uniform and with uneven achievement times; a set of law provisions socially and politically correct in some country, could be ineffective or even self-defeating in other contexts.

The open questions on how to manage the all are many, even if it is accepted that the RL (and SE) practice, directly addresses demands along the supply chain for lower impact in pollution and consumption. The remarks, up now summarised, show that:

- in the *affluent* society, the behaviours' conservativeness does not support economical return, due to the very low repaying cost of the natural capital exploitation, as compared with the human (and financial) ones;
- in front of our living in finite earth surroundings, the growth sustainability ought to simultaneously deal with the *four* capital contributions (natural, human, financial and technical), with balanced value chains;
- in the *thrifty* society, the rushed unavailability of spendable riches would drastically

endanger the people life-quality, unless innovative methods and businesses could counterbalance the materials scarcities.

The reverse logistics happens, in fact, to be sound business: it helps defending the environment, obtaining secondary provisions and widening occupation offers. The hierarchy of consumables (durables *vs.* disposables, etc.) shall be left out, and the economic *vs.* ecologic issues will propose how to manage the reverse logistics business. To start, the EU environmental policy moves from the three conditions for the all-comprehensive systematic backwards flow of the chosen areas: end-of-life vehicles, ELV, and waste electrical-electronic equipment, WEEE, under the manufacturers' responsibility and free take-back. The economical incentives are looking at a series of measures (and ambiguities), which include:

- tax on virgin materials, paid at provisioning (qualms on set apart amounts, to refund the natural capital withdrawals);
- subsidies on recycled stuffs, received at recovery (doubts on the realistic 'secondary to primary materials' poise);
- landfill charges, paid for waste dumping (if 'excessive', avoided, through wreck drop and illegal dump);
- deposit-refund sum, set apart at purchase and returned at disposal (opacity in fixing authorised costs and administrative charges);
- visible fees, lumped in product prices, to balance recovery costs (validity deferred to the reverse logistics competition ruling);
- free take-back, included in manufacturers' costs (effectiveness tied to the overall supply chain eco-consistency management).

All measures are affected by the inability of giving *fair* values to the «natural» capital assets. The reference to market prices has meagre consistency, because, as they come out from instant transactions, the overall decay is neglected (the natural capital never take back the original qualities and quantities). Besides, the example measures distinguish averaged actions (the first three), which apply on the generic provisioning market, and special purpose attainments (the other three), which are directly linked to each definite supply chain.

The full impact (short/long terms, direct/indirect issues, etc.) of the measures requires careful analyses, strictly linked to the information value-added path. The mix of many options exists today, with loose weighs assigned by the EU partners legislations, so that, when these externalities are summed, the charges hugely vary from one to another country, with market biasing effects. The reverse logistics is intriguing domain, at the moment, where the ambiguities have critical effects. It is fundamental to find out shared and enthralling convictions capable:

- to stimulate, in all countries, the higher eco-consistency of the consumers' behaviour (the *consumers* include the goods producers and purchasers);
- to discourage (or to forbid) the use and the selling of the highly penalising tangibles, with resort to measures linked to each specified supply chain.

The former suggests, e.g., more balanced policies, without the amazing choice of typical durables, on the contrary, exacting concern on consumables (beginning by disposables). The latter, first of all, requires establishing objective metrics, by which assess the productivity of the engaged natural capital. Looking at the unit manufactured items, the efficiency has to be rated in terms of *tangibles* efficiency or natural *resource* productivity (like to *human* efficiency or *labour* productivity), with delivery value chain encompassing the lifecycle, from material and energy procurement and construction, to operation and maintenance and to disposing and recycling.

The reverse logistics becomes the supply chain integral part, affecting the full tangibles productivity on condition to enable eco-consistent processes. The focus on given durables (at least, instrumental items, such as cars or home appliances) is useful, emphasising their potential replacement by *functions* delivering (namely, mobility, domestic chores, etc.), each time that the *service* economy, established with clients' satisfaction and suppliers' profit, is more eco-conservative. The bet is to address the tangibles as capital assets (not passive supports) and investments, whose return has to be maximised (by life extension, losses avoidance, wear-out removal, value-added operations, etc.).

Out of the durables, the focus on the disposables (toys, furnishings, garments, textiles, footwear, etc.) needs appropriate co-ordination. The fair-trade protection in the whole European market cannot do without common obligations on product usability and end-of-life, even when the local habits seem to be impediment for unified goals. To try understanding how to modify the items value chain, let refer to some example statements that require deepening and revising:

- at the point-of-sale, the trading price does not include the eco-consistency charges for usage, maintenance, refurbishing and replacement; a different indenture would result odd, but is necessary;
- the time to failure of a complex device is shorter than the one of any of its components, as high reliability ought to characterise the life-span of the whole, and repairing and revamping are not scheduled;
- the consumables replacement is winning option to expand gross domestic product (clothes refit is becoming extravagance and their repair, marginal duty); and the likes.

If the supplier has the contract obligation to support *fit-for-purpose* life-span and to assure *free* take-back of disposed items, the pricing will include such added competition features between the enterprises. The information flow leads to the knowledge-driven entrepreneurship, empowering the consumers, and involving them in eco-consistent choices, from the design phase of every product, to be put in the market. The amount in *intangibles* becomes competition feature, discerning the enterprise's value added. At this point, we can understand that the engineering feasible issues become, as well, actual solutions.

2.1.6. Extended Enterprises and Supply Chains

The industrialism has undergone the throughout change from mass production to customer-whim satisfaction, understanding that the economical return comes from the outside market and that the inside shop shall look up the product-process pair. The economy of scope is the result of methodologies, which instruct how to deal with the *externalities*, in view of shaping the entrepreneurial *internalities*, to assure proper bottom-up business response, to the in market-driven demands.

The economy of scope brings in the «robot age» flexible automation, and the productive break-up, to concentrate each unit on its core business. The change is pushed to outermost deployment, when the driving externalities add up and turn mandatory. The economy of scope methodologies remain pertinent, however with widespread driving requirements and enforced entrepreneurial accomplishments. The technology changes, moreover, are provided by computer engineering, which makes possible the networked organisations, through the innovative web-built and internet-supported facilities.

The new organisation paradigms permit to look after better performance at the corporate and at trans-corporate level. Suitably structured companies include:

- the *distributed* enterprise, organisation with operations in more than one geographic location;
- the *concurrent* enterprise, organisation enjoying multiple production shop-floors, obtained clustering series of companies around a project, to arrive at the critical size and competence, guaranteeing the delivery of complex products; the business operates under unified ruling, and avails of shared and integrated infrastructures, notably, at the networking level;
- the *extended* enterprise, group of companies that work together and act as single business entity, to satisfy the chosen market objective; it embeds the leading manufacturers and all the suppliers and vendors, who contribute to create, distribute, support and call-back the *product-service* supply, under lifelong responsibility, to customers' benefit and environment protection.

The first corporate level is fostered by local opportunities or efficiency (e.g., clients' proximity). The second is typical of huge and tough orders, exceeding the capacities of each company, so that links establish to shape out joint-ventures or the likes. The third is emerging issue of the new delivery lifestyle engagements, and leads to varying facilities/functions clusters, depending on the in-progress supply chain requests, each time shaping the productive *internalities*, according to the business turned up *externalities*.

The *extended* enterprise is the industrialism contest, fostered by the *product-service* delivery, with, certainly, complex entrepreneurial accomplishments, even if the backing technological innovation is nice conquest of the existing computer engineering. If the attention is purposely turned on the technical prerequisites, the three above listed organisations extend towards trans-corporate set-ups:

- the *net-concern* grouping, the networked infrastructure aimed at a suitably assessed business opportunity; the joining characterises by the *clustering* specialisation, and the capability of offering value-added services;
- the *virtual* enterprise, a net-worked set-up among partners that support the *product-service* delivering, by assembling the up-dated provision/service capacities through (mandatory or voluntary) co-operation agreements; the *virtual* enterprise will not distinguish from the *extended* enterprise in terms of the timely incorporated (business-driven) functions and facilities, but in terms of company structure and related ruling by-laws; the administrative, legal and management rules are, however, domain of evolving statements.

The five definitions provide example explanatory situations, and several other choices are possible. In the technical literature, the number of options depends on the preferences of the interested person; in any case, more than the literal details, the multiple issues of enterprise networking is worth especial notice, together with the ability of blurring the *externalities/internalities* boundaries, showing why the infrastructure becomes winning option when the outer business scopes are used to drive and modulate the inner production capacities.

Anyhow, keeping with the above definitions frame, the industrial companies trends show that the *internalities* to *externalities* transit goes on by several ways and intermediate steps, having the *extended* enterprise establishment as coherent issue. The following steps are recalled:

- the simultaneous engineering *product-process*, **2P**, integration, aiming at scope economy by the adaptive exploitation of *intelligent* manufacturing facilities (with focus on optimal *internalities*);
- the service engineering, SE, provision to grant lifecycle conformance-to-specification, by *product-process-enterprise*, **2PE***, integration, exploiting proper PLE-SE aids to manage supply chain *externalities*, chiefly, through the drawing up of proper *voluntary agreements*;
- the reverse logistics, RL, provision to achieve the *compulsory eco-targets*, by *product-process-environment*, **2PE**, integration, with resort to recovery rulers and facilities/functions assembly into the *virtual* enterprise, fit for the backwards track mandatory duties;
- the lifestyle, SE and RL, provision, to comply with the enacted eco-rules, by *product-process-environment-enterprise*, **2P2E**, integration, including the needed facilities/functions into the *extended* enterprise, capable of the all fulfilments, as single business entity.

Of course, these are example steps, and more intricate set-ups can be devised. At the first two steps, the feedback, from the *externalities* to the *internalities*, does not (in principle) entail the environmental bylaws. The updating of the *process* or of the *enterprise* follows economical-driven decisional approaches. With the next two steps, the entrepreneurial business cannot develop unless the *compulsory eco-targets* are fulfilled. The joint *environment-enterprise*, **2E**, adjustment is the best-fit decision, leading to the *product-service* delivery, warranting the on-duty eco-consistency and enforced call-back, with resort to functions/facilities assembly in the single productive entity.

The producers' responsibility discrepancy, between lifelong service provision, and end-of-life recovery duties, explains why the latter accomplishments require, in any case, solutions, should be the manufacturers ready, or not, to cover the new business. The *virtual* solution affects, to tiny degrees, the company's *internalities*, making easy to respond at the regulations, even when not ready for more effective changes, through fully incorporated facilities, nonetheless reaching the mandatory *externalities* by resort to co-operating helps. Looking again at the sketched frame, further definitions are recalled:

- *lifecycle*: the collective set of phases a product or system may go through during its all life: design, procurement, production, operation, up-keeping, call-back, disposal, recovery, decommissioning;
- *product lifecycle management*, PLM: the tool for managing the data about a product, as it moves through the lifecycle: materials provision, on-duty needs and end-of-life disposal; it covers actions, such as: management of engineering and processing data, operation maintenance and conformance assessment, take-back recovery, recycle and reclamation duties;
- *product-process-environment-enterprise*, **2P2E**, design: the paradigm shifts in business, aiming at lifecycle (recovery included) effectiveness, granting natural resources preservation and eco-impact control, following voluntary agreements and/ or mandatory targets (focus on *externalities*); the up-dated productive organisation is achieved, using the function/facility market, to pick-up, negotiate, incorporate and manage the basic manufacture/service modules, which enhance the in-progress value chain;
- *reverse logistics*, RL: the business opportunity, lately subject to mandatory rules, for

the recovery (reuse, recycle) of end-of-life items, in compliance of enacted bylaws, with the related impact monitoring and data vaulting accomplishments;
- *service engineering*, SE: the business opportunity, principally driven by voluntary agreements, along the product life-span, to guarantee the item enjoyment, with conformance-to-use assessment, and related monitoring and certification accomplishments.

The ecology and allied demands are the industrial organisation new challenge: it is not enough the manufacture effectiveness to be competitive, it necessary to grant lifecycle efficiency. The supply chain *externalities* happen to be the critical requirement of the product successfulness, and the suited PLM are standard parts of every delivery, with the due PLM-SE and PLM-RL files. Now, the worldwide class corporation is compelled to cover the two emerging business opportunities: *service engineering*, SE and *reverse logistics*, RL. The bottom-up market-driven efficiency is not enough, and the lifestyle, SE and RL, provision is the emerging compulsory accomplishment of *legal* supply chains.

The coherent outcome aims at the *extended* enterprise, facing the all inclusive *product-process-environment-enterprise*, **2P2E**, design, incorporating, in a single business entity, every supply chain accomplishments. This means the full revising of old paradigms, due to the heterogeneity of the functions and facilities, needed by the productive organisation. The manufacture technologies cannot, anymore, be addressed as standalone means, and are to be inserted, to become instrumental for the overall *lawful* supply chain achievements.

2.2. TOWARDS THE KNOWLEDGE ENTREPRENEURSHIP

The *extended* vs. *virtual* enterprise dilemma blows out, when asked to perform the ecology accomplishments, e.g., compulsory reverse logistics targets, together with the technologies supporting aids, especially, the net-concerns potentials. This section provides an overview of these potentials, thus, suggesting technique-based developments, while the sustainable growth effects are dealt with by the following third section (at least, still, within the known «industry» paradigms).

The net-concerns are exterior views of the conversion towards the knowledge entrepreneurship paradigms. By themselves, they already bring in sharp scientific divide, by respect to the traditional reductionism, calling to assess the complexity into ensemble figures, to fully understand how the whole works. The facts show that to weigh up «how individuals behave» explains the joint conduct, only if the parts interaction is known. When the interactivity dominates, the complex systems are specified by synthetic features, while the analysis is almost useless, providing un-conditioned properties.

The tight relation, among the knowledge society and its pertinent information infrastructure, suggests twin conditions: human capital dominance, the know-how critically affecting the entrepreneurial activity starting and moving on; technical capital centrality, with stress on net-concerns, to manage, monitor and control the supply chain, for whole eco-footprint certification. In the average, the knowledge entrepreneurship fixes wealth generation shifts, with intense changes in the value chain, rather than total uprising. Indeed:

- the market of the IC technology follows industrial economy logic, simply assigning fundamental role to the technical capital, with the human one as enabling support (more than the financial), and the natural one as potential entity under continuous surveillance;
- the infrastructures acquire especial value added, being instrumental help, assuring the

shared, or, at least, the accessible support of the «knowledge» patterns, leading to new ways of wealth accumulation, notably, by means of net-organisations and complexity commitment.

These shifts involve some noteworthy facts, such as:

- the industrialism entails "investors/entrepreneurs *vs.* workers/attendants" modified relationship, with resort to companies, setting in the shadow the financial *vs.* the human capital contrast;
- the market puts in forefront new contrasts, e.g., "knowledge/technologies *vs.* personal expertise", namely, the distinction of the technical capital by respect to the individual inventors and users;
- the acceptation of the industrial economy gives little chance to the growth sustainability, unless that the supply chain visibility assigns plain urgency to the eco-conservativeness lawfulness.

The knowledge entrepreneurship operates the declared shifts, in all industrial areas, old or new. The special emphasis on the «new economy», simply, denotes striking examples, where the agreed changes and facts give insight on the dualism biding the human capital to the trade of knowledge and functions, adding to every material supply chains. These changes are, in the following, studied as «industry» shifts, requiring widening, before reaching «knowledge» consistency; namely, the analysis is still accomplished within the known industrialism practices.

2.2.1. Information Infrastructures and Networks

The information enhanced supply chains and the product-service pair are the distinctive marks of the knowledge society, with the common support in the net structures. These give evidence to the «global village», by which every individual, in each moment, is world-wide connected and can receive or send messages. The option, it is clear, has several purposes, such as the followings:

- co-operative processing of knowledge for problem solving and design;
- transmission, manipulation and management of archives and data-bases;
- retrieval of specialised information, or establishing selective contacts;
- plain entertainment, exchanging news or enjoying recreation and hobbies.

The structures pertain to the new technologies, showing many stakeholders: users, net-workers and developers, service-providers, equipment manufacturers, software producers, access managers, etc., all actively engaged in conceiving new opportunities and applications. These show how the intangible provisions, directly and indirectly, co-operate to the value added increase.

The aimed options help modifying the users' habits all over the supply chain, to add transparency to the allocated goods and functions and related eco-impact. The consumers start to recognise the differences between competing alternatives, and the visibility becomes the spur for conservative choices. The net organisation permits to cover three areas:

- the inner cluster, to link the enterprises collaborating to manufacture and to manage

the product-service;
- the allocated outer lines, for the purchasers' assistance and conformance assessment at the point-of-operation;
- the selective bonds, for the overseeing bodies' access and the eco-burden certification and accounting.

The information structure characterises the highlighted *extended* enterprise. It is the unifying context, accomplishing a set of characterising duties, namely:

- co-operative development of the technical construction files, with resort to virtual prototyping, to fulfil lifecycle overseeing, call-back and recovery included;
- decision support during manufacture, by simultaneous engineering aids, aiming at the scope economy for buyers satisfaction extended to and eco-protection;
- on-duty observation for control, diagnostics, maintenance, repair, etc., by conformance-to-specification checks, for functions assurance and impact assessment;
- monitoring help for the supply chain eco-management, granting the data transparency of provisioning, maintenance. recovery, recycle, reclamation and dumping;
- off-process ascertainment tool, to appraise the resources consumption and the remediation requirements over the all (forward and backward) supply chain.

These five duties help explaining why the net organisations are crucial for the eco-sustainability. The information infrastructures, anyway, provide quite wider chances. The possibilities enabled by the communication nets, with inter-activity aids to foster the stakeholder's co-operation are impressive opportunities of the knowledge entrepreneurship, with potentials, today, not fully disclosed and, in the past, unknown. The internet (hardware set-up) and the world web (software outfit) are aids assuring productivity in old jobs, and allowing start of new businesses.

The benefits of interactive communication nets, out of the function of linking people placed anywhere in the world, come from their capability to grant the job «visibility», enabling the direct control on the organisation efficiency and on the real usefulness of the transaction. Both aspects were in the shadow in most of the bureaucratic accomplishments, leading to hide productivity, as the cost is, simply, transferred to the citizens, through fiscal measures. The knowledge society, from this standpoint, is qualifying progress, at least, at the operation transparency and instrumental neutrality.

Looking at that side of the information infrastructures, it worthy pointing out the objective fairness of visibility, joined to the monitored data spreading. The technical potentials are one thing, and the actual issues, a different one, so that the recalled progress faces obstacles, if nested on old settings, with no restructuring. Special measures might help. The three-party producers-users-controllers eco-infrastructure profits of being law-driven compulsory request, without any frozen antecedents, which can be built with little or no bias against innovation, coming from protecting insiders and habits. Using the conventional theoretic approach of the market economy, having joint social-and-ecologic safeguard restriction, the resulting value chain characterises by two facts:

- the information infrastructure is itself made by industrial products, that, as such, increase the gross internal product of a country with the noteworthy property of intangibles predominance, even when they contribute to, e.g., government policies aimed at simple options (with caution on the conflict between marginal utility and original brain-works);

- the associated value chain brings forth synergic effects, if the net concern enables co-operative communication, processing and vaulting duties, in order to perform the required eco-charges, balancing the advantages of the consumers (manufacturers and users), against the protection of the other people and of the surroundings, by objective charges.

The computer technologies relevance comes out from induced effects, more than direct ones. These effects are common to the all knowledge entrepreneurship, leading to the restructuring of the apparatuses, into horizontal, lean settings, out of accomplishment duplication. As second stage, the eco-protection duties, further, profit of the process transparency, assuring feasibility to the control on the whole supply chain. The focus on «visibility» and «induced effects» is featuring aid, due to the «knowledge» patterns, when duly employed, to enhance the effectiveness, and not to aim at redundant, un-productive (or privacy intrusive) falls-off. Thus, the information structures ought to be assessed on their leanness, before saying about the enabled advantages: the operation transparency and instrumental neutrality are fundamental aids to such purpose.

2.2.2. Entrepreneurship Know-How Evaluations

The intangible technical fortune of a company corresponds to non-material and non-financial resources, which assure return on the strategic horizon (beyond the yearly costs-and-benefits report). These profit sources evaluation is, not only, a challenge, because often random fluctuations superimpose, but is an impasse, as well, because of the fuzziness in the choice of taxonomies and metrics, having general and sound applicability. The facts, already affecting the earlier industrial economy, become critical with the «new economy», due the reached percentage amount. As general rule, four entries to the intangible capital assets are considered (not strictly limited to merely technical contributions):

- ability in structuring and organising the internal activity: this leads to the enterprise set-up and goodwill, with included starting and co-ordination of the production processes;
- ability in expanding the market share and affirming the trade-mark: this is the result of the business striking root, with balanced general and sectional authoritativeness;
- ability in up-grading and innovating the technology know-how: this is the corporate attitude indicator to be at the competitors front, providing better quality/price offers;
- ability in setting network synergies and exploiting externalities: this shows direct/reverse logistics profits, in provisioning, distributing and call-back the products-services.

The classification has example purpose, and other entries might be considered and differently grouped. The assessment of the four items in the balance sheets is useful, helping to better evaluate the enterprise, but generates drawbacks, when misleading accounts are performed (deceitful, for investors), or reserved data are disclosed (advantageous, for competitors). Still, the *old* economy understatement has to be revised, especially to admit the technologies and externalities entries.

The organisation know-how has great relevance. Yet, the corporation culture is tacit entry, not easy to single out and to transmit in coded form; evolving, the enterprise accumulates implicit knowledge, solidifying the procedures. The entry assessment resorts to qualifying internalities figures, built on empirical features, linking: workers spirits and productivity rise, or operation skilfulness and training investments, or success motivation and responsibility allotment, etc.; the existing evaluation methodologies, however, bring to wavering and not easily comparable data.

The second entry shows known issues. The purchaser fidelity is nice metrics. A trademark is winning heritage, showing the company influence (*Coca Cola*, etc.). Know firms exist, also, in the «new economy» (*eBay*, *Google*, etc.), having prominent positions, by finding successful offers. The accounted value does not, chiefly, depends on the allotted pledge; still, the low acquisition costs of clients, often, gives back difficult conservation, whether market-anticipatory competitive policies are omitted.

The third entry best typifies the knowledge entrepreneurship. The joint capital asset is critically connected with the intellectual property protection and registered patents practice, forbidding the access and exploitation by other businesses, unless after payment of the royalties. Besides, the innovation cannot be fully covered by patents; e.g., the software copyright suffers unlike reading and coverage. As well, the patent deposition is two-side arm: it might be better to avoid it, to not leave the information spread out. The accounted value, mostly, is comprehensibly fuzzy figure, even when deferred to the market assessment, by the demand and supply law, as compared with the sharp rules, in the material items trading.

The last entry is critical innovation. The knowledge entrepreneurship fosters evolutionary abilities: the adaptivity to emerging technologies and the flexibility in acquiring new competencies are winning factors, blurring the internalities to externalities boundaries, towards the fourth entry. The company trading potential highly depends on well-established reputation. Today, the «new economy» allows mitigating the proposition. At first, the net organisation setting brings intangible contribution for the business start-up, both, for qualified resources acquisition and for targeted offers proposal. This leads to the firms' proliferation, with selective survival, when the intangible assets do not up-grade. The net support has, also, falls-off in the expansion of the corporations' relationship, making easy varying geometry structures, partners changing, experts temporary help, venture capitals fast assembly, consortiums build-up, quick access of broad stakeholders, etc., up to design/manufacture/trade set-ups adaptation, following the local and temporal emergencies.

The frame is made more tricky at the *extended/virtual* enterprise level, where series of partners interact, and the economy of scope requires leanness and agility. Here, the business has a ruler (not the main producer, in the *virtual* case) and a set of co-operating companies. The performance assessment of the cluster permits to regulate ruler and partners policies and behaviours. The entrepreneurship current evaluation, thereafter, requires set of checks, exemplified by the following pre-set block of assessments, to verify the project suitability, the collaboration usefulness and the partner aptness:

- at the project range:
    - *knowledge dissemination*: the information flow and decision-support tools acknowledgement and evaluation are managed, in order to appreciate the networking effectiveness;
    - *operation programming*: the business is fixed, establishing the partnership strength and acceptance interest, to establish set-up details, duty schedules, node responses, audits plans, and the likes;
- at the concern range:
    - *business participation*: the negotiation and setting of the listed intangible capital assets are individually run, to account for the intra-organisational requirement;
    - *co-ordination acceptance*: the mission assessment, advices and guidelines of the planned collaboration frames are dressed, to guarantee the pertinent integration sharing;

- at the node range:
  - *involvement obligations*: the task allotment, listing the receiver/supplier and inner/outer perspectives, specification is applied to verify all the involved nodes competencies and provisions;
  - *audits/reviews schemes*: the audits/reviews schemes are acknowledged, to establish the periodic and emergency measurement schedules and on-duty tests.

At the single partner level, the examination covers a set of accomplishments, used as self-assessment check of the on-going business profitability. An example prospect includes blocks, each of four tests, at the mentioned three ranges.

- At the node range:
  - *assessment aids*, the facilities set-up evaluation avails of decision support tools to deal with the all receiver/supplier and inner/outer perspectives;
  - *impact certification*, the eco-consistency is assessed, and the conformance-to-use checks are assigned to (third-party) accredited bodies;
  - *issues monitoring*, the co-operation (cost, time, conformance, quality, etc.) is granted full visibility, and the predicted compatibility is shown;
  - *shared metrics*, the nodes standards are taken from agreed measurements list, with the accord to establish, revise or cancel out the shared metrics.
- At the concern range:
  - *control and govern*, each node operates in autonomy, with decisions and issues communication, to allow on-process interaction;
  - *deployment trend*, the ability to create *value* goes behind the performance measurement, by training/learning aids, based on net-concern up-grading;
  - *duty organisation*, clear-cut partnership manages allocation, co-operation, achievement, etc. tasks, giving goals and clout to develop the business;
  - *integration modes*, the facility/function market is dealt with, to find out, negotiate and assemble the partner cluster that grant enhanced deliveries;
- At the project range:
  - *policy and objectives*, the business consistence and appropriateness are analysed as overall outcome, using market-driven cost/benefit figures;
  - *strategy choice*, the performance figure is available for the strategic resetting, ensuring involvement (and assent) of all (active) partners;
  - *tactical setting*, the current facilities/functions steering is done by the business ruler (or reference trader), according to the chosen scopes;
  - *up-grading planning*, the net concern is thought with evolving structure, through new partners and facility/function incorporation;

The weighing and ranking of each (twelve) self-assessment tests depend on the application, and will modify, when the policies and/or objectives change. The overall figure is important to assess the networking value added. The assessment procedures are above all important for partners operating within *virtual* enterprise, to fulfil the analysis by respect to the competition surroundings, in view to modify the cluster «instantiation», by more effective functions or facilities incorporation. The evaluation of the entrepreneurial know-how becomes instrumental to derive and propagate the new forms of industrial organisations.

2.2.3. Marketing Functions and Facilities Spread

The spread of the intangible function and service market, which characterises the knowledge society, has multiple peculiarities, partly, following established trends, partly giving rise to new opportunities. As for the assessed bents, we have the emergence of the human capital as enabling support of technologies; as for the innovative options, the synergies of complexity provide the richer outcomes of the networked organisations.

The value chains turned towards intangibles modify the work market, as well. In lieu of the earlier opposition of the workers, carrying out salaried jobs, against the proprietors, assuring the investments for the transformation means ownership, now, disjoint layouts establish, with complex productive set-ups, and superposed directorial, managerial, executive, administrative, etc. tasks, to assure the business deployment, independently from the financial capital origin, or, at least, under the (widely) loose control of it.

The separation, of the capital possession and company management, creates extra-roles at diversified duty, qualification, responsibility, etc., and this brings the power (and profit) to distributions that relate with the layout complexity. The term 'wealth' does not, any longer, lead to the actual holders strict identification. The dual process brings to detach the 'knowledge', from the pertinent possessors. In both cases, the legal frames became visible, to fix the proprietary rules, with important restrictions and reservations to balance the third parties and to protect the corporations' rights.

The financial and technical capital autonomy, by respect to the factual owner, implies proper exchange rules, with cautions, when defining the transactions legal aspects; e.g., the governments guarantee the facial value of currency; the contract clauses rule the protection of the software, and the like. The term «complexity» is standard challenge, leading to several falls-off, to describe the interconnected lay-outs, with diversified constraints on the individual parts (or nodes), which possess specialisation and competency. The qualification of each node is local property or personal achievement, as it was the case with the craftsmen skilled mastery, since the know-how means training and technological up-dating. The «complexity», in addition, addresses the structure characteristics, in which the local operators have equivalent replacing, due to the segregated 'knowledge' standardisation.

The discovering of the common behavioural features of the linked (physical or social) setting gives insight on the productive/servicing organisation, showing the interplay of concepts such as network, tie, division, emergence, interaction, and the likes. Now, complexity is abstract feature, if described by shallow knowledge: the flight of birds does not reach its ordered shape because of a structural law or of a leader command, but it emerges from the inter-relations among individuals, which follow simple and informal rules, such as the positioning by respect to the neighbours. However, the complexity and networking are, both, domains of study, still to be suitably acknowledged.

The «to de-materialise» axiom, maybe, cannot avoid market «complexity», so that extended explicit regulations become standard must. The good 'knowledge' is object of transactions between people, with onerous trading or free exchange, by rather especial ways: for instance, the supplier does not stop possession and, chiefly, he can perpetuate trading, indefinitely multiplying the same 'knowledge'. The other side, when the production costs are analysed, the incremental constraint appears: the 'knowledge' block is acquired, on condition to possess the adequate cultural background. The facts are industrialism options, which, maybe, replicate the old transition of «industry», originally abstract concept, and, today, tangible achievement. They, nonetheless, distinguish the «knowledge» paradigms by ways towards innovation, making the «cognitive» breakthrough the support of the new «to re-materialise» axioms. The present study starts looking at such prospect.

The information technology represents striking case, with exemplary course in the past.

At the beginning, the high-potentiality computation hardware had prices making, in the practice, impossible the wide-spreading of the cultural background. The investment costs dramatically decreased in progression, and concurrently the individual interest and expertise popularised, with no need to accede to localised and exclusive facilities. The developing communities training gap disappears, so that the high-top corporations decentralise their activity where wages (and life-costs) are lower, or enrol experts already trained with lower investments.

The course is generalised, recognising the industrial economy trends, which move from high-intensity financial investment structures, to new ones, grounded on high-intensity qualified-knowledge organisations, if the human capital training is major requisite. The outsourcing «function» market becomes winning practice, leading to *distributed* enterprise set-ups. The present worldwide *global* scenery, in fact, is affecting the financial, not the human market.

With the knowledge entrepreneurship, the intangibles value chain, which join or replace tangibles, is deemed to be new staples, outperforming the manufacture goods market. The *extended* (*virtual*) enterprises become the typical organisation, grounded on the following actors:

- distributed active proprietors of facilities/functions, interested to become qualified providers of their manufacture/service abilities;
- business devisers or promoters or rulers, requiring the acquisition of given facilities/functions to make the entrepreneurial project run;
- market regulators, with access procedures, remote negotiation and drafting aids, etc., to assure the demand and supply matching;
- appropriate facilities/functions brokers and information infrastructures, to help orienting the buyers/sellers choices and integration.

The *extended* enterprise aims at the entrepreneurial lay-outs, timely up-dated to satisfy what requested by the in progress product-service delivery. The *virtual* enterprise looks at similar aggregation of facilities/functions, with, however, not fixed ruler (out of the main manufacturer), motivated by the business project. The co-operation joint-venture is equally focused, being triggered off by the business specific demands. The *extended* and even more the *virtual* enterprises are highly dependent on the facility/function market. Both will expand into really efficient entrepreneurial issue, if such market exists at proper size and efficiency. The net aid permits achieving effective organisation, clustering co-operating manufacture and service partners, in view of the common business project.

2.2.4. Knowledge Entrepreneurship Efficiency

According to the several time repeated specification, the *extended* enterprise is organisation with the dominant trader, allied with all (or some) of its suppliers, in view of optimising the given business opportunity (to the customers' satisfaction and environment protection). The main features of the business integration, at the intra-organisational setting, deals with series of accomplishments, done, typically, by the configuration manager (broker or agency), using adjustments, e.g., at the, above singled out, three ranges.

■ At the node range:

- *efficiency*, the entrepreneurship value is affected by the information speed and

fullness exchanging/managing among the business partners;
- *involvement*, the alliance aims at co-designing, co-producing, co-selling, co-supporting the delivery to buyer satisfaction and eco-protection;
- *partnership*, the net-concern nodes are, or not, linked hierarchically, dominated by market objectives, and co-operation abilities;
- *technologies*, the core-competency compatibility allows the organisation effectiveness, by net and decision tools, to make the integration work.

■ At the concern range:

- *collaboration*, each one treats the other as business partner, allowing and accepting the other requests, in view of the chosen business;
- *duration*, within the project scope, the nodes share unified vision and work to fulfil the product-service delivery;
- *integration*, the business ruler (configuration manager) resorts to brokers' help for facilities/functions negotiation and incorporation;
- *ruling*: the decision-making aims at best-use of the partners' competencies, with account of the common goal, defined at the cluster setting/resetting;

■ At the project range:

- *communication*, the information sharing, on-line and in real-time, needs to assure the autonomic units in-progress collaboration up-dating;
- *drivers*, the supply chain complexity, with multiple point-of-service jobs, requires manufacture and provision clustering comparable sophistication;
- *size*, the setting occurs among the assembled nodes across the value chain, dynamically adapting the partnership;
- *structure*, the fitting will take the form of net-concern, where each partner is a node (with one or more input and output links);

The affiliation dynamic allocation, integration and up-dating are wholly linked to the on-going product-service delivery. The business does not end at the point-of-sale; it expands at the points-of-service (for control, maintenance, conformance assessment, restoration, etc.) and includes the take-back of the end-of-life items (for recovery-reuse-recycle targets and waste dumping duties). The net-concern new business opportunities largely go beyond the traditional manufacture duties. The productive break-up or out-sourcing is not safe, quite the reverse, it might be risky when non proprietary technologies are involved, and the accomplishments are enforced by law. The net-concerns, then, lead to competencies addition, to be built around (closest as possible to) the supply chain, for the buyers' satisfaction, the environment protection, and the business effectiveness.

The shift from the net-concern and *virtual* setting, to the *extended* enterprise, requires revising the business integration at the organisational level, grounded on the supply chain unified responsibility, face to the «product-service» mandatory targets. As steady state achievement, the *extended* enterprise has a company ruler, managing the production facilities and functions, in view of optimising the given provision and the eco-protection duties. This means that the above listed business integration main features need to still again revise the deeds of the entrepreneurial motivations at the three ranges intra-organisational level, through clear-cut ties.

- At the node range:

  - *involvement*: the service provision is standard, once the facility/function competencies are properly included;
  - *partnership*: the three-party binds suppliers (product-service dealers), the buyers and the certifying body (for the results overseeing and recording).

- At the concern range:

  - *integration*: the whole liability admits business rulers out of the main manufacturer, only if operating under strict accreditation;
  - *ruling*: the provider's commitment has voluntary agreements with clients and compulsory achievements with governmental authorities;

- At the project range:

  - *drivers*: the stipulation covers the product lifecycle, for conformance-to-use service and end-of-life take-back;
  - *structure*: the organisation is fully integrated, confining brokers/agencies help for facilities/functions negotiation and incorporation;

The apparent ambiguity, in the *extended* either *virtual* enterprises settings, is brought back to modify the features at the entrepreneurial motivation, by respect to the generic intra-organisational requirements. Today, it should be pointed out, the legal, permissible, administrative, directional, etc. aspects especially affecting the net-concerns, and, on the whole, the knowledge entrepreneurship, are far from fully explored, and we prefer to differentiate the *virtual* from the *extended* setting by intra-organisational requirements, rather than by lawful instructions. The idea is to consider the *virtual* settings, preliminary or temporary lay-outs, before the effective integration of the needed facilities and functions.

At this point, if suitable legal frameworks are established to make «assured» the lifecycle producer/dealer responsibility of the delivered items, the confusion of the two wordings does not create problems. It, mostly, becomes formal feature of the partnership (the supply chain manager, being, or not, the delivery designer and manufacturer), without lawful consequences on the buyers and third parties. Nonetheless, the deeper understanding of the prospected views brings to different end, when the factual achievements, expected by the competing alternatives, are defined by the intra-organisational ties, say:

- *extended* settings, with structure-embedded drivers and «externalities» covered at the intra-organisational level;
- *virtual* time-dependent instances, with «externalities» presided over by non-integrated bodies or agencies.

The objective measurement of the *extended* enterprise performance, hence, is important rule to assess the effectiveness of each alternative, explaining in which direction the sustainable growth could better be addressed. The topics, actually, deserve detailed analysis, and hereafter only introductory comments are given, to provide insight to the underlying conditioning helps, say, mostly, the information effectiveness enhancement, achieved by networked organisation, with or without structured settings.

2.2.5. Lifecycle Product-Service Supply Chains

The knowledge entrepreneurship deployment is deemed to be standard issue of future industrialism, foreshadowed by the «new economy», and undoubtedly forecast by the «externalities» to «internalities» interlacing of the lifestyle supply chains, where the *virtual*, either the *extended* settings become forced solution, in view of presetting the facilities and functions, to make the delivery feasible. The above analysis aims at defining the metrics, to compare the related performance, in terms of intra-organisational ties, since the legal, administrative, etc. features, today, indifferently permit facing the already enacted bylaws demands.

At this point, if the scenario of product-service supply chain, with the lifelong eco-protection, under mandatory targets, is true, the value testing becomes critical instrument to try understanding a bit more the industrialism to come. The virtual enterprise performance measurement deals with node and network properties. The node is the unit facility having the responsibilities for formulating, detailing and distributing the information concerning the directions and requirements of the co-operating facilities. At the lowest level, the effectiveness depends on networking virtues (out of the whole product-service conception and ruling).

Depending on the partnership lay-out, the control is hierarchic or distributed, always exploiting trimmed scorecard, having the information *receiver*-to-*supplier* cross-flow, with the local-to-global perspectives planning. The local tier arranges with the *two-port* cross-flow and the *inner-outer* scheduler, which balances the intra-organisational vision. With focus on the *receiver*, the viewpoint stresses on the project requirements. The case is reversed for the *supplier*, which has to deal with the subsequent nodes, assuring their right satisfaction. In reality, every local tier holds a many-to-many relationship.

The performance needs to comprise all the measures made between the set of *receivers*, and the ones between the set of *suppliers*, keeping the overlapping risk that, mostly, affects the complex structures. The performance measurement at the tier cross-flow depends on the local viewpoints, and shall be cleverly balanced, by the intra-organisational prospects, namely, giving account of:

- the *reductionism* fixes the local tier behaviour, giving the *feeder/receiver* schedules, conditioned by the *inner/outer* perspectives;
- the *inner* perspectives: the performance depends on local habits, still the standardisation requires to operate within the shared measures list, and to express the assessments after cross-examination;
- the *receiver* schedules: the assessments are joint fulfilled among the node and each of its suppliers, and are finally averaged, over measures drawn from the standard performance measures list;
- the *supplier* schedules: the bonds represent the integrated output interface, jointly built with each client, using the shared performance measures list, from the raw data, found in the cross-flow;
- the *outer* perspectives: the delivery shall adapt to the pooled requirements, drawn from the shared performance measures list, and is affected by how many nodes participate in the net-concern.

Each node is expected to organise its own inner setting, while amalgamating the receivers/suppliers requests, and generating the schedules for the aggregation duties. The measurement system should allow:

- the business ruler (dominant trader): to intervene when the nodes run into problems; to ensure the consistency of the overall assessments; to define policies and to decide corrective actions; to negotiate contracts; etc.;
- the team partner: to establish value delivery flow; to create mission-driven process/function adaptation; to explore measurements for self-mending; to ensure behaviour accountability and recording; to allow the allotted tasks renegotiation; to help, defining trends and detecting out-of-bound threats; to retain responsibility for calling and running the reviews and executing any decisions; and the likes.

The partnership dynamic allocation, integration and up-dating are specified by the (twelve) self-assessment adjustments, at the *virtual* enterprise setting (three) ranges, or by the further (six) intra-organisational enhancements, at the *extended* enterprise trimming. The basic self-assessment checklist will vary for the steady setting/fitting, either: instance, net-concerns and context. The expected issues will arrange according to entrepreneurial prospects, with, again, three operation ranges (see the previous paragraphs) and twelve accomplishment entries:

■ At the project range:

- *brokerage aid*: the intermediation among on-the-market facilities/functions can be carried by a domain broker or forced by the governmental agencies;
- *facilities market*, the core-competencies to be incorporated by net-concern are accessed from the facilities/functions market, after specific enquiry;
- *facilities assembly*, the final assembly can identify the stable *extended* enterprise, either address the current supply chain *virtual* organisation;
- *re-setting rules*, the supply chain servicing and end-of-life call-back imply incessant up-dating of the partnership, to include the pertinent facilities;

■ At the concern range:

- *information system*, the scope and organisation are ranked by competitive comparisons, using benchmarking indices;
- *business ruler*, the business opportunity is made-up to entail worldwide leadership, having vision of the sought market share;
- *net management*, the productive lay-out, support services and business processes are sustained and ruled by the transparent partners co-operation;
- *strategic enhancer*, the timely core-competencies are ranked by synthetic figures in the innovative domains and by complex ones in the assembly;

■ At the node range:

- *local leadership*, the responsibility requires defining the reference dealer, even if specific/shared charges are managed according to contract rules;
- *notification duties*, the environment protection, is subject to the enacted mandatory targets, and the issues are notified to the overseeing authority;
- *operation achievements*, the on-duty monitoring and conformance-to-use assessment are co-operation demands, to make transparent the responsibility ascription;
- *customer satisfaction*, current and final approval is ruled by (voluntary) agreements protocols, in extension to the enacted compulsory targets.

The above list collect examples checks, grouped in the top-down logic (of the project). To understand the performance measurement entrepreneurial checklist, it should be emphasised that the node, in this case, is seen from external viewpoints, when looking at the individual partner and requiring to measure its effectiveness and reliability by respect to outside (not only by respect to the other net-concern nodes). The listed twelve entries, again, can present specific weighing and ranking modulation. The overall figure permits to evaluate the organisation efficiency (by respect to competitors).

The project motivations deal with intra-organisational performance capacity, being this, however, preliminary step to evaluate the given business opportunity implementation. They shall move to the higher level context, to assess the actual return on investment, when the supply chain requirements *externalities* become the productive organisation dominant parts. Then, the balanced scorecard of the local tiers (with *two-port* cross-flow and *inner-outer* scheduler) is affected by the conventional *internalities*, properly adapted and re-issued to take into account the net-concern's peculiarities, at the intra-organisational level, and this explains the *extended* enterprise merits, by respect to the *virtual* enterprise effectiveness.

Of course if no worldwide competitor comes up, the *virtual* enterprise solution is acceptable issue, steered by govern-like agencies (with the related bureaucratic and administrative burden), or promoted by more agile business rulers, primarily, with broker-like involvement. Example situations are mentioned at the paragraph 5.3.7. There is room for progress at the entrepreneurial level efficiency, once the integrated **2P2E** approach is completed, driving up to the *extended* enterprise set-up, because the «business ruler» is operating as principal designer/manufacturer, openly fulfilling the dominant dealer duty and integrating the all supply chain job, since the product-service design phase.

The outlined performance assessment, uncaringly, addresses the *virtual* or the *extended* enterprise, assuring full compliance to every compulsory eco-duty. The only disparity is represented by the fact that the first self-assessment checks series refer to pre-set constraints, where the **2E** externalities cannot be 'optimised' (they are only deferred to control agencies). Through the last performance measurement checks series, the explicit **2P2E** integration is managed by the dominant ruler, with account of all *externalities*, while merely the legal corporation, pool, consortium, conglomerate, syndicate, etc. setting is left to factual decisions (timely aiming at innovative structures, as soon as the law will better settle the overall frame). At this point, the preference, to look after fully formalised *extended* enterprises, and not simply to «agile» *virtual* instantiations, is understood.

2.2.6. Registered Certifiers and Controllers

The earth resources bookkeeping is spur to establish monitoring schemes and objective metrics. The eco-restrictions are enforced by the, more or less deeply, negative yields of the natural capital. These data have quite different footprint, if they are the outcome of averaged rough estimations, either, they are obtained by the direct assessment of individual case histories. In the first case, little cynicism or egoism suggests to remove the personal drawbacks, and to charge the faults on somebody else, so that the remediation is required from the global village generic citizen, and not from the given protagonist. In the second case, no excuse exists.

Unluckily, the municipal/governmental authorities, either look for individual income, either for type-behaviours or standard citizens aggregated data (this is the way to charge the home garbage service in Italy, etc.), and the virtuous habits are not rewarded, so that the public spirit finds difficulties to emerge. Then, summing the lack of interest in rewarding conservative behaviours, to the fear of excessive intrusiveness in the private habits, the governmental bylaws obtain poor issues for the eco-protection. The resort to third-party registered actors should be considered valuable alternative. These certifying bodies operate:

- interacting with the *collaborative* network support, established within the *virtual/extended* enterprise, to manage the *product-service* delivery, with visibility on the lifecycle data and proper security restrictions;
- verifying the monitoring and control network integrity, and providing the conformance assessment records by in-progress statements, as estimates for tangibles consumption and pollution impacts *refunds*;
- overseeing the *virtual/extended* enterprise accomplishments for eco-charge payments, with progress balances, based on unified responsibility, binding the (two) *consumers* parties indentation;
- guaranteeing data vault and privacy protection, with proper specifications in the case the certifying duty is transferred to a different body and/or the joint consumers parties modify the binding contracts;
- operating within an accreditation scheme, notified to local (supra-national) authorities, and confirmed by international establishments, having world-wide acknowledgement.

The third party certifying bodies' involvement needs proper rules, enacted by the national authorities, according to the EU directives, and harmonised, to assure worldwide equivalence. The emerging business will profit of the co-operative net organisations, according to different schemes:

- on one side, the product providers can be spurred to keep in charge the all *service*: supply flow, lifecycle conformance and disposal duties, so that the trade regulation would depend on single indentures;
- on the other side, independent enterprises could profit by safety rules and environment acts expansion, to become *service* dealers, with technology oriented qualification and infrastructure-based organisations.

The two schemes require focusing on the design phase, moving the enterprise profitability to be critically dependent on concept choices. The former identifies with the *extended* enterprise solution; the latter, with the *virtual* setting. The final outcome suggests reaching three achievements:

- the marketing of *products-services*, with proper collaborative networks, to support the clients requests;
- the involvement of *qualified* operators, assuring *point-of-service* and *end-of-life* conformance guarantees;
- the overseeing of *third-parties* certifying bodies, to record the eco-impact and the tangibles actual productivity.

The *third-party* certifying bodies are necessity of the overall-comprehensive market surveillance, which is unrealistic, if committed to governmental agencies, compelled to monitor the traded goods totality. The effective collaborative and overseeing networks develop with the already mentioned topology:

- the inner cluster, to link the net-concern's nodes, running the supply chain PLM-SE

and PLM-RL maintenance/recovery data flows;
- the pertinent links, to support the *point-of-service* communication, among the *virtual/extended* enterprise partners and the individual buyers;
- the selective data channels, to give access to the certifying bodies, under proper security protocols, of the *on-duty/end-of-life* impacts.

The *extended* enterprise solution can aim at optimised issue, by feeding back the supply chain data, to redesign the *product-service* offer. The shared, varying-topology, information set-up is unifying context, sliced into layers, with:

- the inner level (products ideation and construction), within the *extended* enterprise information system;
- the outer level (lifestyle *product-service*), for the data management at the clients' satisfaction and eco-safeguard;
- the overseeing level (delivery sustainability and tangibles charges), ruled by the *accredited* certifying bodies.

The image is coherent with the *controlled* collaborative net, linking suppliers and clients, for monitoring and improving duties, that make the *product-service* course transparent to the accredited certifying body, for checks and conformance assessments purpose.

The three layers net-concern profits of the *free-market* competitiveness rules. The *extended* enterprise is world-class company, after negotiation and merging of co-operating partners, chosen among suited function/facility alternatives. Several *certifying* bodies contend in the market, drawing up overseeing agreements with the consumers (manufacturers and purchasers), to be revised or switched among other parties, under proper continuity conditions.

The environment management gives rise to an expanding business. New *fair* trade rules are required, assessing the *service*, permanently linked to the *product*, with the eco-footprint recording. The lifestyle delivery needs to be standardised into integrated design scopes, with momentum on all details, up to recovery, since recycling is critical, with time/energy consuming dedication, and little interest in the resulting items, unless goal-oriented targets and incentives are fixed by law.

The above three parties' eco-management is effective issue, to grant objective assessments and privacy protection. The direct supplier-user convention can result affected by partiality, likewise the user-certifier indenture. The three-party linkage with registered (by means of accrediting schemes) bodies, entitles fast arbitration, in case of dispute, replacing endless law suits. The charge systems could develop to be consistent with, both, the EU and the USA approaches to regulation:

- with mandatory targets: the taxes basically apply to the manufacturers, and the checks on wastes and emissions could forbid given items use and shall require the exhausted parts take-back;
- with voluntary agreements: the eco-consistency is encouraged by charges modulation, mainly, allowed at the point-of-service, to keep the permanent recording over the consumers' behaviours.

When the setting achieves total fairness, comprehensiveness and transparency, the goods value chain will deal with modified market (supply and demand law) price, due to the eco-charges incorporation. The fact is not neutral. The standards objectivity and assessments transparency are basic measures to achieve fairness and uniformity. The market competitiveness develops unbiased, when worldwide conditions establish, having resort to:

- pervasive means, for the lifecycle monitoring, cumulating provision, on duty and recovery data;
- unified standards, for obtaining exhaustive measurements with the legal metrology validity.

The pervasiveness is relevant property, assured through computer engineering. The objective assessments require standards grounded on legal metrology rules. Together, they involve the knowledge entrepreneurship deployments. The three parties' net-concern builds up, as competition-driven service:

- the joint-consumers' side lets out (on contract) the overseeing execution and the related conformance assessments, but can, any time, change the certifying body;
- the authorities are entitled to rule the legal frames, but without any direct involvement in data keeping and recording, accomplished under proper secrecy by parcelled-out services.

The *extended* enterprise set-up allows implicit collection and administration of the eco-charges, because the manufacturers' lifestyle responsibility is requested to be all-inclusive, so that the on-duty and end-of-life tasks have single respondent. To that purpose, the ecology business takes shape, whether:

- the purveyors cover the all *supply-chain*: materials provision, production, lifecycle up-keeping, backward recovery; the eco-duties are dealt with by clustering several firms within the factual alliance of co-operating multi-sectional interests companies;
- the users buy *products-services* to profit of the *functions*, with reliability figure close to one; the payments include the conformance certification at the points of service, and the end-of-life call-back, after the tax collection against the tangibles depletion;
- the supervisors assure *third party* assessments for the today and tomorrow environment and society protection; the registered certifying bodies report to governmental authorities and use objective standards, having access to the *product-service* delivery lifecycle data.

The entrepreneurial «complexity» is eco-protection necessity. It cannot leave in the shadow that, with enforced regulations, the governmental ruled overseeing is the only alternative, with the consequent bureaucratic burdens. The all is here discussed on merely technical motivations, and further socio-political insights are needed, to better figure out the factual organisational outcomes.

2.3. ENGINEERING ECO-SOUND BUSINESSES

The industrialism evolution has undergone several technology-driven changes, with the in progress advance of the *artificial* energy support, from the centralised power supply, to the distributed motorisation, and to the modulated actuation. The *artificial* intelligence support has, likely, contributed to the improvement from the fixed, to the flexible automation, helping to establish the manufacture settings of the market-driven economy of scope. The affluent society ending, with the growth sustainability demands, brings in new changes, now leading to the ecology-driven economy of scope, where the civic consciousness, to preserve the earth resources and environment for future generation benefits, appears to be impending request. The outlook of the previous sections has, primarily, followed the trends along the interplay between the *financial* and the *technical* capitals. In this last section, the focus is more specifically on the *natural* capital, and related falls-off in the *human* capital. To start, the original four reasons of the economy of scope are up-dated, to deal with the changed context and to provide the new motivations:

- the growth sustainability: it is not enough to manufacture quality goods to grant wealth, as the natural resources reduce to waste and pollution;
- the civil benefit: the average buyer is forced to face severe eco-protection rules, and his satisfaction turns to more conservative and reliable offers;
- the governmental regulation: the manufacturers' responsibility to put in the market new products is extended to their end-of-life free take-back;
- the technology changes: the computer engineering permits overall supply chain visibility and throughout exploitation of the net-infrastructures.

The economy of scope, so, suggests the way to more complex entrepreneurial set-ups, capable to include in the business project the many eco-accomplishments each time enacted. Thus, even remaining within the known industrialism schemes, the ecology pushes to restructure the productive systems, out of the old «industry» patterns, because of the impending socio-political requirements.

2.3.1. Free Market and Economic Global Consent

The apparent health of the world financial markets follows the plain statement that the men should be allowed free trade conditions, to operate in self-utility and to look after increasing their benefits. The personal wealth progresses, if no outer obstruction or political hindrance is erected and each one is fully permitted to act in his own interest. The individual citizen's achievements build up the wealth of his all nation, in the Adam Smith view. The political liberty and the ownership rights are direct prerequisites (and privileged consequences).

The evidence comes from the old industrialism successfulness, as it was run by the UK at the revolution setting, paradigmatically assessed by the, previously recalled, "*long global assent*" or, again, recently, by the US "*short global assent*". The two "*global assents*" promoted worldwide spreading of the financial market, through synergies by removal of the protected position income (which rises local penalty), and by scale effect efficiency, to the multi-national players advantage.

The market self-regulatory ability, typical of spontaneous achievements, in the end, results in the establishment of world competitors, on the go absorbing weaker and inefficient actors, towards business oligopoly formation, capable to promote and to manage the finance operations, with optimal profit. The scenario will lead, perhaps, to the *hyper-market* prospects, later outlined as hypothetical issue of the worldwide trade organisation, ruled by multi-national corporations, surpassing the regulations enacted by local governments.

In parallel with the recent *global* deployment, the *no-global* opposing political movements arise and worldwide scatter, with the common protest against the free market excesses. The quoted dissent, as for the *Washington Consensus* document, is known, and the objection of 'local' opposed to 'global' is easily understood. It results more difficult to find unifying propositions, towards effective alternatives, as it will be, further, discussed in the following.

The ecologic concern is the most clear common *no-global* concept. The boost towards rejecting whatever might have origin in industrial competition becomes basic axiom, with contradictory outcomes, also, in what would affect the technical capital. As a principle, the *no-global* approach means to be against the non-shared innovation, as the elitist benefit disguises an 'arcadia' order, acting the safety role (and fighting, e.g., the «aberrant» genetically modified produce).

The position leads to other issues, such as the dissent on patents and royalties, thus, motivating the «no-logo» groups. The *global* and *no-global* abstract readings are too extreme, to fix coherent economical systems, with future practicability:

- on the *global* side, the economical presumptions have, e.g., full confidence in the market capability to find out alternative intangible resources, allied trade organisations and safe reclamation process, which allow removing or counteracting the pollution downgrading;
- on the *no-global* side, the ecological conservativeness has, e.g., disciples, which think in paradox, to join the present quality of life preservation, the resource consumption and environmental pollution removal, and the full rejection of the industrial practice inheritance.

The demographic and financial factors add to the ecologic and social ones; the communities and governs accountings cannot anymore ignore the need of the four capital assets balances, without exorcizing the industrial activity, and related falls-off (stocks depletion and pollution escalation), cumulating the out-of-balance off-sets, undamaging the not involved people and future generations.

The *no-global* movements have the merit of the ecologic alarm, and request to take into account the impacts on the surroundings, as constraint for the people to come, giving evidence that an inter-generation pact should exist, to safeguard the earth life. Today, the global village cannot be disregarded. The active *post-global* way to face the growth demands means to acknowledge the ecology constraints, and to investigate these problems, in view to find out solutions.

The way might be viewed, as dialectic fruition of balanced ecologic/economic issues. To remain in the industrialism frames, the *post-global* approach explores example patterns, in order to establish the priority to efficiency, e.g.:

- the role of the ownership rights on the natural capital, raw materials and products, to promote and build-up new wealth (preserving the protection of the intellectual property *vs.* the «no-logo» claims);
- the worldwide market management of the (goods and services) ownership rights, with identified people or legal body, holding the subjective interest of preserving and increasing them;
- the bookkeeping habit of the engaged natural capital assets, to monitor the in progress yield, with individual commitment for the unbalances and the third people and future generation safeguard;
- the technology-driven inventions, allowing to widen the tangibles stocks by reliable and effective new developments, given by «to re-materialise» *artificial* processes.

The bird eye view on the *global*, *no-global* struggle shows (odd) binds with *green engineering*, fostering technology product and service advance, to market-driven requirements (to win new buyers', facing *global* competition), and official bylaws (to satisfy the *local* on-duty and recovery rules). The facets, appearing in the on-going industrialism, lead to puzzling facts, such as the following ones:

- the industrial development is effective with the local consensus, based on the individual commitment towards wealth, supported by the protection of the personal socio-economic and politico-legal rights;
- the *local* growth is vitally linked with series of rules and acts that sponsor competition (*free-market*, etc.) and benefits (ownership rights, etc.);
- the industrial expansion is enhanced by the "*global assent*", or worldwide trade accords, pulling down tax barriers, eliminating local subsidies, etc.;
- the *global* growth takes principal benefit from the economy of scale, when the winning competitors reach oligopoly by brand's limited market share;
- the overall picture is consistent with stable and ordered organisations and steady political institutions, possible with governments having the national loyalty of democratic parliaments;
- the wealth build up is deeply affected by the financial capital management, more than by the human and technical capitals; the natural capital is totally neglected, as instrumental means.

Contrary to the listed facets, the *no-global* claims are in full conflict, without, however, similarly practical recipes. The *green engineering* intersection is modest aid. Basically, the lifecycle visibility *value* increases, when connected to fair trade contexts or reported to mandatory targets: general principles are issued (e.g., by CERES, Coalition for Environmentally Responsible Economies). Some company includes *green*-guidelines as standard design constraints, to be allowed to use eco-labels, testifying the conformance to compulsory goals or the achievement of high conservative records. The overall fuzziness, anyway, shows that the industrialism cannot continue along unchanged paradigms. Thus, the present study is worth, to look at how far the current paradigms permit to progress. Indeed, the above listed facts are not puzzling *per se*, but because they pass over introducing the «natural capital» productivity, and the eco-footprint bookkeeping.

2.3.2. Co-Operative/Conflicting/Overseeing Concerns

The *global*, *no-global* dilemma has no issue. The *green engineering* junction provides only palliative supports. Better ideas come out from the ecology-driven economy of scopes prospects, with *extended* enterprises, subject to lifestyle rules. The build-up of bylaw frames, rather than obstacle, is spurring to technological sustainability, giving visibility to competing solutions achievements. During the "*short global assent*", the value creation was typically due to the technical capital, with the further potential of the natural capital provision transparency. The end of it is a break, not the defeat, towards the knowledge society (and related ecological achievements).

To look *positive*, the nation-state hegemony crisis helps finding breaks to the above listed facets. The multi-national corporation is, in theory, free from single government rules, and optimises its financial return by productive break-up and job out-sourcing, thus, transferring competitiveness to low wages regions, with the shared proprietary know-how. Of course, the knowledge is an odd good, no more protected, if shared.

This means that the knowledge society pioneering set-ups are fated to loose their competitive advantage, unless the intellectual property is guarded and (when duplicated) charged. This, possible for the advanced organisations, is rejected by, e.g., the «no logo» people. The growth stops, deprived from technical capital. The *global, no-global, post-global* path shall go worldwide to the end. In the future, it will be difficult to distinguish «global village» blissful districts, because the eco-conservativeness cannot leave out merry islands.

The so forth listed ideas address the knowledge entrepreneurship, supporting all tools for the deliveries smart use, say, *extended* enterprise, mixing apt facilities and functions. The technical details lead to the said three-partner setting:

- the manufacture organisations, with the intelligence-driven re-thinking at the design departments;
- the deliveries users, enjoying lifecycle conformance, environment impact and end-of-life service;
- the certification bodies, with the monitoring and control tasks for the eco-consistency checks.

The engineering activity evolves, reaching conditions of *scale*-economy in the field of *knowledge* processing, with the market of intangibles surpassing the one of tangibles, including the incorporated services and reporting ecological balances (at the same time, the materials processing reaches its highest efficiency by *scope*-economy, through lifestyle integrated design).

The certifying bodies are essential infrastructure innovation, as the *fair*-trade handling prerequisite, with safeguard of sellers, purchasers *and* ecological assets (for the future generations benefit). The explicit active role of the third party can be contentious, when the *extended* enterprise already strives, with coherent know-how, for developing systemic innovation. The voluntary agreements scheme, this way, assures products smart use, through proactive (native and on-duty built) eco-managing; the obligations are internal fact and integrated design keeps awareness under structured and unified frame. This, clearly, is ideal opportunity, achieving highest effectiveness, due to intrinsic leanness. Though, the absence of conflicts on the consumers (producers and clients) side could turn inconsistent the control, unless the overseeing is committed to a third party.

The facets of the possible *post-global* industrialism (overcoming of the *global, no-global* dilemma), might lead to the example requirements:

- the wealth build up develops, balancing ecology and economy, with focus on natural/technical tie, rather than financial/human antagonism;
- the worldwide trade consensus needs to include binding agreements on the natural capital management and restoration;
- the nation-state sponsorship and safeguard shall carefully embrace also the protection of the overall mankind and future generations;
- the population quality of life ought to apply transparent bookkeeping for withdrawn resources, dumping and pollution;
- the worldwide loyalty has to replace the current voter's egoisms, with the *altruism* roles, going beyond the parliamentary democracies.

These *post-global* facets leave in the shadow lots of demanding facts, starting from the nation-state policies and habit, to report to actual (living) voters, and not to non-voting citizens and to (still not existing) future generations. The all frame, thus, should first be understood, incorporating betterments and integrations, once the industrialism we know and related affluent society end is acknowledged. We have to distinguish the political roles (e.g., local authorities fixing eco-constraints, with worldwide validity and acceptance), from the industrial actors duties. These, specifically, are discussed hereafter.

The conflicting patterns are, profitably, surmounted by the joint co-operation and overseeing rehearsals, where the different parts understand that united efforts, only, might lead to avoid the impending defeat. Besides, the knowledge society is, as far the up-now signs show, the continuity of the current industrial society, with the switching to staples into intangible provisions, more than into manufactured goods. The «new economy» suggests opening hints, and the *extended* enterprises, with lifestyle clients and overseeing certification partners are additional practical recipes of what the future could look about, through *solidarity* pacts.

The continuity gives rise to the environment management business; *fair* trade rules are required, acknowledging the *service* (each time) linked to the *product*, with the eco-impact transparent recording. It is hard to think that the all could be ruled by supplier-user indenture, better than by three parties' connectivity, having registered bodies, entitled of monitoring the *solidarity* bylaws and of arbitration in case of dispute. The charge system has consistent regulation:

- with mandatory targets, the taxes apply to manufacturers, with checks on wastes and emissions, given items use forbidden, and exhausted parts call-back enforced;
- with voluntary agreements, the items' eco-conservativeness is encouraged by taxes differences, mainly, conceded at the points-of-service, to keep the permanent recording of the users' behaviours.

When the frames achieve total comprehensiveness, fairness and transparency, the goods value chain will differently deal with the market (supply and demand law) price, due to the mandatory addition, with anticipation or dilution of the eco-charges. The bias is huge: the objectivity (of standards) and the transparency (of assessments) are basic measures to achieve uniformity and fairness. The market competitiveness will grant unbiased worldwide conditions, with resort to:

- pervasive means, for the lifecycle monitoring, cumulating provision, on duty and recovery data;
- unified standards, for obtaining exhaustive measurements with the legal metrology validity.

The pervasiveness is relevant fact, assured by computer engineering. Several metrics could be developed to reach smart assessments. At the moment, the *post-global* active engagements are more hopes, than obvious prospects.

2.3.3. Resource Management and Recovery

The repeatedly proposed topics show that the affluent society, produced by the industrial revolution, based on goods ceaseless exchange, manufactured depleting earth resources, transformed into pollution and waste, cannot last. The progress requires new business paradigms, aimed at measuring and taking the eco-footprint under control. The industrialism issues devise, basically, two ways:

- to pursue a careful waste policy, with relevant eco-charges and mandatory recovery targets;
- to collect eco-fees for raw materials consumption and to impose reverse logistics resurgence.

To address minimal pollution and dumping is, only, an attempt, to run more conservative supply chains. The eco-decay limits the *industrial* growth, in time, unless upgrading ways are found, circumventing entropy: by information value-added chains; bio-mimicry (say, restoration by life processes emulation); or other technical capital driven alternatives.

The sustainable growth, thus, in addition to regulated dump, shall control the tangibles consumption, aiming at circulatory flows in the manufacture activities, with expansion of the reverse logistics domains. The engineering areas will, once again, face important restructuring, according the recalled issues:

- preliminary level, the «to de-materialise» issue: in the value chain, staples are in intangibles, with emphasis on the computer engineering aids;
- advanced level, the «to re-materialise» issue: in the value chain, focus is in tangibles restoration, with centrality on energy/resource stocks refilling.

The restructuring already shows first signs, when future offers in competition are put in the market, stressing on the lifestyle provisions at clients' enjoyment and eco-conformance, rather than on the spot approval at the point-of-sale and no care of the eco-footprint.

The enhanced resources control is new challenge. The purchasing of goods will require improved suppliers' qualification, due to the lifecycle responsibility (now, limited to the free take-back of some end-of-use durables). The marginal company, with no eco-soundness, could be required to draw up onerous insurance contracts to warranty the supply, making impossible its independent survival. The practice moves to revise concepts behind the manufacturers' effectiveness, when the responsibility is not bounded to spot output (provision of a service, function or performance), but requires lifelong commitment and end-of-use remediation, to fulfil the enacted recovery (reuse, recycle) targets.

Today, the facts are not fully clear. The gradualness suggests to progressively enabling improvement opportunities, i.e., products-services, assisted by *extended* enterprises, and networked entrepreneurial options that grant the timely needed tasks and competencies. The scenario means:

- from the entrepreneurial viewpoint: the «externalities» create the business organisation, tailored to support the lifecycle servicing, and to assure the compulsory recovery accomplishments; the «externality» effectiveness is competitive advantage;
- from the user side: the «externalities» are mandatory to enjoy what he has bought; the exploitation of the product is allowed, if the eco-requests are carried-on, according to the established rules;
- from the governments' regulatory frames: the «externalities» are necessary prerequisite to put in the market products, having certified eco-footprint of the borrowed natural capital and disposal safety.

The idea of *virtual* enterprise comes in as factual choice, having the underlain assumption of efficiency by agility, i.e., by avoiding redundant lock-ups, while reaching effective leanness by simple networking. As recalled, the «externalities» are seen by the EU environmental policy, with two scopes:

- to establish stringent conformance-to-use requirement for monitored eco-impact; there through, implicitly, to foster voluntary agreements towards the product-service market;
- to enact compulsory recovery targets at the product discarding, combined with economic instruments based on the users' free take-back plot, under the explicit

manufacturers' involvement.

The «externalities» are not embellishment. They are necessary manufacturing business step, and represent basic achievements to make the activity lawful, after incorporating the enforced *natural capital* managing rules. The *resource manager* will play the fundamental role for the company competition, with coverage of the provisioning, on-duty operation, and end-of-life recovery. It is clear that, for the supply chain lawfulness, the enacted eco-requirements need to be fulfilled by the resources consumers: manufacturers and users (the first ones, in addition, having the overall responsibility of the traded products-services).

The duty allocation vis-à-vis of the authorities is, by now, not fully fixed. It is, perhaps, useful to identify the fundamental facts:

- the provision aims at a client, interested to enjoy it in full eco-compliance, and he has to pay for what it is really delivered for his enjoyment;
- the provision has a design staff, charged to grant a fully-inclusive offer, to satisfy the buyer, through actually enjoyable products-services.

The *resource manager* is design staff function, planning the eco-attainments. Alternatively, the co-design staff should exist and co-operate with the producer, responding vis-à-vis of the authorities, to fulfil the enacted eco-requests, on the user's own (setting, typically, promoted by the *virtual* enterprise operation). Both set-ups show that the eco-features are necessary prerequisites to put the product in the market.

The *extended* enterprise, responsible of the product-service, established after integrated design procedures (with eco-scopes included), coherently reaches full effectiveness. In default, a pace-wise trend has to be followed, notably in the UE, for the series of mass-produced durables, with compulsory recovery targets. When the resource manager is not routine accomplishment in the design staff, the *virtual* enterprise can be devised. The vicarious project ruler has to fulfil the basic duties: he lists the set of dangerous/polluting materials not to be included, and specifies the generic criterions to reach eco-safe lifecycle issues; moreover, with the users, he is required to select the certified operators, consistent for on-duty conformance assessments and end-of-life recovery.

Today, the situation is far from the steady state issue, with manufacturers too much limited on the production *internalities*. Anyway, the future trends are well defined. The *extended* enterprise shall use the resource manager, at three ranges:

- for general duties only, looking at the occasional lifecycle assessments, in relation with the already specified criterions;
- for the mandatory recovery (reuse, recycle) targets of end-of-life products, to be accomplished by the *free* call-back process;
- for right business expansion, planning and accomplishing the service jobs by voluntary agreements interventions.

The *resource manager* is manufacture function, not a man only. The function, with resort to integrated design methods, **2P2E**, helps closing the gap between the *internalities* (product/process fusion) and *externalities* (environment/enterprise matching). If formerly meeting buyer satisfaction meant to sell a product with highest performance/quality ratio, now the companies need to, moreover, supply lifelong service and maintenance and to accomplish disposal and recovery duties. The resource manager function, in a coherent set-up, has in charge three areas:

- the *product lifecycle management*, PLM, with technical commitment on the product lifecycle span;
- the *service engineering*, SE, provision, by "extensions" (today) based on *voluntary agreements*;
- the *reverse logistics*, RL, duties, by "obligations" enforced (for given items) by *mandatory targets*.

The supply chain control, maintenance, restoring, etc., are PLM-SE aids. The end-of-life recovery, reuse, recycle, etc., are similarly covered by PLM-RL aids. With the said approach, the duty responsibility frame, as steady issue, is consistent with the *extended* enterprise set-up, if the integrated *product-process-environment-enterprise* design leads to shape out the most efficient business lay-out, up-dated at the specific *product-service* stage.

The legal and economical rules, by which the sustainable growth is enforced, are, today, at preliminary phase. The fully structured *extended* enterprise is on the way, as technological innovation. The net-concerns are powerful enhancers, and permit to address flexible arrangements, where unified responsibility is adaptively faced, once «vicarious» resource managers assure effective ruling and backing, by broker helping the choice and the incorporation of the required facilities. The all leads to hypothesise a way, where the eco-consistency is faced by new knowledge enterprises, capable to manage the resources preservation:

- to provide preliminary acknowledgements of the *eco-conservative* issues, recognising the underlying growth enhancing requirements;
- to suggest the viable features for *service engineering* and *reverse logistics*, embedding data visibility and collaborative networked lay-outs;
- to outline the business *knowledge organisation*, involving complementary stakeholders and facilitating decision-making supports.

The critical nature of the *eco-conservative* frameworks depends on two facts:

- the waste, scraps and pollution are, chiefly, left to municipal dumps, with community costs, too much spread by loosely-assigned taxation schemes;
- the raw materials are paid the market price of the supply-and-demand law, with no fair concern on earth stocks depletion and future generations.

The EU, already, clearly requires collecting taxes of scope for landfill access and pollution remediation, and progressively aims at establishing refunds on the natural resources withdrawals and dumping spread out (see, the Kyoto protocols). This is equal to establish a *cost* of the actually engaged *natural* capital, exactly as we are accustomed to do, with the loans of *financial* capital and the employment of labour or *human* capital. The payment for *natural* capital assets, merely, means to apply the common practice, of repaying the loans, through interests, and to give back the borrowed amount by lawful redeeming plans.

2.3.4. Lifecycle Service and Reverse Logistics

The comparison, between the listed *global* industrialism (six) facets, and the *post-global* similar (six) rethinking, shows the big two changes: on the ownership right to fully dispose of natural capital pieces; and: on the free market through the up-dated supply and demand law.

The actually exploited *natural* capital pricing depends on entangled motivations, which entail: balancing the handled tangibles and intangibles (e.g., *artificial* energy and intelligence); weighing up the entropy increase; assessing the drifts from the safe-life condition (earth warming up, etc.); and so on; the all, showing the today industrialism tiny soundness.

The further exploration, based on obsolete principles or hypotheses, does not seem to be of great help. The factual reference to the knowledge entrepreneurship potentials, face to the sustainable growth demands, appears better choice. More precisely, the discussion that follows, aims at the two *externalities*:

- service engineering opportunities, for the eco-optimised supply chains of the product lifestyle design;
- reverse logistics accomplishments, for efficient recovery tasks, supported by networked organisations.

The topics develop as the main resource manager task. The duty responsibility frame is specified at the steady and at the transient phases. Stress is on the design-with-recovery-in-mind ideas, as they condition the scenarios to come. Concepts in resource management are given, dealing with *voluntary agreements* for delivery restoration, and with *compulsory targets* for resources recovery; these are argued as design pre-setting and operation options. This stands for addressing the *post-global* facets on promptly achievable scopes; the industrialism adaptation issues, thereafter, define on these requirements.

The goal is to blend intangibles with tangibles, to achieve specified targets, by means of lifelong indenture, binding purchasers to sellers. The computer aids are enabling means, to keep the *products-services* under the producers' responsibility at the points of *sale*, *service* and *disposal*. In that trade set-up, the value chain is plied up, at lifelong provision duties.

The approach profits from the *product lifecycle management*, PLM, tools. These are built at the integrated design phase, combining *product-process-environment-enterprise*, **2P2E**, data, in federated-model architecture, such that the competing organisation, material, process, operation, restoring, call-back and recovery behaviours are tested and compared by virtual prototypes, since design.

The federated architecture links the models in hierarchical manner, as already pointed out, to show regular views of each aspect (manufacture goals, operation performance, eco-conservativeness figures, etc.) at different levels of abstraction and/or lifecycle steps, to verify the achieved properties, to make decisions and to choose effective solutions. This way, the producers exploit *virtuality* at the design, to deal with the delivery lifecycle outcomes, responsibly guaranteeing the supply chain eco-consistency.

The inclusion of *artificial* intelligence, and data sorting/processing abilities, of lifecycle monitoring, knowledge build-up and action putting forth, etc., allows fit-for-purpose setting at the users' whims, merging concept choices and technical processes, into the unique functional precepts of the (actually delivered) product-service. The issue is consistent with de-materialising the «new economy» staples, based on the value added 'inherent intelligence', namely, *credited* native quality, *assessed* on-duty reliability, *certified* eco-consistency or any other acknowledged property, jointly testifying the users' benefits *and* the manufacturers' profitability.

When *actual* value added builds up, no risk exists; the supply/demand law is factual means to specify the market price. The reality is a bit tricker. Most of the accomplishments are law enforced, and the final fees combine administrative and technical requests, with little evidence on how the costs are set up (the regulation charges are hidden or totally arbitrary). Then, most of the time, current references are the administrative fees, in lieu of market prices. This means that we need to revise the *free market* axiom, so powerful with the past two "*global assents*". The question is rather ticklish, and the industrialism effectiveness is critical point.

The trends, up now followed by the parliamentary democracies, move in the welfarism direction, performing income transfer towards the less fortunate people, by taxing in proportion to the individual revenue. The macro-economics measures permit to expand the governmental interventions, mainly, to employment support, without, however, too relevant market biases, to not infringe the "*global assents*" trade conditions with the other nations' partnership.

The ecology-driven revisions add macro-economics bias, with critical issues on the nation-state competitiveness, unless agreements on equal worldwide rules exist, or barriers with relevant duty-fees apply. Besides, the solidarity progressive taxing is little linked with resources protection and dumping avoiding, leading, on the contrary, to poor environmental benefits. The rise out of explicit lifecycle and reverse logistics costs seems, thereafter, imperative demand.

Further, the approach to regulations is controversial, and, e.g., it differs in the European Union and in United States. The dissent on the Kyoto protocols is well known, but also for standardisation, differences in opinion are not marginal.

In the US, due to the public corporations efficiency and competition reliance, preference is given to market-oriented frame, for providing reference standards to meet contemporary and market requirements. In contrast, the European approach is aimed at the national bodies' bureaucratic consensus, to enact standards, based on metrology principles, joined to fundamental ethics considerations.

Now, eco-design deserves increasing responsiveness, and the *product-service* pair involves co-operation liability for enhanced legal environment preservation. The eco-safe development is community survival demand, and is transferred into enacted bylaws about resource efficiency, materials consumption, energy saving, product stewardship, on-duty conformance, recovery/revamping options, end-of-life plans, old crocks disposal, and so on.

The ecologic ethics weight, maybe, has little force, since the guardianship of future life-quality is less evident than the solidarity approaches, with benefits of living voters. The domain weakness, possibly, comes out from the low interest of enterprises and citizens, to struggle in favour of long term thrifty behaviours. The EU has tried to foster interest, in view to promote acquaintance, participation and involvement, particularly, for the climate change risks.

The overall framework, however, remain rather abstract. The average citizen and company is happy to think that most demands are somebody else affair. Quite the opposite, the environment management affects every day life, with monitoring accomplishments, starting since design, to become aware of the eco-burden, from stuffs procurement, to in-use falls-out (pollution and serviceability included) and to trash or worn-out items dismantling.

On these facts, the «industry» patterns need deep revisions. The design tasks are fulfilled when, besides the manufacturing plans, the lifestyle assessments are stated by performance figures expressed through acknowledged metrics, covering the broad-sense *fitness for purpose* rules at citizens' satisfaction, including future people benefits and natural gifts defence. High conservation needs to be standard design scope, with momentum on valuable details, notably, the lifecycle service and reverse logistics regulation. Suited engineering solutions exist or are readily developed; their deployment depends on the enacted *solidarity* rules.

2.3.5. Compulsory Targets and Audit Reports

The government regulations, well prised in the welfarism, address the macro-economics areas, where conflicts might affect international trade, separately from the micro-economics domains, where the internal *free market* is left to foster the competition falls-off. The governmental regulations in the ecological field, to be effective, require addressing the individual behaviours, to look after every specific squandering practice and habit.

The success depends on the civic spirit and mindedness of the all people, and it is highly dependent on the objectivity and credibility of the passed bylaws. The all requests huge infrastructures, to be managed with competition effectiveness, and not with bureaucratic routines. The trustworthiness and independence of the rules, thus, are crucial, but even more imperative, the efficiency and competence of the law-enforced business flows.

For instance, for the end-of-life vehicles, the EU member states are required to establish:

- effective *certification-of-destruction*, *CoD*, procedures, linked to the local automotive registration systems;
- transparent selection of the *authorised treatment facilities*, ATF, where the backward chain formally establishes;
- proper visibility on the dismantlers' quality, shedding capability, shredder residues treatments; process-data reliability, etc..

The *mandatory targets* acceptance is affected, from the technical viewpoint, by downstream efficiency (reuse tracks, after-shredding, sorting processes, etc.) and, from the economic/legal viewpoint, by the upstream carmakers awareness. Besides, the situation is, here, largely favourable, because:

- the automotive deals with registered goods, and the overseeing and control operations can exploit existing structures;
- the carmakers are leading industrial context, having widespread networked after-sale and maintenance services.

The situations involving low-cost consumables and non-specific disposables are by far more complex. This means local/governmental actions to endorse right facilities and supporting means, and to establish pertinent monitoring/controlling outfits. Most of the EU Member States face these issues by bureaucratic concern, rather than by factual effectiveness, maybe, because their unwariness of *virtual* organisations potentialities, to manage dynamic supply chains, with joint visibility and competition. Happily, the eco-rules, grounded on *compulsory targets*, today, limit to the ELV and the WEEE durables, and smooth start might be devised.

If the ecologic regulation widens, turning into industrialism to come reference, the governmental actions ought to be oriented on three issues:

- the resort to supra-national agreements, to institute worldwide standards in terms of mandatory targets, legal metrology and control figures;
- the choice of equivalent tax and/or indenisation transfers, not to encourage unlawful behaviours, altering the economic or ecologic fairness;
- the promotion of competing certifying bodies, as widespread achievements of the environmental protection culture and inurement.

The first issue runs into wide lacks of consensus. For the second, a few clues might come from the EU eco-policy, notably in the ELV and WEEE cases. At the third range, the current position has to quickly evolve. Further remarks look after the present EU scenery on oriented economic measures (see paragraph 2.1.5) and on certifying structures (see paragraph 2.2.6). The topics are reconsidered in the Chapter 5 (paragraph 5.3.7), discussing an example «robot age» project.

As general rule, the EU offers diversified interpretations. The deposit-refund (Sweden) and the visible fees (Netherlands) ways resort to third parties, financed by first buyers with a *deposit* (established by the agency) or *fee* (linked to visible costs), which takes charge of the recovery. Both approaches are questioned, as the carmakers responsibility is by-passed, with interfaced downstream external bodies (possibly implying dismantlers/recyclers state-aid, in the agency case).

The EU scheme aims at *free* take-back. Assigning the full charge to producers, the final owners are relieved from any charges at disposal. The financial resources are directly transferred form manufacturers to dismantlers/recyclers, with no third party biases. The incentives, to improve the recovery (reuse, recycle), involve the producers, leaving out bureaucratic cost or scope tax. The scheme is, mainly, used in Germany, where the carmakers compete, making monopolistic drifts unlikely.

The third party interposition creates visible charges, laid on local buyers, and allows inter-industry incentive transmission, by fostering the process innovation on the supply chain (especially, the backward track). The *extended* enterprise is effective, and includes overseeing bodies; it brings to simple issues:

- the producers shall internalise all, or most of, the reverse logistics costs;
- the targets, officially fixed, are enforceable, equally binding all parties;
- the controllers monitor the lifecycles and provide transparent records;
- in case of non-compliance, the member state (directly) applies regulatory and administrative measures to the manufacturers.

The German carmakers prefer autonomous negotiation of the recovery fees, at least, for newly registered cars. Similar position is taken by the French carmakers, even if established on regional networks. The scheme obligation can follow stable progression, with, possible unequal cost distribution, reflecting the market power of the company, more than objective item's recoverability. This, nonetheless, is standard fall-off of the competitive markets.

Looking at the certifying duties, the testing jobs and joint accrediting frames are quite broader structures by respect to the existing national standard bureaus. The tasks cannot be accomplished by governmental organisations, but need to be done by public companies, chiefly, not involved in innovation, as the institutional duty is monitoring and recording.

The third party certifying is essential business, making possible the *extended* enterprise concept to reach smart outcomes, also, at the privacy protection level. The social (economic, ecologic, etc.) rights, as already recalled, may suffer local restrictions, with the caution, not to upset the citizens' feeling, by undue intrusion in their life. The parting of governmental authorities, fixing goals and standards, and certifying bodies, monitoring the individual supply chains, is suited way to grant privacy protection, by vaulting disaggregated data, and transmitting only the aggregated eco-indices. The *free-market* paradigms, in the emerging scenery, has to change, as the costs build-up needs to include the natural capital refund, which has to face the ticklish question of the resources downgrading into pollution.

2.3.6. Global Village: Networking Assistance

The *global, no-global, post-global* path brings in more shade than light, if the industrialism future is forecast. The market mandatory regulation, manufacturers' responsibility, citizens' privacy eco-monitoring, etc. are severe hindrances, which oppose to the wealth expansion according to the earlier industrial paradigms, thus, to the life-quality progress. The survey of engineering supported eco-measures is, now, turned to some more sophisticated opportunities, offered by networking, in which the «complexity» virtues show the leeway, towards still not fully assessed outcomes (notably, for the social information infrastructures build-up).

The above remarks show that the eco-consistent enhancements, driven by the emerging environmental policies, request dramatic engineering, technology and management revisions, to promote changes, such as:

- the principal contribution of value added, shall be in intangibles, providing wealth build-up free (as much as possible) of *natural* capital consumption;
- the conservative behaviours need to aim at service engineering *voluntary agreements*, granting the delivery lifecycle conformance-to-specification, entrusted to *extended* enterprises;
- the (co-operative, conflicting, overseeing) net-worked aids have to manage the EU *mandatory targets*, to achieve the enacted eco-protection targets.

Perhaps, on these premises, the mankind wealth will not significantly expand, and the quality of the future life will decrease. The *global, no-global* quarrel is the symptom that the sustainable growth demands are little understood, and the *post-global* synthesis is deficient. The EU environmental policy, enacting mandatory targets, with enforceable time and quantity thresholds, is very relevant, as for the consciousness promotion. The gradualness might be appreciated, not to alter the life-quality levels by sharp shakes. The objections come out in front of lacks of consistency, e.g.:

- to address landfill limitations, and not starting by disposables (no ban on *use-and-dump* practices !);
- to enforce recovery (reuse, recycle) targets, without energy balances (no *energy* side-effects records !).

The trend towards sustainability is being discussed from the point of view of the product-service dealers operating, having acknowledged *resource* productivity metrics, *under* overseers and controllers, representing the third party by respect to clients. The three party's links, most likely, are critical prerequisite for the success of an eco-consistent *service* economy.

Very nice option is, also, the «complexity» preservation and exploitation. The «complexity» investigation allows figuring out sets of noteworthy features, fairly useful to understand the knowledge entrepreneurship.

The scientific method, in the western world, aims at the ever-lasting relational ordering, which leads to nice laws, assessing the natural facts by causal schemes. When successful, the resulting frame collects «deep knowledge»; otherwise, only «shallow knowledge» is obtained by the empirical observation. The attention on this fact is «complexity» issue, and it allows discovering a set of emergences that regularly appear, when dealing with interlinked organisations.

The networking «shallow» science brings to identify some keen opportunities, such as the "small world" option, namely the surprising efficiency, which allows the individuals connect. The "small world" chance appears in different situations, notably:

- when local hubs, connected to the neighbouring nodes (clusters), are also liked by long-distance (weak) bridges, playing the *representative* role;
- when the local communities of nodes are linked by scattered long-distance bridges, adding options to the (centrally) cohesive *monolithic* societies.

In both situations, the "small world" effect interconnects each node to another one, by quite low number of steps (the "six passes to target" paradigm), exploiting the long distance (weak) links to make the transfer from two, moreover, separate communities. As already discussed, the "small world" nets derive their behaviour by adding few random bridges among well-structured patterns, either well chosen hubs, acting for regularly linked clusters. The lesson shows the advantage of the weak links, to improve the cross-connectivity. This is important requirement for the stability of local communities and enterprises; the ordered behaviour, mainly, is given by the links assuring the social commitment, provided by the embedded privileges of the integrated structures.

The networking is reality of the «new economy», permitting the *global village* deployment, at least, as computer engineering fall-off. We already have the clear evidence that alternative net organisations are achievable, with notable similarities and differences. In both cases, nonetheless, the offered options allow quite helpful restructuring tools.

Following a bottom-up process, the generation of a "small world" net leads to the *representative* set-up, hierarchically optimising the inner reactivity, to answer the market requests by evolving technologies: the high performance is assured by minimal redundancy.

The *monolithic* set-ups follow from a top-down process, conceived to recover the stability after external attacks, with resort to the makeshift bridges. The links assure good connectivity as an added option, while the local aggregation among equals makes easy to surrogate the function of the neighbouring nodes; actually, the downgrading is immediately apparent, but the collapse is avoided by cheap measures.

The «shallow knowledge» allows to assess the behaviour of complex systems, from the world wide web, up to the financial businesses or the meteorology. The approach is becoming basic option to find out the hierarchies and regularities into local (strong tied) ensembles, further linked by weak bonds of cross-acquaintance, address contiguity, content similarity, shared reliance, goals consonance, and the likes, mostly leading to the "small world" behaviour of *representative* (with hub organisation) or *monolithic* (with random makeshift bridges) set-ups.

The idea, to accept «complexity», without the reductionism short cuts, permits lots of new achievements, basically issued by computer engineering studies:

- software delivery, with chiefly programming or processing value;
- computer aids for integrated design, to deal with the eco-requirements;
- devices assuring operation transparency and on-process servicing;
- multiple-user outfits, for monitoring and checks accomplishments;
- interface and diagnostic protocols, for the real-time monitoring;
- pro-active maintenance aids, for machinery and plants life-extension;
- consulting support to the reverse logistics, for recovery upgrading;
- net resources, for certifying bodies overseeing and eco-fees accounting.

The list assembles a set of known businesses, and can widen to other similar facilities and procedures, exploiting network abilities, characterising the *extended* enterprise world and the *product-service* delivery market.

The survey, focusing the sustainable growth domain, needs to considers the service transfer with clients' interest. The cost transparency is background fact. When the charges asked to supply the administrative service are (perceived to be) very high (as compared to the benefit to the community), the apparatus shall be revised to thin the organisation and to rethink the loads.

The overall net structure has to become the technical support of the *solidarity* accomplishments, for the eco-protection of the generations to come. The above recalled capabilities provide preliminary hints on how the process transparency is exploited, once the suited *hyper-democracy*, or other equivalent political set-up, is built, to manage the *global village* of the future mankind.

The EU environmental policy is perceived severe in the scopes, and redundant in the procedures. To make the revision easier, the following steps are useful:

- to foster the horizontal management, giving the task visibility to all the stakeholders;
- to empower the back-office, e.g., charged of one-stop-shop-access mode services;
- to endorse side policies for promotion and share of the accomplishments social benefits.

This way, the control, instead than hierarchic, is left to the transparency of the actually performed procedures and on the genuine worthiness of the results, or, in other words, the administrative machinery is pushed to restructure itself from the outside, with promoting role in the servicing and active web-communication. The effectiveness through transparency is restructuring challenge of the administrative accomplishments: the *global village* merits the opportunity, once the net concerns technologies are correctly explored.

Chapter 3

3. THE KNOWLEDGE PARADIGMS

The industrial revolution, along with the lines up now recalled, originates the recent transformation-driven economy, based on the intensive exploitation of the earth native resources, to put in the market large quantities of material goods, to the benefit of wide amounts of potential purchasers. The characterising features can be differently defined; still, the following facts deserve attention:

- the accessibility of large *artificial* energy, inside huge manufacture shops, where to process the raw materials with planned efficiency;
- the availability of large work-force, to be integrated in the productive flow according to "scientific" organisation patterns;
- the nation-state capitalism, to obtain appropriate governmental baking and people competition, enjoying parliamentary democracy;
- the (economical) global assent, to ensure the worldwide market spreading, with return of the (big) financial investments.

These facts appear linked to precise places (UK at first, later, western world) and civic frames (individual autonomy, free market, etc.), so that the industrialism effectiveness should not to be thought generic ending, rather special achievement, not necessarily reproducible in different set-ups. The study, up now carried in that frame, shows, however, that such industrialism requires deep revising, if not total rejection, face to the ecologic emergencies, not permitting the prosecution of the affluent society, built on these old «industry» paradigms.

The idea, now, is to look at the industrialism as changeful reality, being able to have alternative origins, and to follow different paths. Relevant changes have been faced, leading from the economy of scale, to the economy of scope, initially driven by the affluent society lifestyle, which, yet, moved to look at the computer engineering applications, giving birth to the «robot age», as technology-motivated phenomenon, rich, moreover, of the «new economy» distinctive innovation mix.

More dramatic changes are requested today, because the growth sustainability shows its clear limits, opposite to the industrialism practice to draw raw materials, up to limitless quantities, and to transform the all into rubbish and pollution. The affluent society is, forcedly, reaching the end, and the industrialism shall undergo utmost renovation, to enable eco-safe paradigms. The hypothesis, now, is to look at a started breakthrough or fully revolutionary divide, after which the *cognitive* (or any other naming) set-up establishes, to permit further growth. By such way, the «knowledge» paradigms have been recognised as the enabling support for the new work organisation. Basically, the changes can be viewed, distinguishing:

- the already available technical innovation, suitably expressed, introducing the «robot age» way into the problem solving habits;
- the conditioning socio-political framework, showing that the «knowledge» paradigms are worthy chance to be explored;
- the forecast of actually realistic sceneries, and related assessed constraints, to give the technology-to-sceneries link.

The topics are reviewed in the three sections that follow. As an all, the trend is in the lines of assessing new *durable* development prospects, once acknowledged the limits of the «industry» paradigms, we have experienced. At the moment, we cannot decide whether we are at the beginning of a new revolution (exploring the benefits of *bio-mimicry*), or forcedly we shall address decreasing goals, towards thriftiness. At any rate, the changeful «knowledge» paradigms already introduce, by themselves, discontinuities in the industrialism, not to be underrated, and the chapter aims at deepening the recognised features, at the mentioned three levels: the «robot age» technicalities, to deal with *flexible* automation; the conditioning *global village* socio-political framework; the devised *future scenarios*, which are possible outcomes to be faced. The purpose, of course, is not to suggest forecasts, rather, to outline reasonable frames, pushing to acknowledge the critical situation now reached in the civilisation progression, which cannot be surmounted, unless after dramatically modifying the men lifestyle.

The subsequent two chapters refer to mostly personal experiences, occurred in the engineering activity dealing with robot technology developments. The results give, for sure, limited outlooks, being restricted to the sub-set of projects factually carried within an academic research group. They, nonetheless, allow to offer fully vivid description of what the «robot age» technicalities already means. The all is limited to introductory overviews, to help fixing some technical potential, when suggesting the prospected conclusions of the last chapter. The final chapter looks, again, at the *future* scenarios, leaving aside the technical «robot age» plain views, to outline more hypothetical prospects, where the «cognitive» options might play the winning role.

3.1. PARADIGMS IN ROBOTICS COMPLEXITY

The «robot» is taken to signify the technical switch form the industry-driven, to the knowledge-driven work organisation, and opening, in the prospect, the new achievements of the *knowledge* economy and society or «knowledge capitalism». The word «robot» has recent and literary origin, being coined, from a Slavic root meaning "hard labour", to represent sorts of human-like aids, capable to fulfil the set of heavy duties, relieving the man's efforts and fatigue.

The robot, accordingly, is deemed to replace the front-end workers, fulfilling all the handling and dexterous actions, requested by the assigned duty-cycle. To that goal, the following definition applies:

- (industrial) robot: task-oriented equipment, which can be programmed for accomplishing a variety of duties, with due autonomy, having appropriate intelligence of the work progression and of the conditioning surroundings.

The main characterising attributes are:

- task-orientation and programmability, explaining the robot *versatility*;
- (suited) autonomy, showing the extent of task-execution *flexibility*;
- proper *intelligence*, say, interfacing and data processing capabilities.

The traits: *intelligence*, *flexibility* and *versatility*, are typical of the man, and make clear why the «robot» classifies as «human-like aid». This disagrees with the old abstract attribute «industry», when zeal and commitment were the mostly prised virtue of the engaged workforce. It is, however, consistent with the change from the *fixed* (mass-production flow-shop), to the *flexible* automation (varying-mix job-shop) of the economy of scope. The (industrial) robot comes out, thus, as the technological answer of already assessed (market) demands.

On these premises, originally, the robot «complexity» is quite low, limited to the strict tiniest levels of *intelligence*, *flexibility* and *versatility* compatible to fulfil the desired tasks. In parallel, technology-driven advances appeared, and the robot starts to be the equipment able to interact with poorly structured surroundings, to accomplish self-teaching and to act with enhanced autonomy.

In the following, the «robot» is taken as symbol of changes in the manufacture world, and intermediary from the industrial capitalism we know, to the knowledge capitalism of the new productive organisation. The change, out of technical drive, is involving education and social facts, at several levels. For instance, the dialectic struggle 'production means *vs.* labour' disappears, but also old-fashion enterprises stability needs to be revised, at least, on two points:

- the centrality of the «knowledge» possession and qualification, for every individual employee;
- the fundamental role of the «complex» net-organisations, for the overall information infrastructure.

Both make clear how the competitive advantage rules have to modify, in the knowledge society, since the situation of each country, moves the capital assets, to lower the financial relevance (as the investment entity decreases), and enhances the technical function (for qualified operators and for net-fitness).

The infrastructure «complexity» needs to be assessed on the three *intelligence*, *flexibility* and *versatility* levels. The first is, primarily, dependent on the computer engineering additions. The following two are cross-linked and, basically, rely on how the technologies are transferred in the operation surroundings, to balance the hardware (the *set-up*) and software (the *fit-out*) requisites, to fulfil the enterprise policy.

3.1.1. Productive/Service Supply Intricacy

The idea is to start by the robot innovation in the structured surroundings, say, the manufacturing automatic set-ups, and to aim at forward logistics supply chains up to customised-quality delivery. The integrated control and management of the process/service operations shall account of the physical (the set-up) *flexibility* and of the logical (the fit-out) *versatility*, namely:

- the effective *set-up* of facilities, *tailored* to the operation mixes included by the enterprise strategic policy;
- the proper *fit-out* of schedules, *adapted* to operation agendas within the enterprise

tactical planning.

The off-process and on-process adaptivity is efficiency request, to be tackled over at the proper level:

- the *setting* involves the structures, components, facility-configuration and control, **CFC**; the set-up of **CFC** frames presents as everlasting activity; a choice provides reference for identifying current process set-ups all along every facility's life;
- the *fitting* deals with the information aids, monitoring, decision-manifold and management, **MDM**; the **MDM** frames fit-out, by acknowledging the infrastructures' state and trend, offers data for the on-process improvement of the efficiency.

In the traditional manufacture plants, the incorporated facilities and functions, broadly, classify in two classes:

- cluster of multi-task cells: operating by group-technology, and requiring permanent optimal logistics up-dating;
- chain of special purpose units: processing varying mixes, by adaptive rigs, job-enrichment, etc., and re-routing flow.

The economically effective solutions scenery comes from many stand-points. The return on the (off-process) setting and on the (on-process) fitting for govern-for-flexibility issues, is, mainly, assessed by computer simulation. The choice of the appropriate set-ups refers to typical factors:

- investment costs, with paying off within enterprise policy's horizon;
- process performance, to grant technical specifications and trade goals;
- delivery plans (or time-to-market), according to customers' expectation;
- quality figure, fitness-for-purposes and conformance-to-specifications.

Time-to-market and delivery quality are imperative for the enterprises aiming at remaining, or becoming, worldwide competitor. With market globalisation, the company cannot be sure to propagate its *protected* trade segment; the fast supply, at customer-driven quality, is critical request.

Once the market policy acknowledged, the company looks at most effective resources setting, avoiding downgrading due to exaggerated work-organisation sternness. The *flexibility* is distinguished by range and horizon, into hierarchical information lay-outs, so that:

- at the business project range, the overall planning is performed, according to the enterprise policy, over the established *strategic* horizon;
- at the facility co-ordination range, the selected delivery mix is scheduled, for the maximal productivity on the proper *tactical* horizon;
- at the supply chain operation range, the breaks (at unexpected or planned occurrences) are overridden within the *execution* horizon.

The organisation intricacy is visible, compared to earlier habits. The previous *scientific* work organisation was based on the job-allotment paradigm and on the three-S constraint "*simplify, specialise, standardise*", optimised off-process, with specification of each operation, so that no ambiguity could be left to the front-end operators. The final delivery is granted to be released within tolerated quality, on condition that nothing is moved off the pre-set plans.

The *intelligent* work organisation is based on including the decision manifolds on-process, in view that, through adaptivity, the optimal running set-ups and fit-outs are continuously maintained, making use of the three-R options: *"robotise, regulate, reintegrate"*, so that:

- to robotise means flexible automation by the facility/function versatility to support fitness-for-purpose innovation;
- to regulate is concerned by proactive up-keeping, to uniformly reach the process conformance to specifications;
- to reintegrate explores product and service amalgamation, for the market-quality enhancement through complexity.

The cluster of multi-task cells are basic set-up, so that the installed equipment technological versatility can face changing situations, provided that the front-end logistics satisfies the in-progress product-service flow. The three-R paradigm, in lieu of *scientific* job-allotment precepts, enables *intelligent* task-assessment rules; thereafter:

- robotics is understood as the ability of giving decision supports, based on the on-process knowledge intensive surroundings;
- regulation shows on-line commands, for adapting/restoring the processes, depending on the (changing) products-services;
- reintegration grants steady quality delivery by conditioning facts visibility, after inhibition of affecting biases and drifts.

The *flexible* automation develops combining materials and information flows, to run in parallel, overseen by the control flow. The three-R paradigm allows the «complexity» dynamical allocation, because:

- the real-time monitoring and analysis generate time varying elements and the adaptivity never leads to frozen plans;
- the duty schedules evolve with the process and high level tasks are enabled for preserving the enterprises' effectiveness.

The understanding of flexibility effects is used to choose the **CFC** and **MDM** frames. Through iterated modelling and simulation, structured databases are built, back exploited to orient the choices and to re-design the enterprise by patterns of structure. Reference data continuously evolve, providing the structure-and-action patterns, so that the enterprise keeps changeful re-set/re-fit into positions, tracking the economy of scope conditions.

For facility/function upgraded fitting, the decision loop starts by managing the flexibility issues, over the selected horizon (strategic, tactical, execution), with the right enabling logic: inventory decision; planning decision; dispatching decision; schedule decision; and the likes. The structures and horizons cannot be separately acknowledged. On these premises, suited supply agendas are planned, specifying:

- the productive/service policy (items mixes, jobs planning, supply logistics, activity dates, etc.), and the reference tactical schedules;
- the value chain and sequencing forecasts, according to customers' requests and environment protection, with transients managing criteria;

- the maintenance and restoring plans, with chosen reference signatures, risk thresholds, reliability figures, and so on.

Results, properly established at the design phase by modelling and simulation, provide the uniform basis to compare (for each **CFC** frame) the economic benefits enabled (and acknowledged) by means of the related **MDM** frame. The computer aids embrace the capability of recursively running the setting, testing and fitting steps, according to the outlined decision loop.

In the last quarter of the XX century, the manufacture engineering makeshift changes are, at times, startling. The timely IBM president, in the eighty's years, happened to comment [1]: «*We are on the threshold of a new era in manufacturing technology, and there is a critical need in industry of people, who can make full use of that technology and enhance it in years to come. There can be no factory of the future, unless there are universities of the future, educating those people now. The challenge is so important and the opportunity so timely, that IBM is making manufacturing technology the focus of the largest single-programme educational donation in its history*». The consciousness that a breakthrough was there is clear, and the «knowledge» paradigms are, since the beginning, understood to be «robot age» milestone, deeply linked with a new culture.

3.1.2. Business Modelling/Simulation Features

To understand the business performance depending on *flexibility*, the relations between supply chain variables need to be stated in actual running conditions and assessed by figures giving the *productivity* as this result s from on-duty data. The efficiency, rather than stand-alone quantity, is highly entangled issue, combining: enterprise capacity, process effectiveness and cross-coupling effects, where:

- enterprise capacity, given by the product/service amount, supplied on the strategic horizon; it is not static: any products/services mix (instance of delivery requirement) establishes an enterprise capacity; whenever the mix changes, so will the enterprise capacity; for given strategic horizon, it is assessed in function of: delivery schedules (market requests, etc.); process plans (acquired jobs, work time cycles, etc.); resources amount (facilities, tools and operators, etc.); or other up-dated information on the set-up;
- process effectiveness, shown by data with reverse impact on productivity: facility/function/labour availability, scrap/waste percentage, set-up delay, etc.; their dependent nature makes partial figures useless; for efficiency, the coupled effects need to be mastered; the effectiveness is improved by: intelligent planning; self-adaptive match; company wide quality; proactive maintenance, and similar other data on the current fit-out;
- cross-coupling effects, stated by the losses in capacity or effectiveness due to combined occurrences that make unfeasible the beforehand established schedules, such as: resource shortage and process blocking; the example event: 'supply shortage and bottleneck propagation' explains such effects: the items feeding stops have: downstream outcomes, with the delay in the duty and delivering; and upstream upshots, with the flow break off and the undone jobs stacking.

---

[1] J.R. Opel (1983), Perspectives in computing, Oct.

The flexibility effects on real enterprises cannot be evaluated by *reductionism*; limited example cases are, at must, included as validation benchmarks. Computer aids are, thus, general reference for the build-up of the basic knowledge, and are exploited as consultation support to devise, select and assess consistent options, in order to fix/re-set the **CFC** frames and to fit/restore the **MDM** frames.

Expert simulation is useful means to gather actual evidence of advantages and drawbacks, which the govern-for-flexibility issues offer, by joint management and control. The plants running conditions are tracked and their performance assessed, with resort to functional models that combine structure descriptions, with human-like judgmental ability. The knowledge coding expands on several layers, keeping oriented scopes, namely:

- the facility description infers the causal relations (structural models) and judgmental frames (behavioural modes);
- the functional modelling leads to generate the algorithmic and the heuristic blocks for the virtual-reality checks;
- the testing and evaluation are performed on actual business plans, varying the govern logic on the strategic, tactical and execution horizons;
- the development leads to set the **CFC** lay-outs and to fit the **MDM** frames, through iterated decision, over the singled out contexts.

The code includes two sets of blocks: the first generates the facility dynamics (structural frame); the second provides the judgmental logics (behavioural frame). The computer code assures the testing of alternative set-ups, by simulation. The engineer exploits the option since the conceptualisation stage, once the functional models are established on parametrical bases.

The govern modules support flexibility effect assessments, when the plans are enabled by decentralised control and supervisory management, as the case arises, along the strategic, tactical or execution spans. Changing goals, moreover, require *dissimilar* expert codes, to accomplish, e.g., the planning, diagnosis, govern, etc. tasks, each time based on programming aids, which, regularly, share knowledge-based patterns, for the easy coding of the decision logic and the 'emulation' of the human expertise. The knowledge-based programmes are examples of the *artificial* intelligence methods. Their complexity characterise by two facts:

- multi-agent philosophy: awareness is focused on the supply chain activity (product-process delivery), assuming co-operating evolution of front-end agents, within message passing communication environment; the decision logic is effective to govern real-time concurrent processes;
- functional characterisation: the expert simulation allows "prototyping for evolution" by iterating the "evaluation-modification" cycles; the decision logic emulates the area experts reasoning patterns, to allow assessing the performance rank of each tested alternative.

The knowledge-based systems are useful means for engineering applications based on empirical milieu. The ability of encoding *information* by «knowledge», rather than plain «data», is basic drive. The «knowledge» is data, with attached relational context. In the software, one distinguishes:

- declarative knowledge, mainly, the enterprise *structural attributes*, at the input interface; the business *process information*, at the output interface; the updated information is stored in the current memory for further use, to make possible the procedural firing;

- procedural knowledge, basically, the *behavioural aspects* of the business facilities; the 'rules' are common coding to express the *methods* (how to accomplish actions or to take decisions), chosen for declarative knowledge processing and maintenance.

Object-oriented languages are good option. The objects, by 'attribute, method and belief' coding, grant unifying programming; the joint relational frame allows inheritance and makes feasible to define generic elements, re-usable, when useful; the structure induces incremental descriptions and classification.

The knowledge-based codes are nice engineer's aids. The software develops as multi-layer construction, with vertical links between the relational, generative and information layers. The all simulation software organises by federated-model architecture, to help investigating new details, each time the request arises.

The *relational layer* embeds pre-processing modules, as friendly interface for model selection and agenda setting, to help defining the manufacturing facilities structural attributes and the behavioural properties of the related governing logic. The availability of expert simulators is obtained by referring to specialised data-bases connected, through management blocks, with, both, the generative and the information layers, so that virtual-reality checks are run, with the control flow full transparency.

The *generative layer* contains the solving-capabilities; it propagates the causal responses by algorithmic blocks and acknowledges the consistent suppositions by the heuristic blocks. The decision cycle, aiming at achieving flexible automation concepts, is prettily conditioned by iterating the emulation/simulation loops.

The *information layer* performs the restitution duty, through post processing modules. The user can call for graphic display; the process-information is shown as relevant facts sequence, by situational specifications, or is processed to provide the performance evaluation, as compared to competing schedules.

Together, the three layers are the simulator kernel. The conceptualisation and the acknowledgement layers are, further, used as interface of the outer learning loop, closed by the intelligent governor, in charge to adapt the enterprise control and management, namely:

- at the *conceptualisation layer*, the user has access to the structural models and to the related behavioural modes; the technical versatility automatic exploitation can be used, to enable co-operative and co-ordinated actions; the meta-processing abilities add condition lucidity and fore-knowledge opportunity to the (subsequent) relational layer;
- the *acknowledgement layer* supplies process-data visibility; the assessed facts help to expand the describing data and to start the chosen flexibility horizon interim analyses; at this layer, special cross-processors, by means of hypothesised knowledge self-accrediting, enable the expert-simulator, to operate on-line, for adaptive supply chains.

The correct facilities setting is assessed, by simulation, with due regard of the enterprise policy and supply chains strategies. The reconfigurable lay-outs, based on modular units, are options to maximise efficiency. The robotised solutions are prised, when one-of-a-kind production should be granted. The opportunities are hereafter commented in relation to the *flexibility* goals. The remarks help locating the «complexity» analyses in the manufacture paradigmatic context, showing nice problems-solving procedures, having tested successfulness.

3.1.3. Context Flexibility and Achievements

The mass production, aiming at the highest productivity by economy of scale, being characterised by vanishing flexibility figure, z_F, has design stage fulfilled at the conceptualisation and acknowledgement layers, and does not need relational, generative or information layers. On the opposed side, the one-of-a-kind provision characterises by the largest flexibility figure. In between, the figure takes middle values, with higher placing obtained by alternative means, namely:

- to modify the enterprise reach, by off-process operations, out-sourcing or productive break-out, aiming at external facilities/functions, to accomplish the in-progress delivery; the effectiveness depends on the value added core business and on the nested-partners co-ordination;
- to incorporate technical versatility up to the *extended* enterprise setting; this issue is reached by using networked facilities/functions, to reach the product-service autonomy; the effectiveness depends on the co-operation and integration of the production/provision partners.

The simulation is used to decide on technologic facilities, to set the business lay-out and to fit the decision manifold. For the co-operating partners, the govern blocks are directly interfaced to the facilities, with twofold operability:

- to generate updated information for re-set and out-fit duties, if the product-service delivery modifies and new enterprise capacities are needed;
- to enable joint 'management-and-control', to fulfil recovery options at the emergencies and to preserve (steady) provision on the tactical horizons.

The inference capability inclusion makes possible 'anticipatory' management loops, with responses that close, as standard govern-for-flexibility rule, of the on-process adaptive control of the business project. The option is central from two standpoints:

- for assessment purpose (PLM option), to expand the designers' expertise, allowing for interactive enhancement of lifestyle knowledge references;
- for implementation purpose (SE/RL option), to take in value added duties, within the given provision, by enhanced capability and wherewithal.

The lifestyle design is, now, enforced duty. The PLM option, simply, means efficient data-bases management. The SE/RL option means on-process inference abilities; the set-up shall be standardised to be easily adapted to varying operation or environment requests. The expert governor module is connected on-process to 'oversee' the supply chain and to 'manage' the service/recovery plans. The typical components of the expert module (as usual) are:

- the 'conditioning logic', encoding the procedural knowledge;
- the 'data-memory', with declarative knowledge updated by current data;
- the 'inference engine', carrying out the decision cycle: information-match, conflict-resolution, and rules-firing.

The *management* duty choice resorts to learning loops, aiming at assessing the flexibility cross-coupling effects on efficiency. The effects depend on the *govern* options, at enterprise capacity range, in view of delivery schedules and due-dates. The 'efficiency' is evaluated after collection of huge data, by simulation tests.

The '*govern-for-flexibility* vs. *net-concern operation*' upshots are assessed, to upgrade the business efficiency, fully enabling the technological versatility with due account of leanness. For capacity allocation, the control-and-management of flexibility is made up of decision rules in relation to, e.g., capacity setting, logistic flow, function/facility allocation, and production/service plan. Each rule offers alternatives, regarding how the company has to be updated, to meet the requests, using integrated design practice to adapt *product-process-environment-enterprise* by stepwise betterment of the incorporated functions/facilities.

The performance figures and the flexibility coupled effects are empirical data, obtained by experimenting on real companies or, through functional models, by computer simulation. The effectiveness is enabled by economy of scope, as above said, with *leanness* check-up assessments. The view is, perhaps, reductive, as the progresses are consistently stated by the knowledge-intensive *extended* enterprise, through integration of market, design, manufacturing and finance features, with account of the (enacted or devised) eco-rules.

To reach more comprehensive assessments, issues are stated by collecting and comparing standardised indices and by tracking and acknowledging established methods. *Extended* enterprise concept is basic help to combine product-process *internalities*, with the market and ecologic *externalities*, into unified data frames, travelled by knowledge through computer engineering aids, along the net backed structures. The ensuing lay-out supports the *intelligent* organisation and controlled partnership; mainly, two philosophies apply:

- hierarchical structures, based on distributed problem solving abilities;
- multi-agent clusters, with message passing management.

The communication model is centralised, in the first case; distributed, in the second. Intermediate combinations have, also, been developed. On-line flexibility is enabled either ways; the multi-agent architecture enhances the re-set and out-fit flexibility, as independent entities are defined, to acknowledge the different jobs (the latter philosophy, essential in the virtual enterprise contexts, is common tool for quickly reconfigurable infrastructures).

The *flexibility* analysis, to embed in the enterprise policy, rightfully addresses sets of factors, depending on series of hardware/software quantities either through operation contrivances or through planning tricks. The figures are reported with reference to the combined product-service delivery, included into the enterprise strategic span. The *flexibility* figure, z_F, is complex issue, where *agility*, *leanness* and *smartness* play cross-linked roles, to make difficult the *reduction* to elemental causes. The list that follows has illustrative motivation; it gathers example factors for explanatory purposes:

$$z_F = \alpha_V \; \alpha_C \; \alpha_S \; \beta_S \; \beta_P \; \beta_D \; \gamma_F \; \gamma_R \; \gamma_M$$

where z_F is the *flexibility* figure divided into characteristic components:

■ the three initial factors deals with the process *agility*, supported by the technical processing requests compared with the operation contrivances:

- α_V, *versatility factor* of the delivered mix, showing the number of product-service types, which might be processed together, on the given time span: year, month, week, day, shift, hour, minute, etc.;
- α_C, *co-ordination factor* or delivery complexity (involving dextrous fixing, with

combined processing); auxiliary fixtures and cooperating robots may be requested;
- α_S, *sequencing factor* or number of operations to accomplish by elemental tasks (this applies, if complexity is solved into sequences of fundamental steps);

■ the subsequent factors are useful to make comparisons between regular results, to reach *leanness*, after homogenisation of the product-service flow:

- β_S, *range size factor* or items' dimensional class: from micro devices, to finger or hand sizes; then: from easy handling, to heavy weigh parts;
- β_P, *capacity factor* or productivity issue, showing the deliveries amount on the reference time; quantity mostly replaced by normalised characteristics;
- β_D, *reciprocal duplication factor*, namely the inverse amount of replicated fixtures, working in parallel to assure the supply by just-in-time schedules;

■ the last three factors show the planning tricks, to obtain reliable delivery plans, in view of the organisation *smartness*:

- γ_F, *fitly assets factor* or resources' exploitation ratio, providing the activity-time of the facilities directly involved in the principal process, from the initial (conceptualisation, development, etc.) stage, to the current (on-duty, idle, failure, maintenance, etc.) stage;
- γ_R, *uncertainty factor* or reciprocal product-service design-and-provision robustness, to grant conformance-to-specification to the all supply chain by means of the principal process alone;
- γ_M, *delivery margin* figure or overproduction referred to trade agreements, so that effectiveness is supported, while completing the delivery within the due date.

The analysis is example method. The listed figures provide explanatory hints on how to deal with complexity to make preliminary comparisons. The *flexibility* figure, z_F, by itself, requires judging technology, organisation and method terms, and the obtained estimates are, mostly, *a posteriori* assessments, suggested by the domain experts, after *on-process/virtual-simulation* data collection.

The switching from the *scientific*, to the *intelligent* work organisation means to use a rather different entrepreneurial mind, more or less depicted by replacing the «industry», with the «knowledge» paradigms, because the on-process decision keeping mechanisms are offered to mitigate the strategic horizon stiffness, by the due tactical horizon flexibility, allowed by the execution horizon versatility. The *flexibility* figure, z_F, thus, is just a cataloguing reference, permitting to weigh the underneath factors and to balance the individual effects, by acknowledging their on-process consequences.

3.1.3. Flexibility in Structured Surroundings

The dilemma opposing «complexity» and «reductionism» is mitigated, if unit actions refer to structured surroundings. This applies to the manufacturing world, and suggests smooth transition from the «industry», to the «knowledge» frames. The functional ranges of flexibility preserve common characterisation all along the intelligent work organisation. The back-up knowledge architecture presents with hierarchic set-ups, reflecting differences on physical facilities and, as deeply, on logical resources. This leads to specialised aids, such as:

- expert consultants, aiming at the enterprise policy strategic management;
- distributed supervisors, for the facilities monitoring and coordination;
- localised actors, to perform peripheral diagnostic and task sequencing.

The structured surroundings permit to establish hierarchic knowledge frames, so that the enabled solutions distinguish by their flexibility range:

- the upper range aims at the economy of scope effectiveness, resulting in, through simulation-based investigation, "supervising managers", with the balanced utilisation of the strategic flexibility on the organisation range;
- the intermediate approaches the special purpose automation productivity, once proper set-ups are configured, to make feasible optimal schedules, ruled by on-line distributed controllers, along the coordination level, at the tactical flexibility range;
- the lower exploits robot versatility to grant process continuity by execution flexibility managed by adaptive commands; the efficiency is timely issue, due to the process breaks, to face planned or unexpected occurrences.

In this frame, the *flexibility* defines as: number of type changes of the product-service mixes, allowed by the functional versatility of the front-end resources, on each (*strategic*, *tactical* or, respectively, *execution*) time span, fulfilled according to the enterprise policy. The set-ups selection, properly arranged to grant return on investment, can be inferred by using, as performance index, the *flexibility figure*, z_F, analysed according to the previously defined factors.

The strategic *flexibility* is normally based on a few years spans; shorter spans are becoming relevant in front of market-driven changes. The tactical *flexibility* depends on the supply chain: the service side is evaluated on the hour/day spans; similarly for the execution *flexibility*, with second/minute spans. The figures are used since the design stages, for the facility configuration setting/fitting, and the control strategy choice, according to the decision cycle sceneries. The *flexibility* analysis moves from the enterprise policy, orderly exploring:

- partnership configuration, so that different product-service types are made available, after proper specialisation of the functional units or modification of the processing lay-out;
- dispatching updating, with buffers and by-passes, to adapt the supply chain scheduling in front of unexpected or programmed discontinuities, through dynamic redundancies;
- partners integration, so that the different provisions are scheduled to be delivered in sequence, for ordered market batches, after suitably adapted enterprise re-setting;
- agenda-driven supply chains, steering the delivery to involve in-progress product-service adaptation, allowed by the available facility/function set-ups/fit-outs;
- or further similar setting/fitting actions, granting the partners tuning to the product-service provision.

For the business-driven tuning, the tactical and strategic flexibility are settled. The other situations have lower tactical flexibility, with condition switching at the fit-outs or, respectively, at the set-ups. The execution flexibility, further, depends on the ability of facing emergencies by recovery plans, there through, granting to resume the supply chain tasks at unexpected events. Several reference time spans are in use, depending on the flexibility to be evaluated, and the above mentioned example figures are given as typical references.

The structured environment cannot remove the *complexity* enduring intricacy. The throughout investigation of single link in the supply chain, avails of:

- *empirical scales*, assessed (by the representational theory of measurement) mapping the enterprise performance standards, into characterising indices;
- *functional models*, established, after task decomposition, by duty sharing between agents, as reference method for computer simulation/emulation.

Example procedures based on empirical scales are given in paragraphs 2.2.2, 2.2.4 and 2.2.5, discussing the net-concerns effectiveness. The functional models decomposition provides deeper insight, when joint with computer simulation. The procedures are, mostly, clarified with block-schemes or water-fall style diagrams, adding iterative loops within and among the different phases, and performing the interlaced specification of product, process, environment and enterprise.

The flexibility figure, z_F, nine factors are example suggestions, to understand how to organise cost-effective set-ups of the *extended* enterprise, consistent for a given product-service delivery. The factors have unlike relevance; suitably scaled estimates are used, when better guesses are not affordable. The initial selection of the technological resources moves from functional requirements, stated according to the business project, by comparing experimented settings.

The agility guess is *a priori* base, inferred from expected efficiency (average duty cycle, less idle, failure, setting, fitting, etc. times). When the desired volumes out-span the throughput of single facilities, paralleling is used. The equipment can be specialised: individual items track separate paths, to reduce the mix allotted to the single unit and to suppress some set-up/fit-out stops. The concurrence is, also, used for paralleling independent service operations. The sequencing represents the number of elemental jobs (giving account of complex provision duties) to be done by the serial schedules. The productivity is improved, enabling multiple deliveries by concurrent devices. The versatility provides the number of types processed by the facility, at the same time, without refitting:

- on one side, goal oriented (high productivity) facilities deal with special items; fixtures availability evens up the product-service types over the considered horizon (with idle resources, waiting for un-assigned jobs);
- on the opposite side, fully robotised processing sections operate with self-enabling mode, making possible to process time varying product-service mixes, avoiding (or minimising) idle resources and re-setting stops.

The leanness estimate is a homogeneity reference; modularity is nice option, to simplify composite provisions. The enterprise capacity is strongly affected by the processes cross-coupling; the overall productivity is *a posteriori* outcome, but proper planning options shall be exploited for choosing preliminary set-ups. The duplication request is related to the need of high reliability for complex tasks that ought to be fulfilled, after careful redesign, by co-operating robots, either:

- simultaneously, by the combined delivery through provision concurrency;
- sequentially, by changing the operation fixture to fulfil each next task.

The smartness guess directly faces intricacy. The fitly assets figure measures the active work-time portion on the overall-duty span; it helps fixing amortisation plans for frozen assets. It is relative guess, as compared to nominal productivity, and may be appraised, measuring the actual duty cycle duration, or, more suitably, the averaged provision job, over the considered time spans. The uncertainty figure depends on the supply chain quality ranges. The production of very cheap objects could be compatible with increasing the defective provisions. It shall be noticed that the uncertainty figure differently affects the individual deliveries; it depends on monitoring, diagnostics and recovery implements, arranged in advance and on the selected quality tolerances. The margin figure is caution brought in, compared to just-in-time delivering for wide mixes, so that completion of smallest batches is reached in the average, without increasing (too much) the refitting stops. It leads to over-production as compared to some provision agreements, in order to reach completion times, properly inside the overall engagement due dates.

Suitable *normal* scales define, to classify the enterprise adaptivity. With due regard to flexible automation, the scale factors of robot-operated facilities usually reach low productivity levels, as compared to special purpose devices. The higher productivity, however, has to surmount serious inner and outer logistics requests, to grant material dispatching. The extended enterprise architecture, yet, enables data-file exchange (to share the decentralised processing units *information*), and decision-aid opening (to split the local knowledge-based systems *intelligence*).

3.1.5. Lifestyle Proactive Support Provision

The switch, from shop *internalities*, with fittingly structured surroundings, to lifestyle supply *externalities*, is impasse, pushing the transition from «industry» to «knowledge» paradigms. The actual effectiveness is open problem, and suitable opportunities, to go a little further, are related to the ability of exploiting, to the best, the on-the-field-acquired empirical knowledge. Craftiness and training are widely trusted in human overseeing. The translation into automated govern tasks is obtained by specialisation, frosting «complexity» off-process. The knowledge-based approach and expert-simulation are aids to assess flexibility by uniformly combining causal and judgmental knowledge; they are offered to the engineers for making possible:

- the optimal setting of products and processes, according to the canons of simultaneous engineering;
- the best return on investments, by preserving the leanness into intelligent (knowledge intensive) businesses;
- the good protection of the environment, by on-duty conformance and end-of-life recovery enhancement;
- the efficient enterprise setting/fitting, incorporating the facilities/functions, adapted to the product-service delivery.

For the set-up of *extended* enterprises assuring the return on investments, the feedback in capacity allocation decision cycle is built by monitoring the flexibility effects; it provides reference data for the overall procedure. The step measures the effectiveness of competing options by means of functional models, and the use of virtual reality. The most effective alternative is the last step in the decision loop: each choice represents an optimised setting of the flexible capacity allocation and an effective governing policy, with constraints on the business operation, e.g.:

- to upgrade the value chain integration;
- to adjust the product-service offer mix;

- to change the facility/function assembly;
- to modify the on-duty inventory policy;
- to fix the recovery (reuse, recycle) goals.

The series of steps move from preliminary facility/function settings, to match the case requirements. The modelling features develop into the standard: *causal blocks*, to simulate the transformations, undergone by the physical resources; and: *judgmental blocks*, to emulate the govern rules, enabled by the logical resources.

At the later steps, the testing is performed by simulation. The causal inference provides the assessment of performance by means of categorical features (patterns of actions); heuristics is called for to implement govern-for-flexibility procedures. The performance figures are weighed up for the considered business project, with several schedules and operation conditions. Cross-coupling effects are common occurrence for time-varying, low-inventory, lean-cluster set-ups. The effects of flexibility have impact on efficiency, in such a way that company output depends on several factors and not only on the engaged facilities marginal efficiency.

The lifestyle assessment permits to extend leanness, by monitoring the value added by each partners of the *extended* enterprise, for cost-effective integration, aiming at scope economy by knowledge-intensive frame, checking rules, such as:

- to extend the offer mix variability, to satisfy larger consumers' wishes;
- to suppress redundancy and set-apart resources, using recovery flexibility;
- to abolish non necessary functions, exploring decentralised responsibility;
- to avoid investment in special rigs, and to exploit robot's versatility;
- to limit inventory and enable adaptive, bottom-up, just-in-time schedules;
- to exclude competency segmentation, and solve troubles where they arise;
- to enhance customers-driven responsiveness, by minimal time-to-market;
- to institutionalise co-operation agreements, for the on-duty eco-servicing;
- to manage the mandatory recovery targets, with resort to domain partners;
- to exploit people involvement, improving value added by shared concern.

The proactive support provision is flexibility-driven selection. The goal takes principal part to widen the offer mix, and critical role to avoid idle resources. The return on investment arises from sets of precepts, uttering the complexity running objective by 'intelligent' task-assessments, such as:

- *functional integration*, along the business principal process, to support the synergetic cooperation of every incorporated partner/facility;
- *total quality*, for globally conditioning the enterprise organisation to be customers' driven, by aiming at 'fitness for purpose' supply;
- *flexible specialisation*, through partners intensive involvement into wide mix of technological integration and productive break-off;
- *lean engineering*, to avoid redundancy and minimise investment, for the planned supply chain, all over the shared business project.

The efficiency is flexibility (three-R option) outcome, based on specialisation (three-S choice). Achieving flexibility depends, of course, on the initially pre-set layouts and facilities. The return on investments is, anyway, widely subject to the govern-for-flexibility exploitation of the enterprise facilities. The improvement is obtained by iterating a decision loop, which refers to the functional model of the facility behaviour, and is validated by supports based on the appraisal of the node performance and the comparison of current figures against the expected levels of efficiency. Such decision logic is effective, on the condition that every alternative, matching the application case, is investigated. This is feasible through simulation, granting virtual reality display of actual supply chains. Indeed, it is not possible to pre-set and to apply the versatility as enterprise policy and to control and manage adaptivity as current request, unless the effects are measured and the facilities are tuned to the flexibility at the project/concern/node range (see paragraph 2.2.4).

The expert simulation is used at the design stage for the lifestyle evaluation of alternative facilities; it becomes permanent consultation aid, during exploitation, to select or to restore the best choices. The upgrading is achieved by specialised software for factory automation, which embraces artificial intelligence aids. The decision cycle for setting/fitting flexibility aims at economy of scope, by changes in the work organisation. The efficiency differences are found in interlacing the material and information flows, which entails the knowledge sharing and decision break-off issues and mainly leads to the well known issues: *piece wise continuous betterment, diagnostics and monitoring maintenance, lean engineering check-up assessment* and *cooperative knowledge processing*.

The four issues are like views of tasks dynamical allocation, by the intelligent preservation of complexity, into facility set-ups (granting return on investment by adaptive schedule delivery) and into process fit-outs (aiming at highest efficiency, by recovery flexibility). The goals are addressed by recurrent procedures, using distributed knowledge processing plans, to move back on-process the decisional manifold, consistent with flexible automation.

The switch from «industry» to «knowledge» paradigms is mostly important, when we ought to deal with lifestyle *externalities*, by setting/fitting the involved *internalities*. The possibility to arrange structured surroundings is big advantage of manufacturing business robotics, and the opportunity can be used to introduce formal methodologies permitting to preserve complexity, with suitably managing the effectiveness. The next paragraph provides an overview of the options, shortly recalling the, already discussed, modelling characterising features. The remarks of these paragraphs, as already said, apply at the *extended* enterprise, and, generally, cannot be exploited to appraise the *virtual* set-up.

3.1.6. Product-Process-Environment-Enterprise Design

The lifestyle product-service delivery, to address sustainable growth, needs to widen, through integrated *product-process-environment-enterprise*, **2P2E**, design. The methodology means to connect the *product-process* «internalities» with the *environment-enterprise* «externalities», when singling out and implementing the modelling and simulation features, *M&SF*, requested for the advanced checks, in view of the setting and fitting of the devised supply chain. The related knowledge frame moves along entangled tracks, covering:

- *product*: technology innovation, requiring enhanced focus on new offers ideation (*product-service*, etc.), by the self-consistency of the integrated design (applying proper *M&SF* tools to the delivery *extension*);
- *process*: adaptive and lean management of the manufacture steps (flexible automation, etc.), with account of *varying* function/facility infrastructures (managing the proper net-concern *M&SF* opportunities);
- *supply chain*: appropriation of the *point-of-service* and *point-of-disposal* duties, to rule the all delivery under unified responsibility (including all the *product-service*

M&SF requested by eco-protection rules);
- *corporation*: production resources choice and incorporation, to optimally achieve the business objectives, by dynamical organisation setting/fitting (resorting to the suited facility/function *M&SF* inclusion).

As compared with *internalities*, the *externalities* «complexity» level is bigger, since the structured surroundings, properly maintained in old production set-ups, forcedly has to face the challenges of the flexible networking between a series of endlessly adapted facilities and functions. The reference *M&SF*, accordingly, are more entangled. The relevant issue in the *M&SF* deployment is to link the views, developed for dissimilar purposes and by different teams, up to *model federation*, making easy to create high-level representations, which allow to reuse the existing data and casings, and to propagate the knowledge body by seamless continuity.

Aiming at the four modelling and assessment ranges of the knowledge frames, the integrated design strategies achieve the entire lifecycle transparency, allowing the designers to face the «*externalities*» demanding requests, as helpful constraint, from the early conceptualisation stages, and to respond by suited product lifecycle manager, PLM aids, embedding the *super-model*, whose *federated* architecture is the way to deal with «complexity», bypassing the reductionism shortcuts.

The designers become able to detail the candidate prototypal deliveries, from *virtual* factories, for *virtual* point-of-use sets, up to *virtual* point-of-disposal ends, to improve producibility, utility performance, on-duty reliability, maintainability, eco-impacts, recovery falls-off, etc., in real running orders. They become ready to perform these checks beforehand, with proper completeness, to quickly build the effective *super-models*, having the ability to zoom-in at the critical details, and to compare the *virtual* alternatives.

The business project paradigm shifts, to reach profitable *value chain*, are the outcome of the past few decade tools, preserving the methodologies, but with the policies covering lifecycle performance, as the offer necessarily needs to close on conformance-to-use assessments at the *point-of-service*, and on recovery targets at the *end-of-life*. The sustainable growth is recognised to have promoted the divide, moving back the natural resource restrictions in the supply chain range, after that the robots assures the scope economy efficiency at the *manufacturing* range, with the *M&SF* implementation at the *design* phase.

The explicit focus on the four *product-process-environment-enterprise*, **2P2E**, spheres, is critical checks for the business success. The joint **2P2E** design becomes best practice, if the lifecycle, disposal included, duties are standard responsibility of unique suppliers, typically, *extended* enterprises. The *M&SF* account evolves to include the all business and operation management areas within the politico-legal regulations.

The bet is problematic, since the physic-based frames of the engineering and manufacture are replaced by the economic transactions, human and intellectual activities, and socio-politico-legal constraints. Objects and events are specified by texts, frames, spreadsheets or graphic trends; model validation and simulation test lead at data forecast for cost check and value chain appraisal. The approach gives effective means to acknowledge the lay-out, and to provide hints to predict the on-process control and the steer actions. The PLM-SE and PLM-RL are tools, where the technicalities embed heuristics.

The suited «*extended* enterprise» *M&SF* incorporation is domain open to new studies, where the computer aids play critical roles, with resort to declarative and procedural knowledge, to the exploitation of the networked infrastructures, and to the availability of the facilities/functions market. The spheres addition (from the **2P** simultaneous engineering, to the **2P2E** holistic design) is new challenge of the knowledge entrepreneurship.

The productive setting/fitting still exploits (basically) structured surroundings, and the different situations, asserting economical return, encompass:

- the exploitation of wide-versatility facilities, assuring the concurrent offer of widespread product-service mixes, distributed among the net partners;
- the use of provision enhancers, concentrating customised deliveries, by the incorporation of parts or functions provided by experimented suppliers;
- the resort to modular facilities, with by-passes and buffers between special purpose units for the adaptive scheduling of moderately varying delivery;
- the establishment of multiple partners, carrying independent **2P2E** duties, to reach pace-wise tracking of eco-issues with customers' satisfaction.

These are example situations, showing the «complexity» level of the *extended* enterprise integration. The focus on the *product-service* provision allows devising design methods and practices, especially suited for the knowledge intensity. The *super-model* architecture needs to aim at granting the visibility over the points-of-service, in connection to on-duty conformance-to-specification checks, and at the point-of-disposal, to comply with the enacted recovery rules. To that purposes, the major innovations develop along complementary lines:

- to establish appropriate knowledge frames, which include the full lifelong scope-economy, through up-grading (service) orientation;
- to exploit modelling and simulation features *M&SF*, which provide all the pertinent backward track (reverse logistics) product-views.

The two lines are supported by the *super-model* architecture, having:

- varying-geometry boundaries, to expand the useful views that specify the «*product-service*» peculiarity;
- embedded simulation-emulation tools, to allow on-duty virtual tests for the behavioural experimentations;
- co-operative infrastructures, to support problem-solving issues, open to the co-operating net-partners;
- automatic propagation of changes, assuring the consistent updating of the super-model data-frame;
- straightforward evaluation of alternatives, trends, risks, etc., grounded on reliable relational schemes;
- easy producibility, affordability, etc. analyses, performed using intelligent decision supports;
- fast and accurate exploration of lifecycle occurrences, to establish friendly concept-to-business figures;
- balanced assimilation of the reverse logistics functions, in full compliance with the enacted recovery accomplishments;
- efficient archival of globally accessible knowledge-bases, readily available in the current activity;
- ubiquitous service through the shared net-worked infrastructure, enabling effective, inter-operable tools;
- and similar other **2P2E** approach options that the computer aids provide for co-operative analyses.

The **2P2E** way shows to what extent the knowledge entrepreneurship settings affect the traditional manufacture domains, compelled to operate with intangible parts fuzziness, to deal with *factory internalities* and *lifecycle externalities*, using sophistication, perhaps, even greater than the one tried-out by the quoted «new economy», due to the strong dependence on the sustainable growth and the eco-impact assessments.

The all shows that the *product-service* working out ought to exploit the **2P2E** design, with twofold up-grading, to improve the business native effectiveness by assembling the right facilities/functions, and to widen the chosen actors reach up to the eco-protection. The involved *super-model* set-up shall, thus, offer well assessed sets of opportunities, such as:

- *product-service unified data-frame*: the delivery of *products-services* is primary business achievement, and lifecycle knowledge is fundamental requirement, since the earlier design stage;
- *integrated data-flow management*: the parametric, hierarchic, cross-linked, delivery and business models ensure that decisions are right away made, achieving entrepreneurship-wide impact;
- *distributed, flexible operability*: robust communication and shared data-base joined to processing facilities help establishing teaming relationships, to face the emerging requests whether they occur;
- *plug-and-play interoperability*: all technical and economic modules have to be seamlessly compatible and self-adapting, to become operational right away, without idle spans and integration cost;
- *total connectedness*: all acknowledged actors are linked by communication infrastructures that deliver the right data at the right time, whenever they are generated and required;
- *fully enabled product-service transparency*: the **2P2E** tool, linked to suited science-based and experience-driven knowledge, fully shows the decision schemes and achieved performance.

The unifying *super-model*, for the conditioning knowledge coherent planning, distinguishes because of the specialised features, requested to deal with *products-services*. This is considered by several research initiatives all over the world, at different levels of complexity.

The low level is the simultaneous engineering approach, with *product-process* fused data-frame. At middle-level, the product-process-enterprise and/or product-process-environment, **2PE**, are developed, by the pertinent knowledge inclusion, as «new economy» driving fact, to deals with the company abilities in the *function* management, or with the authorities' acts in the *eco-footprint* assessment.

The tracking of the **2P2E** way is challenge towards eco-sustainability, opening the business to *externalities* (measurable falls-off in terms of environment, to deal with the all supply chain, from material provisioning, along delivery lifecycle, up to end-of-life disposal and recovery), used as permanent input, to reorganise the corporation *internalities* (actual facilities and functions, moved in foreground to fulfil the required product-service provision).

To such goal, the *externalities* to *internalities* interplay has starting approach in the modelling and simulation features, *M&SF*, methodology, and driving reason in the integrated **2P2E** design. The *extended* enterprise deployment is the coherent issue, with productivity enhancement achieved through the knowledge paradigms, along the *de-materialising* track, and the «complexity» management.

The integrated **2P2E** design, as said, is simultaneous engineering extension, in which the joint product-process, **2P**, design made possible the cross-adaptation of products, with client-oriented qualities, and processes, with issue-driven scopes. The methodology develops at different conditional patterns, so that, at the end, it will be useless to distinguish which one, the product or the process, comes out as the leading inspirer of the improvement. In parallel, the **2E** build-up expands into *environment externalities*, referring to *enterprise* embedded functions/facilities, so that the lifecycle service provision is achieved by small or no entrepreneurial up-dating, as for the productive organisation, exactly as the earlier client-bent of the product quality was reached by the already offered process flexibility.

The integrated **2P2E** plan offers benefits, by the interplay transparency, tying the *environment* data (compulsory lifecycle eco-targets) and the *enterprise* data (improved productive set-up assimilation). The reference knowledge expands on less structured surroundings, still the business efficiency resorts to setting/fitting pace-wise upgrading, oriented towards the entrepreneurship organisations, which master all unexpected occurrences.

3.2. MANKIND LOYALTY AND WORLD CITIZENSHIP

The knowledge-driven work organisation is today necessity, and to explain it as technology-supported achievement does not give the right account of the actual changes in the mankind life. In the present section, the survey is moved to look at the more general *externalities* of the mankind steps forward, to try to describe the emerging necessity as civilisation defence or fortification means.

The outlined sceneries are based on the scheme, to look at solutions consistent with the earth survival, keeping on life propitious conditions, face to the pollution lowering and to natural stocks rebuilt. The scenarios consider the solidarity with the generations to come, fundamental demand, not to be postponed to the quality of life of the present citizens, through the *altruism* mind.

The step-wise «complexity» incorporation in the «knowledge» paradigms is technical defy towards efficiency. It coherently leads to the advances, such as the ones recalled in the previous section, up to the *extended* enterprise set-ups. The «knowledge» paradigms play big roles in the economic and ecologic contexts:

- with technological enhancement issues (net-concerns, etc.), and innovative business opportunities (relational corporations, etc.);
- through instrumental operation outcomes (eco-impact visibility, etc.), and on-duty aids («natural» capital bookkeeping, etc.).

The knowledge society makes possible wholly new growth processes, along deployment styles out of the western world conventional habits of the «industry» paradigms of the *reductionism* planning. The *complexity* preservation paradigms provide the innovative development rules, at once, along the «to de-materialise», and, soon, the «to re-materialise» alternative. This does not entails the man habits opening out; in fact, e.g., the relational business further addresses the cultural and social needs satisfaction, as obvious response to expressed demands. It involves innovative entrepreneurial set-ups, due to the ecology pushing moments, forcing to modify the «manufacture market», based on the «supply and demand law», into the «knowledge market», where the exchange does not, anymore, affect *material pieces*, to be in the *single* possession of one or of the other contractual partner. In a short future, furthermore, the *cognitive* breakthrough will explore blending the *artificial energy*, *intelligence* and *life*, into the transformation economy adventure, where the inventive processes by-pass the shortage of non-renewable earth stocks, by the resort to the bio-mimicry promises. In the present paragraph, the mankind deployment styles are reviewed, along the outlined adventure prospects.

The western world culture strongly affects the outlined concepts; the analyses, however, try to figure out issues, going beyond the assessed schemes, and to start being more coherent with the «robot age» changeful knowledge, since these views provide more eco-consistent frameworks. The «market» concept, e.g., instead of stressing on the *competition* picture, proposes its settling *rational* side; it is, thus, viewed as the reasonableness outcome, to be preferred by the wise men, to opt the profit sharing, in lieu of the selfishness. The alternative concept is coherent with the civilisation progress through *solidarity* (or *altruism*) schemes, because these are the only safe and reliable protection of the mankind future.

3.2.1. Membership Principles of Men Communities

The men communities' organisation requests bilateral linkages: of the person, interested to the membership; and of the ensemble, fixing shared obligations. The attribution of personal or family belongings and the exchange of stuff or animals are preliminary facts requiring regulation, and, thus, the market, before even the money, is essential start for formal agreements, to be converted into legal frames, when the partnership begins to understand the need to distinguish between friends and enemies.

The market we know is recent acquisition; its widespread issue even less clear attainment in many societies. Concepts, such as *free market* and *price-market*, are, perhaps, only theoretical suppositions. We can hardly figure out how trading was carried in the ancient Rome time. The goods exchange involved limited durables (housing, etc.), unique consumables (foreign food supply, fabric provision, etc.), covering, in the average, low quantities of items, compared with the self-produced foodstuffs and commodities.

It is not evident how the ancient towns, like Rome, could organise and pay the provisions required by the overall community. The food board (*annona*) existed, based on landlords taxes and colonies charges, carrying the import commerce of the basic produces (corn, wine, etc.) and the storing up for timeliness distribution. In current situations, the money had little usefulness, as the payments happened in kinds or against the execution of functions and jobs.

Economics was science to come. The ancient scholars and philosophers could have been engaged into ethic discussions, about the *fair* reward for accomplished tasks, or the *right* price of exchanged objects, or the *just* pay of on-duty soldiers. In that time, the slavery was common practice, and there was no reason of wages, to be bestowed or received, or way to discuss about their amounts.

The financial capital, as well, had little impact, being restricted to rare cases, where the payments in kinds could have been tricky. The loan at interest was not ethically correct, because the people forced to borrow money was in big poverty, otherwise he could have found alternatives. The idea of venture partnerships, to obtain useful returns, was, obviously, accepted, but the benefits were appreciated as shared gains (rather than financial rewards).

Without wages and interests, no self-consistent price theory establishes. The prices were loosely defined by the production costs, but, in the slave-run factory, the work was hidden quantity, with no productivity assessment. Then, the only question could have been, whether *reasonable* or *iniquitous* prices were raised. The answer should have been different for the *utility* value of many current items (bread, beverages, etc.), necessary to the everyday life, and for *exchange* value of classy goods (diamonds, masterpieces, etc.), almost useless to satisfy basic needs.

On such bases, the ancient societies are of little help, when the wealth creation and the life-quality institution are examined. Two points, however, merit especial attention:

- the citizens duties and rights, ruled by the legal loyalty responsiveness;
- the individual ownership protection, clearly fixed by the Roman law.

The latter is, mainly, consequence of the former. Both, most likely, are highly affected by the restricted citizenship quality (not all people are Roman citizens), and by the habit to accept the competition and to recognise positive rules, enacted as the way to protect the personal safeguard and belongings.

The abstraction inclination, further, is important feature, pushing towards the law forms and legal sureness, more than towards the native spirits and empirical evidence. The economics (the *dismal* science) is the gloomy arrangement, to give *a posteriori* confirmation on how to manage the affluence or destitution demands. The well reputed «market value» (or «supply and demand law») provides instant balances, on condition to remain in stable steady conditions. No guess on future trends exists; on the contrary, one is forced to accept as true that the «supply» will remain constant, out of any possibility that the offer might be stopped. The *free-market* automatic balance is, perhaps, contingent issue, depending on rather odd conditions, and the justifying «law» is forced to suffer too many exceptions, when dealing with the actual price forming mechanisms.

From the «industry» time, we have been living in the «manufacture market», with the staples in products obtained by transforming raw materials (later dropped in rubbish and pollution). In parallel, the «market value» came out as the palpable quantity, giving factual evidence to the planned profit, which is reached through the manufacturing business, when the earth stocks are unlimited. The sustainable growth checks show that the «supply» cannot increase. The instant balances with increasing «demand» are fictitious, whether the tangibles costs are hidden value, leading only to the thrifty future of dismal shortage, if no alternative is discovered by the men's invention.

The only relief could come from the «knowledge market», perchance partially outlined by the «new economy», with the staples in intangibles: the information, which can be sold without loosing possession, making easy its indefinite growth. The «market value», this time, becomes more subtle affair, requiring legal rules, out of the empirical «demand and supply balance» evidence.

Yet, the alternatives behind are rather thorny, with lot of spikes pushing afar the rationality of the devised propositions. The apparent formalism shares ideas in legal metrology, with thorns, when the metrics are assessed. At fault, the religious or ethical commandments are important to rule the relations between the citizens, educating to equity, fairness, honesty and integrity, but of little help in fixing the simple standards (such as the supply/demand balance snap-shots).

At the Rome empire end, really, the Europe faced periods when the concepts, such as legality and ownership, became shadowy. In the Middle Age, the religious anxiety was constant orientation, so that the financial and trading activities did not escape the general prescriptions, this time, with punishments or rewards entailing the life after the death. The European Middle Age ends with the *mercantilism*, and the build-up of the political and economical orders of the «capitalism» and related (anthropic) market-conditioned institutions. The «market value», in that period, is still concept to come, where the trade fragmentation made meaningless looking at steady supply/demand equilibrium settings.

3.2.2. History Outlook and Capitalism Phases

The «capitalism» can be described by many ways. Looking at the origins, we find the period of the Italian communes, with relatively evolved social privileges and claims, but without clear central political authority: the empire was dissolved, the barbarian rulers were uncouth and the church was hybrid substitute. The issue was found in bottom-up organisations, assuring:

- the citizens duties and rights, ruled by direct democratic participation;
- the entrepreneurial protection, entrusted to guilds and trade companies.

The democracy defines as «government through discussion», and requires the tolerance or acceptance of the different viewpoints, and the discussion support or learning awareness from other people. The capital company, the other side, is the mode to achieve critical size, to projects going beyond the personal reach, putting assets together, and managing the ventures, again, taking direct part to all decision (basically: one head one vote, alternatively: vote in proportion to share).

The company sharing resorts to four typical capitals: "mastery, backing, work and materials". The technical expertise, the financial contribution, the personal job and processed supplies are all needed. Without one of them, it is fully impossible to take profit of the company. From then, the «capitalism» evolved, and we already have distinguished three phases:

- city-state capitalism: this characterises as "*mercantilism*", when the trade companies were maximally absorbed by the transportation burden, and the engagement-to-profit bargaining risk was tackled by the merchant;
- nation-state capitalism: this is centred on "*industrialism*", moving, in the raw materials transformation into value added goods, the major business, and profit, requiring suitable workforce and governmental aid;
- trans-national capitalism: this is yet-to-be scenario, according to the *global* views, where the staples are on the intangible exchanges, and the growth, if any, could, possibly, come from the *hyper-market* efficiency (should the ecology alarms be postponed or neglected).

The three phases have intermediate breaks, showing the earlier steady course end, without properly switching to the later path. Example issues are:

- the Netherland trade organisation, after the Italian city-states, and before the UK nation-state deployment;
- the US hegemonic ruling, reinstating the nation-state, to the continental range, with wider political backing.

The picture is schematic, and the outlook is prospected only to give support to example statements, used to explain the rise and fall of the «manufacture market». Two terms require some attention:

- capitalism: anthropic ruling of interpersonal relationship from below; the primary fact is the personal decision to divide profits and losses, gathering critical amounts of «capital» assets, under shared responsiveness and split governance (the vote assured at each head, or *caput* in Latin);
- democracy: government from the people, through discussion; the existing example refers to the city-state, sufficiently small to make possible that all citizens gather together; alternatively, parliamentary democracies, where the representative function could be object of transparent check and free removal.

The definitions provide, maybe, partial views, and are used to fix the devised frame. The analyses of the capitalism refer to four assets: the «technical, financial, human and natural». At early stages, the capitalism could have been meaningless: with primitive technologies, tiny riches, human resources confused into slavery, the natural capital is only free asset, and it happens to be mostly used for payment in kinds. Today, it is similarly meaningless to make «capital» assets balances, but omitting the «natural» contributions from them.

With the craft shop, centrality is on the mastery: in the cottage industry, every operator was taught on the work to carry out, by example and training, shifting the skill to new craftsmen. The industrial revolution gave attention to the production means, and centrality turned on the financial investments, because of the extended availability of unskilled workers. The switch back to the technical capital suggests that the «industry» paradigms are plainly addressing mankind development plots, induced by socio-economical reasons, when sudden contrasting ecologic obstacles arise that cannot be set aside.

Are we at a new breakthrough ? The industrial revolution is identified western world excellent upshot, specially promoted by the UK nation, and done, because of the supporting political reality. Indeed, three backdrops establish:

- the parliamentary democracy, with involvement of the (oligarchic) upper-class, highly motivated by effective *nation building* actions, which made straightforward the switch, from Amsterdam to London, of the basic trade and financial businesses;
- the nation-state capitalism, based on reducing the internal taxes, balancing the governmental budget to keep stable the money, forbidding the foreign merchants, unless they kept locally their money, fostering the formation of venture companies; etc.;
- the worldwide colonial settling, through the extraordinary expansion of the commercial fleet and military navy, to have the permanent overseeing and control of the foreign economical opportunities and market chances, to fix dominion ties with the local rulers.

The citizens' loyalty to the national government is outcome of the established rights and duties transparency, so that each one appreciates the advantages of his active partnership to the common good. The parliamentary entrustment keeps full visibility, even if the direct participation is, no more, possible at the nation level. Besides, the change is not remarkable. The bigger city-states were already ruled by oligarchies, with subsets of inhabitants enjoying the status of *citizen*.

The government through discussion, when all influential citizens can, directly or indirectly, oblige to receive account of their ideas/wills, is crucial prerequisite of building value; whether biasing privileges or hidden inefficiencies establishes, the feedback sets up, to eliminate them or reduce to negligible levels their effects. If the tolerance on contrasting views and positions and the acceptation of revising the bylaws to mediate the interests and conflicts of the oppositions become good practice, the legal settings assure effective government conditions, and are, now, recognised as crucial prerequisite of the industrial revolution success.

The next two backdrops are well acknowledged and no comments are needed. They are reasons for the nation-state capitalism rise, and, also, of its fall. A further remark might be useful on the definition of democracy as the *government through discussion*. In fact, the *people* democracies are known prospect, also, justified by voting, through blatant plebiscites. The big diversity is on how the political ruling class is selected and up-dated; if there no free discussion and bottom-up removal, no chance exists to aim at efficiency, and wastefulness or private interest, quickly, establishes.

The vigilant and permanent control on the on-duty government, exerted by all voters, has, however, big hindrances, if decisions, which contrast the immediate wellbeing, should be taken. This is the case with the sustainable growth, when the restrictions on the natural capital are requested, to safeguard the third people and future generations. The government through discussion with the current citizens is ticklish question, and entails a mix of self-evident and documented explanations, before reaching acknowledged agreements, on protecting the whole mankind, up to our descendants.

Then, what can be said about the trans-national capitalism ? or on the yet to be *hyper-market* ? The observation of sudden socio-economical reasons, not to be set aside, is not sufficient to justify the «industry» paradigms desertion. The ecology motivations give evidence on the paradigms non sustainability, but this is far from suggesting worthwhile mankind development plots. Thus, the study has to look a little more on the development trends, to find out possible hints.

3.2.3. Rise and Fall of the National Economies

The nation-state economy is current institution, through which the wealth of, more or less extended, communities is managed. The paradigmatic reference in to the UK (and "*long global assent*"), but the necessity to overrun the narrowness of the surroundings has brought to the USA (and "*short global assent*") settling, and to the EU gathering, with, in short prospect, new sub-continental powers, such as China, Russia or India.

The transfer of the hegemonic position to larger national-like contexts is the result of combined reasons, which partly recall the descent of the city-state role, due to the necessity of larger backing, and, in addition, comes from the interplay of hard-technologies saturation and soft-technologies diffusion, or of «industry» *vs.* «knowledge» paradigms, opposing «manufacture», to «function» markets.

The size motivation turns up from the labour market width, to set up of suited amount of workforce at the industrial level, and, to a larger extent, from the back military support, to contrast competing governments and to maintain the imperial domination. The mercantilism closes, when the merchant corporations cannot any longer remain self-caring, and require nation building aids to stay alive. The fact is, perhaps, to remind, when hypothesising the *hyper-market* deployment, ruled by cross-border corporations, out of local governments support and control; the past history is there, to put in doubt the paragraph 3.3.1 scenario, too abstractly based on the hegemonic efficiency of the civic entrepreneurial spirit.

The technology motivation is, perhaps, more subtle falls-off of tricky politico-economical changes. The industrial nations, with widespread affluence, during the second half of the 20th century, started off the de-industrialisation process, due the high workforce wage and the multi-national companies' break-up choice. At first, the process means creating new work opportunities in the service areas, a few of low appealing, making the outsourcing meaningless, some other requiring highly qualified workers, especially in the «new economy» areas.

The work division, at the 21st century beginning, shows the «manufacture» domains concentrating in the Asian countries of recent industrialisation, and the western world countries largely fed by lower price goods, with proper benefit at the internal inflation level. The overall outcome is apparently nice, with the steep growth of previously poor communities, and slow increase of the already affluent ones.

Further subtle analyses might bring to discover undermining effects, such as the income allocation penalising the less fortunate citizens (due to unsuited know-how), to the advantage of the highly qualified minorities. The work-restructuring befalls, at times, as impending thread, especially, on middle-age workers, whose positions are out-sourced, and no similar options are made available.

Here, the welfarism plays very nasty roles. At the macro-economics level, it leads to extend the nation charges, for the social amortisation. Then, the country competitiveness decreases, requiring higher taxes, with decreasing efficiency as average national figure. With parliamentary democracy, there is very little chance to get out from the spiral down worsening, since the governments respond to the current voting majorities, and are replaced, should the social protection be taken out or decreased.

The sketched picture is quite rough, but well assessed, so that details can be omitted. At the root, nevertheless, the industrial transformation, from the old self-sufficient flow-shop organisation, based on scale economy, to the recent job-shop setting/fitting entrepreneurship, based on scope economy, is, as well, bringing in ghastly tricks, because of reasons, affecting the companies at the proprietary and the management levels.

The old factories were, most of the time, directly ruled by the owners or under their tight control. The today companies have chief-executive-officers and boards of directors appointed because of their management and administrative merits; the stock exchange is used to place money, not to look at governing the enterprise: if not satisfied, the stocks purchaser switches to different ones, and his satisfaction depends on the market-value rise or decrease.

The share-holders meetings, in lieu of active capitalists taking part, are bare acceptance of chief executives or directors reports, so that the company ruling is deferred at the management levels. The short term return becomes main request, to grant stocks rising trends, with satisfaction of the investors and emoluments for the successful executive-officers.

The sketched surroundings (governs responding to parliaments and executives reporting to share-holders) are both constrained on short term issues, compelled to reach immediate satisfaction, even if the long term success becomes unsafe. The temporary benefits cannot repeat, if the nation, as a whole, and the companies, as primary wealth sources, run after policies without long term prospects.

The division into macro- and micro-economy is further ambiguity reason. The macro-economy, to be effective, shall transfer resources:

- (almost) without affecting the world trade, so that the (temporary) internal adjustments are perceived as neutral by the commercial partners;
- (almost) without long term effects, so that the anticipated spendable riches do not negatively bias the wealth in the times to come.

If this is the case, the macro-economic measures, at once, when enacted, start micro-economic reactions, to quickly establish prices and wages dynamics aiming at suppressing the temporary relief, even before the closure of the (above recalled) political loops. If the two mentioned conditions do not apply, the country is forced to look after protection measures, by duty charges and import/export restrictions, and to widen the national debt, transferring bigger obligations to the future.

The tight linking between «manufacture market» and nation-state capitalism finds different balancing requests, when the «knowledge market» applies on the *global* context of the worldwide capitalism, ruled by multi-national companies. The considerations up now developed show the frailty, together, of the «industry» paradigms, and, this, perhaps, is issue easily imagined, but, at the same time, of, also, well prised institutions such as the parliamentary democracy or corporations management, unless sufficient future obligations are included, aiming at *altruistic* policies, for the future generation safeguard (free from the instant citizens voters and shareholders), out of the fragile *hyper-market* option.

3.2.4. The Global Village vs. Hegemonic Districts

The incentive to aim at the personal benefits and attraction to collect resources at the own individual advantages, even if this means to divest the other persons of their own possessions, is assumed as expected behaviour, and the assessed way to limit it moves through regulation and legality.

The *artificial* constructs, knowledge or money, are civilisation signs, but have to be accepted and practised as useful safeguard of the associated life. The market itself is mark of evolved societies, as compared with raids, forays and plundering of the nomadic hordes. The *legality* of recognized agreements has to be bilateral tie, and this distinguishes *rational*, from *brutish* acts.

The earlier affluent countries' de-industrialisation is, definitely, involved fact, where, at least, two phenomena interlace. To some extent, the number of workers in the manufacturing domains decreases, with the production moved to low wages countries, but the averaged national income increases, since the supply chains are entrusted to centralised rulers, with their return creating profitably businesses. The set-up is instable, and the temporary benefits become wavering issues.

The switch from the «manufacture» to the «knowledge» market does not lead, automatically, to alternative wealth build-up. The concept "market" bears quite a clear-cut import; it corresponds to the space, within which the economical agents interact to achieve benefits and satisfaction, through *durable* outcomes, no mind where the producers, workers, bankers, dealers, consumers or controllers operate. The enabling condition means that no natural or human reasons should be erected in view of biasing the exchanges, with the damage of contenders.

Surely obstacles exist, assuring competitive advantage to definite players. The barriers might be physic, such as the distance, which raises the transportation cost, or legal, such as the tariffs, which distinguish the geographic and/or merchandise sections. The former are considerably lowered by technologies, through enhanced shipment and communication means. The latter are deferred to inter-governmental agreements or to overwhelming abuses.

The *global* concept aims at obstacles elimination. The *"long global assent"*, as basic enhancer of the original industrial deployment, and the *"short global assent"* these times, have looked after finance barriers abolition. The recent one, however, distinguishes, because human blockades are kept, with relevant restrictions on the immigration permission. The people relocation was, just about, free, up to the 20th century end, when the fair wage fixing and workplace lost moved to control acts, with rising strength.

With no people-and-finance barriers, the dramatic wide apart ditch between industrial and poor countries could not set up, based on the productivity abyss, in the manufacture business. The retrenching of thriving districts is, as well, the start of defensive measures, to protect the local inhabitants and to damage the secluded ones.

The *no-global* view aims at non-discriminatory contexts, where, nonetheless, the local protection is fostered as primigenial virtue. The loss in productivity can, even, be commendable vision, since it levels the before risen chasm, with overall equality advantage. The outcomes in lower spendable riches could found cushion provisions, through the income sharing, with less fortunate guard.

The *post-global* view, rather than to look in favour (*global*), or to fight against (*no-global*) the «economic» *global* market, means to live in the «ecologic» *global* market, where the raw materials are rationed, and it is needed to fit quotas in the use of the «natural capital». The uneasiness face to the affluent society practices, already clearly felt by the *no-global* positions, is, nevertheless, to be turned into positive spurs, and to aim at (possible) ways out by «to re-materialise» options; if the bet is won, the *altruism* might be understood as the *rational* issue, opposed to the *brutish* behaviours. The environment protection acts, accordingly, are *rational* answer, not to squander the shared «natural capital», depriving the people to come of their allotted heritage.

The recent welfare (social conquest of the modern parliamentary democracies, obtained through about two hundred years' long struggles of the nation-states) has to become worldwide preliminary condition, being the whole mankind forced to live in the common *global village*. The «social» concern gains special meanings, to correspond with the regulatory frame, permitting the earth inhabitants to share the common sort. But such a society can be deployed, on the condition that the all members divide the common space and resources, with equal opportunities, rights and duties, according to accepted agreements. There is no evidence if that picture is possible, or mere utopia, and such *altruistic* view is optimistic hope.

The utopia comes back to say that the market is sign of smart societies, where *legality* is accepted habit, binding of the members operating within its recognised limits. No hegemonic district is compatible with the «ecologic» *global* market, as the earth is interlinked domain, and the dispensations, which might concern the other capital assets, technical, financial or human, quickly become critical thread, if applied to the natural capital (and the fitted quotas are removed).

The ideas behind the «ecologic» *global* market differ from the ones referred to the nation-state markets, and offer non-resolvable conflict on the anthropic point of view. The exchange of goods and services developed as alternative to wars and to forays, since trading is recognised valuable benefit, leading to *altruism*, and not to raids; now, moreover, the exchange area has to cover the *global village*, in its totality, and the fine protection compromise shall cover the generations to come. The *rational* legality shall become shared agreement, becoming standard piece of the *solidarity*, granting universal civic rights to present and future peoples.

The market, to conclude, comes out from the effort to suppress the destroying antagonism, by the constructive competition. The scope is not to foster, rather to regulate the man rivalry, fixing *legal* limitations. The predatory instinct is present in the mankind history, and reappears with the slavery practice or the imperialism rehearsal. The regulation becomes necessity, when the voracious habits cannot anymore provide competitive advantage. The *rational* behaviour is the one to aim at co-operation, not to *brutish* acts, since they, perhaps, allow local brief relieves, but with overall disruptive effects, affecting the future heritage.

The *post-global* solution, if feasible, can only be reached through the *legality* extension to the entire *global village*. This presumes worldwide regulations, and the interest of the *new* citizens, to behave with *altruism*, having attention on future generations, because the present benefits has to comprehend the left heritage, and the unselfish behaviour is only way for mankind survival. Short remarks on past trends might help understanding, whether the fair behaviour could spread all over the earth, or not. The digression is, typically, motivated by the «technical capital» definition, directly linked to the mankind civilisation progression, there through, to the entangled framework of the related *artificial* enhancement means.

3.2.5. Civilisation Trends and Fair Endurance

The trade regulation was unique constituent of the European nation-states, as soon as the predatory (piracy, hijacking, etc.) acts were banished, recognising that the privateer's behaviour could not grant steady advantages. The issued guidelines and bylaws make drastic distinction between internal and external rules, with the protection of the national citizens (the ruling class or, in our days, the voters), and the careful negotiation of infra-governmental agreements.

With the industrial revolution, the «manufacture» market is winning practice, to expand the wealth, on condition to reach wide spaces to sell the mass-produced items. This means to look at the *free market* establishment, as the prerequisite of the nation's wealth. The demand-and-supply law is fideistic panacea, and the *self-regulating* market the axiom, to be the foundation of economics. The «industry» paradigms transform into merchandising stuff all acquired resources, according to organised efficiency:

- the know-how and related production make the corporation strength;
- the financial backing grants the unswerving investment rewarding;
- the workforce could be engaged or dismissed depending on needs;
- the processed materials lead to the subsequent manufactured goods.

The entrepreneurship is the industrial activity outcome, drawing from the *self-regulating* market suited input sets, and providing proper output sets, with profit achieved from the difference between the output (the proceeds), and the input (the costs). The demand-and-supply law cannot be derogated, because it *scientifically* establishes the equilibrium value, balancing buyers needs and producers offers, at spot situations, guiltily neglecting the safeguard of the potentially damaged. This is the *brutish* side of the *scientific* axiom, taken out of its conditioning context.

The *self-regulating* trade allows to get rid of fairness measures, leaving aside the bylaws, and promoting the widest competition. The welfarism is the reaction, derogating the full *free market*, to protect the less fortunate citizens; it leads to the macro- and micro-economy ruling, obliging the nation-states to the above recalled severe tightrope walking, to balance short/long term issues and inner/outer rules, still leaving in the shadow the «natural» capital balancing.

For better insight of the changes undergone by the nation-state industrialism due to the micro- and macro-economy interplay, the typical features are listed at the two stages. During the highly competitive expansion phase:

- many potential buyers and active sellers are, all together, on the market;
- the producers are comparatively small, with localised interests;
- the competition is on prices, to be quickly adapted, with scale returns;
- the workforce is little aware of union ruling, and has separate interviews;
- the enterprise takes maximal benefit from basic necessities delivery;
- the work market is highly aggressive, with large equivalent offers;
- long economic cycles establish, without remarkable external intrusions;
- the workers face unemployment, with no or little social protection.

The later evolution brings to new set of features:

- concentrated active sellers are supplying many potential buyers;
- the producers are big business, with integrated technical capabilities;
- the competition is on customer-quality satisfaction, with scope returns;
- the workforce is organised into trade unions, with national contracts;
- the enterprise plans specific strategies, to exploit oligopoly issues;
- the work market requires sophisticated jobs, based on domain training;
- the economic cycles are overseen and controlled by the macro-ruling;
- the welfare is widely spread option, with unemployment protection.

From the inside, the changes of the economical frames might show continuity, with, however, some warnings:

- the *global* market expands the number of active sellers in competition;
- the enterprises look at leanness, outsourcing the non-core processing, but, in parallel, merge and acquisition aim at huge financial aggregations;
- the wide offer mix re-opens the competition on prices and added services;
- the trade unions are less strict, but the social benefits widespread charge;
- the companies' policies need to face suggestions with worldwide origin;
- the low-wage countries are impending spur, showing diversified offers;
- the product-service delivery is approach for enhancing clients' fidelity;
- the eco-protection and producers' responsibility are mandatory demands.

The *global, no-global, post-global* parable is, yet, not fully acknowledged on its essential falls-off, and the *global village* frame is perceived as challenge, still, out of the strict bounds on the natural capital. Besides, the higher quality of life, enjoyed by larger populations, allows the worldwide selling of the luxury articles, with the profit, initially, limited to the high-standing proprietary firms.

Only at actually abstract levels, the growth sustainability starts surfacing, with incomplete focus on resource stocks bounded availability, and with local concern on landfills location and pollution. The (mainly) economical analyses permit nice debates, to show the civilisation trends, with the associated trade of goods (from personal belongings, to land ownership), of transformation work (from slaves, to *artificial* energy), of riches (from on-kinds exchange, to financial instruments), of knowledge (from individual mastery, to *artificial* intelligence).

The ecological analyses, when proposed, are weighed against the economical warnings (the ones listed, or similar remarks), to delay and minimise the duties, as if the industrialism could, in the future also, evolve with continuity. More exactly, the warnings are linked to the fair endurance of the industrial countries quality of life, figuring out castling procedures, to defend and propagate the survival of the hegemonic districts (contradicting the *global* axioms).

The difficulty, to recognise the imminent ecological exigencies, moving from the economical frameworks, is further boosted by the old «industry» paradigms, through the transform into marketable items of the all four manufacture business drivers: technology, finance, workforce and materials. The market *self-regulating* ability completes the account, allocating the balance values to each driver, with no need to establish absolute standards, as if the *hic et nunc* estimates are sovereigns, and not affected by earth stocks decrease and habitat worsening. The raw material price becomes quizzical outcome, maybe, leading to speculations, but no metrics on the tangibles loans become object of the *rational legality*.

The «knowledge» paradigms are better acknowledged, becoming basic aid to keep the competitive advance of the hegemonic districts, justifying the high life standards and intensive technologic commitment. The spendable riches, gathering soft-accumulations, start bringing about the twofold outcome:

- the «*de-materialising*» opportunity, highly enhancing the value added of intangibles in the supply chain, to lower the consumption of the material resources in the wealth creation;
- the «*re-materialising*» opportunity, looking at closed-loop material cycles, for recovery purposes, or exploring instrumental cognitive deployments, in view of alternative stocks provision.

The clear-cut outcome is not without drawbacks in guarding the proprietary soft items, readily duplicated at no cost. The converse one is more problematical: when dealing with inanimate stuffs, due to the entropy decay; whether addressing animate items, due to the precaution principle.

The men civilisation *fair* endurance, on the sketched viewpoints, happens to require the continuity from the nation-state capitalism, with progressively larger containers, before at the sub-continental (or continental) range, to finally become worldwide-state capitalism. The deployment means addressing the *global village*, in its totality, and seems to open several unsolved questions on how to fix and to manage the *legal* limitations for the eco-consistency. Rough preliminary concepts are skipped in the last section of the chapter; mostly, the «industry» patterns end is feared, but the «knowledge» patterns alternatives are open challenge. The next paragraph is looking at the «*re-materialising*» track, reconsidering ideas outlined in the paragraph 1.2.4. The last chapter re-assembles most of the concepts, where the *rational legality* is, again, explored as the winning compromise, to suggest the necessity of the *altruism*, protecting the mankind all, descendants included.

3.2.6. Artificial Intelligence and Precaution Principle

The political overview on the *global village* constraints shall register the odd fact of leftist movements strongly placed on conservative attitudes. The left side, up now, was aiming at progress, accepting new ideas and technologies, as useful means to put in discussion previous privileges. Thus the weird *global* views of the short term economical advantages (possibly) enjoyed by the old industrialisation countries, is followed by the awkward *no-global* stance of the *green* groups.

The *no-global* policy means to be against high-speed train and nuclear plants, telecommunication repeaters and high-way construction, etc., because all changes have negative falls-off, as compared to previous accepted situations. The rejection of innovation is serious drawback in democracy (government through discussion), when consent is rejected, as it happens in Italy, by national polls, for the nuclear plants, by local uprising, for "disturbing" rail-roads, by green representatives, for genetically modified foodstuffs, and so on.

The nation-state, already in big difficulties, because insufficient 'container', to face the economic struggle in worldwide contexts, is completely disarmed, if also the political approval is entirely against the scientific and technological progress. The bio-ethics, today the striking metaphor of innovation, cannot be automatically only *positive* either *negative*. Thus, its rejection as totally evil will gather political majorities, without real and understood documentation.

The «knowledge» paradigms means addressing actual facts: we are not aiming at ontological studies: who is the man, the living being, etc.; we are interested into actions on the man body, on the life processes, etc.; the «cognitive» acts stand for bio-processes, emulating or duplicating the life. The anthropic conjectures allow to concentrate, not on the origin of the mankind, rather on what the man is able to originate.

Odd predicaments appear; the leftist movements, lost their usual function, are castled into conservative *no-global* positions. In the contrary side, the *global* mind is confined into the *self-regulating* market ability, as if the *ecologic* limits could be ruled automatically, without awareness of the natural capital squandering (and the earth life conditions devastation).

The outcomes provided by computer engineering or *artificial* intelligence are felt not dissimilar from steam technology or *artificial* energy achievements, and the impending up-turning facts are fully neglected. The attitude is consistent with wary conservativeness. No objection exists, should the «*to de-materialise*» choice be sufficient. Face to the «*to re-materialise*» request, the «knowledge» paradigms needs to be turned actively into «cognitive» acts and bio-mimicry processes. This is the bet, when using «knowledge» paradigms, for sustainable growth, and such attitude leads to the «active» post-global courses.

The *artificial* life researches are novel challenging gamble, and the *precaution principle*, the necessary prerequisite. The conservatism, however, turns the issues into worst-case sceneries, as any action is potentially unsafe. The nuclear plants, chemical fertilizers and pesticides, electro-magnetic fields, etc., all have possible hazardous falls-off, more or less as it happens with the genetic manipulations or the greenhouse effects. The psychology "when in doubt, do nothing" might result paying in political contexts, but with long terms damages, due to the irresponsible riding of the irrational side of the most sloth prejudices.

Happily, *artificial* intelligence turns out to play roles at the «to de-materialise» and the «to re-materialise» levels. The former, warranting the transparency of all processes, is spurring to operate, when "do nothing" means "fall into ruin". The latter, also known as bio-mimicry, resorts to duplicated genetic codes, enhancing the resources proliferation efficiency, as it takes place in the abstract domains of the knowledge build up, but now turned in the tangible fields of *artificial* life and inherent encoding carriers.

The *artificial* life belongs to the highest «complexity» domains. The easy line is, perhaps, to address the entities (cellules, etc.) with self-reproductive potential. The living being is, thus, described as a single or a collection of such entities. The «self-reproductive potential», next, portrays the information (or generating code), out of the support (structure to be copied). Equivalently, the feature corresponds to the knowledge autonomy, non removable quality of the living entities. These, by themselves, are not autonomous; mostly, to survive (and duplicate), they need interacting with the surroundings.

A cellule can reproduce in many environments and made to interact with new entities. This shows the cellular build-ups complexity, and the duty differentiation among cross-linked cells, with the astonishing issues to organise the skeleton or the leg and to assemble the eye or the brain. The experimental verification of the evolutionism follows from discovering the progressive complexity of the living beings, and the concurrent disappearance of species, no more compatible with the surroundings. The animals threatened with extinction, accordingly, satisfy a law of the nature; the species diversification could expand or decrease; the generation of new varieties being not excluded, whether consistent surroundings establish.

The evolutionism hypothesises sets of early bio-chemical processes, leading to cellular life, followed by later environment adaptation and duty specialisation, to yield complex living organisms, up to men. The species stability or reproductive invariance is supposed to set up, so that betterments are acquired and transmitted, while the unessential changes, quickly, disappear. The life, from its lowest level, generates ordered bio-topes, combining materials and energy, into structured lay-outs, and progressively incorporating food to survive.

The «generating code» (or self-reproductive potential) «ordering» property is the «natural» way for the foodstuffs procurement, along the alimentary chain, to nourish further life expansion. The chain partially compensates the world entropy decay, exploiting outer inputs (e.g., sun radiation, for chlorophyll synthesis). The systematic resort to the life «ordering» property is hypothesised to be valuable aid in view of the sustainable growth, through industrial exploitation of bio-mimicry, say, enabling organisational efficiency, into physically consistent processes.

The living entities require energy to survive: the generating codes exist, joined to their (living) supports; and the «self-reproductive potential» vanishes with the death. As an all, the bio-mimicry could lead to processes with high productivity of the engaged natural capital, always with the universe entropy increase (possibly, little affecting the earth resources, if the energy/material provisioning comes from the outer space).

At this point, the research might aim at linking «tangibles accumulation» and «information effectiveness», to devise viable innovation, which will approach the *disjoint* neutral yield of the «natural capital». The «knowledge» paradigms have to be exploited, to make transparent the «natural» capital productivity. With focus on the sustainable growth technologies, several paths shall be tracked: *materials* recovery, *artificial* energy, *artificial* intelligence and *artificial* life.

The *artificial* intelligence does not replace the materials or energy wants. If not entering into the (regenerative) knowledge build up mechanism of the living beings, it can only add value by side opportunities, i.e., performance monitoring and lifecycle visibility. Just the *artificial* life really exploits the «self-reproductive potential», and can be turned to achieve industrial productivity.

Yet, the precaution principle hinders the bio-mimicry track, due to *no-global* political hostility. The safe alternative ways (sun, water, wind, etc.) have marginal prospects. The nuclear fusion plants are option to come. Nevertheless, the struggle between innovation and precaution cannot stop the research of solutions. Besides, the information infrastructures, due to their inherent monitoring capability, are the fundamental aid to provide visibility, with transparency of the emerging warnings, out of unproductive alarmism.

3.3. QUESTIONING LOOKS TO THE FUTURE

The continuity between entrepreneurship based on «industry» or «knowledge» paradigms is factually recognized, looking at the technology possibilities or at the *global* economical approaches, and the above two sections provide bird-eye views of some noteworthy features. We have, in addition, emphasised: that the technical potentials show the way for steps ahead, with outcomes still unveiled; and that the economical frames are not without ambiguity leaving unanswered the impending ecological requests.

The sustainable growth deeply modifies the just technological-driven trends, and rejects the *global* economical approach, unless as a short term gamble. If the «knowledge» paradigms might turn into factual way out, the future scenarios need to be sketched, to weight the inherent advantages and drawbacks. The plans are schematic, and, of course, related to the privileged interpretations, based on the recalled *global, no-global, post-global* description.

Hereafter, three scenarios follow, each coming out as if only one of the three positions could remain dominant. Halfway sceneries might appear, mixing two or even the three positions, so that, instead of alternative outcomes, the all sequence, *hyper-market, autarchic* closure and *altruism* deployment, could develop, leading to the happy end, after experimenting the total uselessness of the previous tracks. It might, as well, happen that no happy end could develop; the prospected analysis is exercise of optimism, and it is dreamed up to look for positive ways out.

The three scenarios, no mind if in parallel or in series, are completed by hints on how the «knowledge» paradigms are primary prerequisite of the *altruism* regulations organisation. The considerations, here, are mostly on the generality of the offered technological potentials, because further socio-political guesses are completely outside the prospected outlook on the future.

The scenarios are, moreover, followed by technical complements, to provide specialised hints on what the information and communication technologies today make standard acquisition:

- the way to acknowledge the «natural» capital unbalance, assuring visibility to withdrawals, repayments and interests, through legal metrics, as it is the case with the

other capital assets;
- the models to describe the industrial manufacture productivity, with overt reference to four capital assets, so that the entrepreneurial profit does not disguise hidden imbalances;
- short comments on technologies, central in the «robot age» deployment, as characteristics of the new «knowledge» paradigms.

The all chapter provides starting stimuli, to the example «robot» technologies collected in the subsequent two chapters. The three prospects are, again, object of considerations in the last chapter.

3.3.1. The Economic Worldwide Free Market

The utopia of communities co-operating by free exchange methods, leading to worldwide increased wellbeing, characterises the *"short global assent"* at the end of the 20th century. It was also experimented by the *"long global assent"*, started around 1840, stopped in 1914 by the first world wide war, and followed by highly segmented and instable settings, showing the earlier nation-state capitalism fall. The two *"economic global assents"* differ, because:

- the *"long"* one made evident that the *reference* container was insufficient, and the sub-continental size becomes necessary;
- the *"short"* one shows that the size provides (interim) *economical* relief, with no coherent *ecological* answers.

To understand the economy/ecology dilemma, apposite remarks on the *global* viewpoint structure are useful. The world economic space has falls-off, bringing to the world communication space. The distance barriers, likewise, drop, due to transportation efficiency and remote-transmission abilities. These are exactly the prerequisites of the «global village», to make all peoples understanding the shared sort (common to the planet earth). Instead, the *"long global assent"* is followed by fragmentation settings, before the *"short global assent"* period, fast undermined by the lack of consistent «natural capital» stop.

The market integration and fragmentation courses are described by assessed economic models, showing how to bias the local competition by customs duties, and how to protect the internal market by technical standards. The global assent presents advantages, leading to averaged wellbeing enhancement, and drawbacks, indiscriminately exploiting the natural capital. Only now, the economy/ecology dilemma starts showing the affluent society patterns illusions. In short, the global village driven by economic reasons is not new, only more affected by information and communication technologies advances. The big breaks are:

- the differences in the ruling with imperialism function;
- the tangibles impending regulations and eco-demands.

The *"short global assent"* has main ruler in the USA, but the empire ending is disclosed by the insufficiency signs, even, of new sub-continental container. The capitalisms, we have been discussing about, require the balance of the nation-state political and industrial potential, against the dominated free-market covering. The latter depends on the *"economical assent"* area and strength. The new patterns are, most likely, axed towards polycentric arrangements, with several nation-states, at sub-continental level (China, Russia, India, Brazil, etc.), able to threaten the USA leadership, in terms competitive industrial outcomes.

This replicates situations at the *"long global assent"* end, when the polycentric set-up, with several European nation-states in competition, did face the infra-two world-wars instability. However, at least, two differences appear:

- the weigh of the multinational corporations and nomadic capitalism;
- the pervasive trans-governmental regulation of the trade exchanges.

The two together refer as *hyper-market*, namely, the political set-up, in which the worldwide market interests outrun the national laws. In short term, the *hyper-market* allows proper elasticity, to drive away the war risks, carrying-over trade agreements and clearing steadiness, ruled by multinational corporations and trans-governmental regulation (and not by democratic administrations). At least, this is an interpretation, not devoid of suggestion (and of supporters).

The *hyper-market* means, notably, the power transfer, from governments, to over-national bodies (companies and agencies). In different terms, the political control is removed from the direct voters, and ruled according to market-driven paradigms. With the product-service delivery, the citizen is required to interact, in all times and affairs, with companies and agencies, having unmitigated access to his privacy, out of any (democratic) representative entrustment.

To aim at effectiveness, the *hyper-market* expands the intrusion areas. When dealing with wealth build-up, the new entrepreneurship considers the value chain, and looks after competitive advantage by the visibility of the up-grading addition (in view to drop aside unproductive costs). In the case of complex product-service delivery, quickly, the value chain, from sequential, becomes an intertwined net of tangible supplies and intangible provisions, needing radical changes, to deal with the information modules and the decision aids.

For lifecycle managing, the multiple and concurrent information acquisition, handling and storing are entangled by real-time mechanisms, to assess, check and choose the *hyper-market* scheme, planned to track the *extended* enterprise set-up. The related economical systems are, necessarily, time-varying plausible answers, to changing constraint surroundings, typically requiring *virtual* architectures; to continuously enable the most effective lay-out, for every rising situation.

The emerging net-concern includes nodes, with specialised duties (according to established classes: suppliers, producers, dealers, clients, certifiers, controllers, etc.), and operation tasks, specified by legal guidelines and indenture agreements. The access to the overall networks requires the strict conformity with hierarchic prescriptions, generally, allowing only sectional entries, with monitored jobs and limited activity consent. Every node is viewed as dedicated source of: information about provisioning demands or supplies; flow switching and task allotment; local govern and choice aid; and so forth. The *rational legality*, here, is self-ruling spur, embedded in the market self-steering capabilities of the competing multi-national corporations (*free* from local political authorities, otherwise their efficacy suffers from those external biases).

Out of its exact hierarchy level, each node is the logistic actor, with allocated duties decentralised commitment, and assigned achievement responsiveness. The all modifies the people attitude in front of traditional supply chains, as the number of active actors hugely increases, making their business duty a specific step of the value build-up. The *virtual* enterprise framework, apparently, is economic answer for higher return on investments; at the same time, each node can play the active role in giving transparency to the processed natural capital, with three-party duty, to accept, to process and to transmit, the data-flow of the identified supply chain, according to the allotted commitment and responsiveness.

The *hyper-market*, in the sketched outlook, is being considered at two levels:

- the economical effectiveness, assuming that the market laws will overrun the political institutions, suppressing the national diseconomies, to look at competition in wider areas, including administrative and clerical duties;
- the ecological awareness, fixing that the visibility on the citizens' actions could be exploited to provide transparent recording of the environmental impact (overall resources consumption and pollution increase).

The transfer to public companies of many clerical and administrative tasks, in the past typical of the governmental authorities, is not without serious outcomes, also, in the taxation systems. The welfarism aims at charges in proportion to the income, to help the less fortunate. The *hyper-market* approach will charge, based on the delivered service, or to the caused eco-impact, loosing poverty shelter aids and infringing the individual secrecy along every-day habits. Moreover, the above second level presents, at least, fuzzy contours. As already said, the eco-protection acts are pushed by the mankind lasting survival awareness; it is hard to discover how this can affect the trans-national companies self-ruling *rational legality*.

The governs de-construction and the privacy invasion bring the capitalism at situations, approaching the ones of the city-states, when the venture companies, happened to be devised to gather technical, financial, human and natural assets up to the thought business levels, even if the overall political infrastructures were not fully developed steady governments. The similarity is merely at a very superficial level; the *global village* features show that the history will not repeat. At least, the *mercantilism* happened to operate in a tiny area, surrounded by un-known spaces, still to be discovered, while the today cross-border corporations manoeuvre in the earth market, with little chance to get resources from the un-known universe.

The switch from nation-state, to worldwide *hyper-market* capitalism is bet, in which the economical effectiveness is made to play the main role. The ecological awareness remains at the subordinate level and no certainty exists on its long term sustainability. The scenario, even if protracted, cannot turn into steady issue, and the subsequent phase has, essentially, to choose between the *autarchic*, either the *altruism* policy. Right now, the *hyper-market* capitalism enjoys the *global* vision preferences, but the sketched considerations suggest that it might only be passing scenario, with limited benefits, if, as in the sketched lines, the ecology/economy interplay moves initial steps (event with low probability figure). From the chiefly theoretic point of view, the *hyper-market* scenario is the brilliant outcome, which joins the *rational legality* of the market fairness (as opposed to *brutish* acts), once removed the political fragmentation, to minimise bureaucratic and administrative costs, by virtuous self-regulation. It is difficult to guess how far such scenario is realistic, in view of the mandatory ecologic restrictions.

3.3.2. The Ecologic Conservative Ruling

The local habits defence, by *conservative* policy and self-sufficiency resource management, is choice, experimented in the past, alternative to the *hyper-market*, basically, prising the *autarchy*. The scenario faces, at least, two messes:

- the fear of unknown outcomes, preferring the «in doubt, do nothing», since every innovation is not without risks, due to *total* precaution principle, up to the *caution headway* remissive thriftiness;
- the castling of the richer districts, with attempts of violence and resistance, to expand and shelter the local wealth, up to world wars deflagration, once the neighbours' subsistence makes aggression desirable.

The two swindles associated with the *caution headway* and *autarchic* policies depends on the lack of spendable riches that the strict conservatism implies. This is fundamental fact, while *caution* or *autarchy* is side choice, not by itself wrong, when explored with balanced fairness. Further remarks are useful.

The four capital assets are used, to obtain revenues, by transformation courses, promoted by the man, for satisfying his needs. The explicit account of the natural capital is noteworthy fact. It is, as well, obvious that, with the present population, the opportunistic economy cannot be realistic. The (so called) biologic cultivation does not provide enough food to feed everyone, and resort to intensive farming is needed, even at the expense of biotype decay.

The appeal of arcadia is recurrently coming in human history, as if the natural, order around the human beings, is something enjoying self-sufficient primigenial consistency. The opinions are, certainly, full of suggestions, and eco-protection by experimented 'biologic' practice is prospected as the ecologic answer, to mankind safeguard. Up now, the farms extension has spoiled virgin lands; the manufacture has transformed raw materials into rubbish, etc.; these are risky evils, and the man action brings in continuous changes, until the decay increases out of control (this, according to several analyses, is the case with the greenhouse effect).

Following that picture, the global village, with ecologic drives, is conditioned by conservative scruples, looking after regulations that *forbid* with the precaution principle, and *not permit* in view of the ecology safety. The warning awareness is rejecting technologies, which modify existing 'natural' equilibriums, by systemic actions, bringing to 'artificial' constructs, not acknowledged before. By the 'total' precaution, the economy/ecology dilemma could only try to replace the «energy decay», due to entropy, by the «information increase», with knowledge build-up.

The problems complexity needs not to hinder the obligation that sustainability has to be charged to the parties having benefits, and not poured on unrelated ones, to keep the impact on the environment under acceptable limits. The scheme needs to receive the widest consent; but this is not the case, today, lacking full visibility on the natural capital use and misuse, so that somebody is free to inhabit a village quarter, without eco-taxes, refusing the *global village* constraints.

The *no-global* vision, out of extremist closures, deserves deepening. The good and evil should have a divide, and clear-cut checks need to specify the meaning of 'permissible' transformation and 'natural' evolution, by contrast with 'forbidden' manipulation and 'frankenstein' evolution. The precaution principle should be the standard prerequisite, without stopping the technological innovation, which, only, leads to growth. This way, sustainability, at the *extended* enterprise level, requires a five dimension change, to promote strategies, which assure:

- pricing based on the return on investment involving resource productivity (vs. labour productivity);
- product-service offers, by unique indenture covering the whole lifecycle and recovery prescriptions;
- ecological responsibility up to the points-of-service/of-withdrawal, further to clients' satisfaction;
- facilities and competencies, with the explicit know-how requirements and supply chain visibility;
- accredited eco-consistency-driven quality systems, with third party control of the supplied product-service.

This five dimensions up-grading well exemplifies what is meant by the resort to the *caution headway* paths, towards growth sustainability with *"ecologic global assent"*. Indeed, coherent situations can be found: the *extended* enterprise makes 'best practice' lifecycle and call-back ruling under unified liability. The *autarchy* (resource managing for the self-sufficiency), as opposed to the *hyper-market*, still, presents the other (above recalled) mess.

With the *hyper-market*, no castling is admitted, and each one is the rival of all the others, with unbiased contest. The governmental authorities' weakness and the multi-national companies ruling, together, are proper ways to contrast barricades and wars. The *autarchy* is preliminary step to support ethnic, religious or political divisions, to foster the local affinities and peculiarities, as opposed to outer ones. The national/district ambitions or homeland/local hopes are gluing options to fight for *vital* advances and conquests.

The *no-global* rage against the market culture, now, goes across the countries, with pirate's fights and privateer's larcenies. Quickly, however, once the barriers become nation/quarter combat features, the war will turn to established schemes, and conventional armies start to fight, quite surely, reaching worldwide war. The threads shall require the involvement of the richest countries, through preventive battles and extensive re-arming programs.

The new weapons could represent big profit opportunity of, having upshots in workforces' engagement. The fear of technology rapidly vanishes, with chemical, biologic, bacteriologic, electronic, nano-engineering, etc., issues under secrecy or concealment protection, and investments, hastily, doping the current economy. In countries, where today the nuclear option is standard reference, the escalation will be obvious consequence, when the enjoyed wellbeing is put under attack.

Possibly, two kinds of wars might distinguish:

- borderline battles, when the combat separates wealthy and poor countries, and the objective is to safeguard the higher quality of life;
- shortage-driven struggles, when the lack of resources (petrol, water, etc.) is spur to conquest and robbing someone else belongings.

The *autarchic* policy, by itself, can be eco-conservative and assure sustainable growth. The big hindrance of the *no-global* way, besides the scientific closure, is the protection of local traditions, fortunes and privileges; in the past, the political closures have always been wars warning sign and hegemonic claims. Right now, the *autarchic* capitalism is hidden *no-global* vision issue, and the considerations suggest that it might be passing scenario, repeating old instabilities. The explicit warning, nonetheless, is useful, and it is here prospected.

At the moment the *no-global* vision plays the return role, in opposition to the *global* game. The approach is typical of political movements of the industrialised countries, where the natural capital unbalance is clearly evident, and the shrewd resources exploitation is desirable option. As already noted, the involved interest groups, basically, position themselves in the left areas, promoting, nonetheless, a highly conservative policy. At the same time, what is even more intriguing, alike standpoints acquire wide consent in the social milieus, where the worship of the assessed traditions is especially strong. The *autarchy* is the quite easy issue of the two groupings coalition, and the depicted scenario might enjoy the converging of somewhat awkward motivations.

From chiefly abstract stances, the *eco-conservatism* denotes fair *rationality*, if all different deployments are *brutish* tricks, not capable of generating real wealth, on the contrary, increasing indiscriminate deterioration, to speculators' profit. The proposition is dogmatic, while the in-opposition *cautious-headway* and *autarchy* programmes do not represent viable solutions, to the populations' welfare request, even if local and elitist advantages might clearly be obtained. At least, the present elaboration of the *no-global* propositions does not permit to figure out effective falls-off, out the antagonistic apprehension face to novelty.

The egoistic behaviour justifies, if the *ecologic precaution closing* identifies with the absolute conservative rejection of *risky* innovation. The certainty of the dramatic decreasing of the life quality could excuse fighting against more affluent societies, with predatory scope. The supra-national institutions and the negotiation efforts are only hope, on condition that the agreements are respected, because the shared advantages *vs.* the impending damages makes preferable the co-operation, at least, as standard reference. This suggests looking at the third scenario.

3.3.3. The Relational Entrepreneurship Bet

The third scenario, after the *global* «hyper-market» and *no-global* «autarchy», looks at the *post-global* «knowledge» paradigms. The idea is to accept, and never to oppose, the facts unveiled by both the viewpoints, and the opportunities offered by the worldwide economy of scale and by the local culture thriftiness. The above scenarios are both driven by the personal utility paradigms; the first one leading to sets of multi-national corporations and financial managing elites; the second one, to retrenching secluded oases, protected from the outer starvation.

The *post-global* age, it should be said, is already started, at least, if we accept the end of the "*short global assent*", and we acknowledge (as it is further noticed in the paragraph 6.3.7) that the *hyper-market* hypothesis is fairly implausible, and the *autarchy* one, too much dubious. This opens the perilous transition period, in which the *global/no-global* opposition does not identify a winner, experiencing a series of troubles, without prospecting ways-out. In the present analysis, we prefer looking at the active *post-global* approach, in which the *global/no-global* impasse offers motivations, to keenly work out solutions, and this is hereafter outlined.

The third scenario means to look at *altruism*, with focus on future generations, because the present benefit has to comprehend the left heritage, and the unselfish behaviour is the only way to make the mankind survive. The *new* citizens accept of not being the owners of any «natural» capital pieces, but only to enjoy them in usufruct, with the obligation to maintain and transmit the capital and to respond of all misuse, through assessed solidarity, paying proper fees for the loans.

These *new* individuals are, in parallel, citizens of the world, say, inhabitants of the shared *global village*, and the local communities' city-dwellers, in which their activity is accomplished. Their nationality might correspond to the language they speak; the rebellion against ineluctability is major scope, and the «solidarity» with the unknown (born and future) fellow man, the basic duty. In lieu of cross-border corporations, we like to imagine transborder inhabitants; instead of «enterprises», institutions planned to make efficient reaching short term scopes, «communities», people gathering around shared ideals and durable achievements. Looking back at paragraph 1.3.3, this happens to be impressive shift in the western world lifestyle, once the relational links play dominant roles, and the *solidarity* with the people to come is obligation, grounded on *rational* bylaws.

The market economy will modify into *altruism* economy, because this is the way, through which the transparency of the individual commitment is weighed to acknowledge the function/service transfer. The relational entrepreneurship allows the provision (free of charge) of information, tuition, know-how, entertainment, education, training, etc., without any shortage limitation, since the same content is easily duplicated with no bound. The governmental authority can be strengthened by the shared distribution and ruling of the social costs, when the relational duties cover larger parts of the *global village* assignments.

This third scenario might be utopia. Still, on one side, the ecologic urgencies are so evident that the *altruism* alternative is the only clear-cut response, easily to implement; the other side, the net-concerns and worldwide web instruments bring in the technological aids, to enable the «knowledge» paradigms. Referring to the *global village*, the appearance of *virtual* options implies high-sophistication value chain management, with distributed physic and logic units, assuring co-operation rewarding. As said, the *rational* behaviour outperforms the *brutish* acts: the axiom needs to be usefully acknowledged, when planning the time to come.

Multi-site, multi-company and multi-task set-ups shall operate, to support the *product-service* delivering, with full transparency of the all performed duties. This makes sense to the *"ecologic global assent"*, providing the credibility framework, given by the total visibility, which avoids hidden privileges. Aiming at the change gradualness, it would be very important to orient the companies towards voluntary ecology responsiveness and commitment, anticipating the mandatory obligations, requested by environmental policies, with consistent worldwide outcomes.

The changes do not happen by themselves; they might in progress develop on the availability of *ambient* intelligence, as nice aid of the *altruism* economy. The build-up of the relational business, however, establishes after wide modifications of the relationships between citizens, in trading acts and civil codes. The changes will affect the tangibles market, as well. Leasing-like indentures for instrumental *durables* (cars, house appliances, etc.) shall modify the control ownership fruition. Restriction on *consumables* might imply the lifecycle sellers' responsibility, to hinder non-sustainable practices (the cotton clothes, e.g., contain today toxic by-products, and remediation requires more energy than manufacture). Prohibition of *disposables* could factually be reached, by very high fees at the landfill dumping. And so forth.

The cosmopolite trans-border inhabitants of the *global village* guarantee their loyalty to the confederated countries they timely live in, permanently respecting the ecological obligations, to protect the other world citizens and the generations to come. The overall political structure is challenge, to be discovered, going out of every consistent engineering domain. It is, here, named *hyper-democracy*, just to put in clear the control of the citizens on the on-duty governs, so that transparent uniformity shall keep the «altruism» engagement of each individual. Further hints on the prospected *global village* government appear, here and there, in this book, but without any real purpose to suggest a clear-cut solution.

The «altruism» *post*-global «knowledge» paradigms are bets, still, unknown, which require double gamble: the trust in world populations consent; the faith in an anthropic principle. The points are not disjoint, as the eco-problems can appear urgent, but with no (or too risky) solutions, to reach shared people welcome.

Aiming at gradualness in the acceptation, the strategy weakness is explained by inherent errors, say, to castle into a few village quarters and to take no care of the outer *global village*. Indeed, the *"economic global assent"* appears to bring, as the past experience shows, to instable set-ups and self-feeding drawbacks, due to an illusory expansion of wealth, which, ultimately, treats the natural capital, as being spendable riches by the populations that, first, take the possession of the all. This is, unluckily, mistake, which undermines the future of the affluent society. In the past contexts, several other errors follow:

- capital allocation errors: in free markets, the capitals flow to reach highest yield, so that, today, the resource-saving disregards investments, until the rise of the tangibles cost will grant huge rewards;
- business organisation errors: the product innovation creates divides by respect to the older production plants, and the distributed inefficiencies are, most of the time,

tolerated as means of risk reduction;
- supply regulatory errors: many community services and utilities are ruled by 'political price' policies, and the common practice aims at minimal-fee social provisions, not at optimising the saving;
- operation directions errors: the manufacturer's instructions privilege the product reliability, most of the times leaving aside the eco-conservative (but legally risky) opportunities;
- sale advertising errors: the product labelling could be mistaking message, as the higher consumption of the refrigerator depends on bad insulation, not at all on higher performance;
- and so on, showing that, both, the «hyper-market» either the «autarchy» solutions, even when apparently fitting, have relevant side effects.

The «knowledge» paradigms, due to native visibility, allows bypassing these, and similar other errors. The *ecologic* position holds the merit to have evidenced, the first, the environmental critical demands, but, in the same time, the fault of rejecting positive anthropic increase, preferring arcadia-like scenarios. Moreover, the «autarchy» is dangerous egoistic castling, quite soon leading to war.

The *"economic global assent"* can fight the quality of life dramatic declining, but the temporary delay will only worsen the subsequent measures. Moreover, the *hyper-market* will result in wealth distribution partiality, creating subsets of extra-riches, to the destitute majorities' expense; the egoistic behaviour, this time, is not castled, but worldwide revolts are quite like to rise. Refusing both scenarios, the *post-global* answer is preferred, presupposing «knowledge» paradigms consistent interpretation, or the *anthropic* optimistic meliorism, which tries to get round the economic/ecologic dilemma by the *altruism* alternative, through, as fist instance, the relational entrepreneurship and intangible supply chains, in view, as new way, of the *cognitive* breakthrough. The third section in chapter 6 reassesses these short remarks, suggesting the *active* post-global approach, based on rational eco-spirit. In the following, more technology-driven considerations are developed, to supply proper insight on the eco-footprint assessment and overseeing demands, together, these two options, being fundamental engineering aids, of the conscious ecologic behaviour in the *altruism* alternative.

3.3.4. Resource Allocation and Bookkeeping

The initial two scenarios are possible: the first, maybe, already on the way; the other, unluckily, experienced in the past. The last scenario calls for huge efforts; it can only be understood, founding the *altruism* on strong *legal* regulation, and on the invention of the new *democratic* citizens' consensus over the worldwide basis. That political outcome is beyond discussion, being, quite abstractly, named *hyper-democracy*, as if the bottom-up control could be extended to protect all the world citizens, next generation included. To make plausible this last scenario, the *post-global* «knowledge» paradigms viability has to be liked to factual methodologies, and these points are shortly reviewed in the following.

The sustainable growth is man's right at universal range, with outcomes to be extended to the future generations, not today, represented by any political party or governmental agency. Then, the democratic consensus or international agreement results deprived of justifications, whether limited to place before the interests of the today citizens, by respect to posterity. The one-issue socio-political approach, put forward by «knowledge» paradigms, leads to protect the generations to come, presently without voice, by compensation ways, such as:

- to create a tax system that consolidates the part of wealth equivalent to the withdrawals actually accomplished from the natural capital, according to a *"deposit-refund"* like arrangement;
- to forbid the natural withdrawals that exceed quotas, physiologically equal to the reverse logistics recovery, or (hopefully) the bio-mimicry stimulated generation, in view to keep the neutral yield.

The first way is formal, since, transforming different capitals, the equivalence criteria are, at least, ambiguous. The second, if coherently applied, faces entropy limitations, and today runs into life quality decrease, towards the thrifty society. It is, moreover, possible to merge the two ways, using the "deposit-refund" choice as first instance, and keenly researching innovative technologies, out of reverse logistics, allowing the active generation of *replacing* resources, to achieve neutral yield of the *inherited* natural capital.

Many unanswered questions exist (e.g., in terms of comprehensiveness), but, as an all, the EU environmental policy seems to aim at the combined way. Thus, the bookkeeping of the tangibles decay and pollutants increase becomes primary demand, with, as side request, the assessment of the restoring onerousness.

Then, the closed-cycle economic/ecologic processes are prerequisites of the manufacture markets to come. The lifestyle design is necessary, with resort to the set of aids, such as suited PLM, with incorporated PLM-SE and PLM-RL tools. The analyses need to be quantitative, to make meaningful comparisons, fulfilling the assessments by recognised standards. The consistent closed-cycle valuation brings to concepts such as the new metrics (or similar equivalent standards):

- **TYPUS**, *tangibles yield per unit service*: the measurement scheme covers the whole materials supply chain, from procurement to recovery, so that every enjoyed *product-service* has associated eco-figures, assembling the resources consumption and the induced falls-off requiring remediation. The assessed figures are, most conveniently, expressed in money, with the arbitrariness of establishing stock-replacing prospects. The point is left open, but, needs to be detailed, to provide quantitative (legal metrology driven) assessment of the *"deposit-refund"* balance.

The figures are effective standard, if aiming at the «natural» capital intensive exploitation. The supply-chain lifecycle visibility is reached by monitoring and recording the joint economic/ecologic issues, giving the quantitative assessment of all input/output materials and energy flows. The current tax system has directly to operate on these data, establishing consumption rates at the input, and pollution rates at the output, to obtain the 'wealth equivalent' of the overall impact (as for the first mentioned way).

The *extended* enterprise suggests a better way, for the visibility of the lifestyle delivery «service» side. The idea is to stress on «function» provision, assuring the clients' satisfaction (independently on products ownership). A given «function» can be assured by several means, with different averaged eco-impacts. The neutral yield objective, if not right away reached, is approached, by aiming at «function» provisions, with vanishing overall eco-impact of the *service* delivering, since the supplier (as an all) fulfils the associated «natural capital» balancing duties.

The (welfarism) tax system pursues social targets, up to wealth equalisation, through the individual's contributory capacity rates (progressive taxes are, as for Italy, constitutional acts). The systems have to be coherently up-dated, for future generations' protection, with charges collection depending, rather than on actually enjoyed income, more explicitly on the tangibles intensity, or the *productivity* of the involved natural capital. This is striking change, as compared to the affluent society habits; e.g., in the USA, and in most advanced industrial countries, lots of trade incentives exist, based on expenses deductibility from the taxable income. When the goods value does not embeds the materials full costs, the recovery does not pay, and the consumption radically increases (fact know as *rebound effect*).

If a metrics, such as the **TYPUS** one, is adopted, the conservative behaviour is quickly fostered. The ecologic bent of the taxing systems becomes enabling spur, to turn the «knowledge» paradigms towards environmental friendly goals. The **TYPUS**, *tangibles yield per unit service*, metrics can, of course, take other forms. The objective is to look after capital conservative arrangements, notably, as for the natural assets. Nonetheless, the merely static conservation is against progress, and the «replacing function» (today) sophistication is explored on its capability of re-orienting the citizen's needs, with like satisfaction. The «to re-materialise» way might be enabled, hopefully, in the future, when bio-mimicry or similar processes are made available.

Even remaining at simplified bookkeeping, the management of the ecologic taxes is, today, rather muddled. The domestic garbage collection, e.g., is, mostly, treated as community's burden, with charging fees linked to nominal parameters (e.g., in Italy, flat surface, proportional, perhaps, to the individual's contributory capacity, not the amount of produced rubbish), so that no incentive to more eco-safe behaviour is prospected.

Also, the goods taxation, in terms of the inborn tangibles productivity ratios, could seem an unacceptable biasing measure, by respect to current market rules, aimed at maximising disposals, to minimise the human work content. Indeed, the tax systems need to collect amounts of money covering the supplied service, and, in the same time, to lower the citizens' fees for the ended duty. If the charges are exceedingly high, the community will cease to be competitive, and the life quality growth is compromised, by unproductive administrative costs.

With the «knowledge» paradigms, the tax regulation restructuring is required, because the socio-political aspects management becomes important contribution to effectiveness. The biggest question is how to distinguish community's, from individual's duties. The averaged taxation of input/output materials and energy flows is comparatively simple, when linked to nominal parameters, limiting the control to out-of-all quantities (provisioning and land-filling); this could be local communities business, with visible fees refund, depending on the efficiency of the local bookkeeping.

Today, the these fees evaluation is quite obscure, due to political biases, and it leads to taxing approaches, which draw from the accumulated capital, rather than in proportion to the *actual impact* (net consumption joint to pollution). The eco-protection, switched into individual consumers' business, is starting point, to look for higher efficiency, on condition that the whole supply chain, from provisioning to recovery, is tackled according to *all-comprehensive* productivity criteria (at the *industrial* effectiveness level).

The image might profit, looking at the *hyper-market* scenario. Here, efficiency is the abolition of the barriers, not only to remove the unproductive administrative charges, but, more relevant, to create unified market situations. In political terms, the (visionary) *hyper-democracy* leads, as well, at depriving of authority the local governments, but, now, transferring it to supranational commissions or agencies, assuring the *rational legality*, as infra-generational pact.

The **TYPUS**, *tangibles yield per unit service*, metrics helps solving the economy/ecology dilemma, because the eco-impact is assessed for the end-of-life items, including reverse logistics recovery. There is the trusty involvement of the consumers' side to optimal resource productivity, exploring the *product-service* deliveries, reaching the highest economy/ecology effectiveness, obtained by the concurrent information flow transparency. Then, the definition of the **TYPUS**-like standards, with legal metrology form, falls among the infra-governmental duties, ruled accoeding to the *hyper-democracy* prospects; not the assessment businesses management, more efficiently accomplished with resort to apt certifying bodies, operating in the *extended* enterprise setting.

3.3.5. Manufacture Business and Productivity

The manufacture productivity is trick, when the generated wealth is obtained by transforming raw materials into pollution. The effectiveness distinguishes the competing countries, at scattered industrial levels, and the conflicting social and political groups, with local or global roots and more or less developed interest in the earth resources preservation. The *altruist* scenario requires acknowledging the manufacturing business, on two prerequisites, namely:

- to recognise that the natural capital use is authorised on condition of fully refunding the withdrawals;
- to assess and to bill the materials costs, with resort to uniform standards, and legal metrology schemes.

These prerequisites have enabling means in the knowledge paradigms, once adequate standards are available, such as the recalled **TYPUS**, *tangibles yield per unit service*, metrics. The manufacture enterprise shall have accordingly up-dated description. With earlier models, the delivered product quantities, Q, are assumed to change, depending on the contributed financial I and human L capitals. Simple relations have been employed, such as:

- $Q^\circ = \alpha_o IL$,

or, with incremental description:

- $\Delta Q^* = \beta_o IL - \beta_I I - \beta_L L$,

when linear input/output models are assumed for the local marginal description of the quantities, or the related increments around an optimal setting. More complex models are in use, to include market entry thresholds (I_{min} and L_{min}), or saturations (I_{max} and L_{max}). Similarly, instead of bi-linear dependence, close to steady state, the contributions lack of symmetry could have resort to modulating exponentials.

Our research requires to make explicit the dependence on the technical K and natural T capitals; if four assets are considered, the basic model-patterns leads to:

- $Q^\circ = \gamma_o KILT$,

or, with incremental description:

- $\Delta Q^* = \delta_o KILT - \delta_K K - \delta_I I - \delta_L L - \delta_T T$,

where all contributed technical K, financial I, human L and natural T capitals are included, to supply overt account of the knowledge and tangibles effects. Again, the tetra-linear dependence assumes to operate nearby equilibrium assets. With optimised choices, the negative appendages accomplish the sensitivity analyses, separately accounting the individual capital contributions.

The earlier dependence on the financial I and human L capitals, maybe, was issue of the dialectic opposition between plant owners and labour. The neglect of the tangibles T factor was surprising, as the manufacturing activity has no output, without materials and energy. The addition of the intangibles K is characteristics of the «knowledge» paradigms, and an earlier entry could have been devised from the industrial revolution (or, at least, from the «new economy» deployment).

The **KILT** models provide reliable image of the delivered product quantities, Q. Lacking one contribution, the balance is lame, and the reckoned productivity figures, untruthful or meaningless. The analyses are useful, when the piling up invariance is investigated against the resort to non-proprietary technologies, or to off-the-market loans, or to work out-sourcing or productive break-up. As with the earlier pictures, the models can be modulated with thresholds and nonlinearities.

The tetra-linear dependence means the assets equivalence, as single input, and their synergic action, when cumulated. The company return is optimal, when the (duly scaled) factors are balanced; the current scaling expresses in money the four capitals (comparison of non homogeneous quantities is meaningless; the output Q has proper value, with the four inputs homogeneity); in the same time, the return vanishes or becomes loss, if one contribution disappears. The loss represents the enterprise imbalance between constituent (know-how, money, work out-sourcing, bought semi-finished parts, etc.) flows.

The partial bookkeeping is merged into global cross-sectional assessments, to provide whole pictures, with visibility on the four capital assets, as inner, either, outer enterprise's function/facility. The basic industrial set-up generates the value chains, providing Q guesses based on market requests, including the welfarism charges, according to the enacted rules, and, from now on, the ecologic fees, e.g., assessed by the **TYPUS** metrics. The four manufacture assets easily apply, without modifying many assessed traditional habits.

The mass economy description deals with the delivery figure Q^o, given by the steady production flow, on the selected strategic spans. The flexible automation (with integrated design, in progress open to time-varying scopes and externalities) leads to ΔQ^*, which corresponds to the incremental delivered product quantities on the tactical spans. The steady description states that the changes in technology, finance, labour and supply can freely be fulfilled, without affecting the throughput (products and services).

In the bi-linear model, the tangibles (utilities or commodities) are attainable without limits. They do not affect the manufacture business; the affordable growth trend is undefined. The changes in technology, knowledge or know-how, simply, rescale the productivity of tangibles, processed along the material flow. The *new* model presumes direct concern of the sustainability bounds for the energy saving, pollution avoidance, natural goods preservation, and the likes. These bounds lead to introduce the concept of resource *efficiency*. The replacement of material goods is cost to society, with non-negligible environmental impacts, which shall be paid by the benefits holders (and not poured out on the future community).

The scaled T-factor measures the «natural» capital use/misuse, with annexed implications. The technology K concern has several fall-outs at design, production and sale phases, and, in the manufacture business, is dealt with, for *trade fairness*, by quality engineering rules. The standard definition of *quality*, «conformance to specifications», binds design (technical specifications files) and delivery approval testing. The quality functions deployment, or other company-wide quality-control, is recognised as way to improve efficiency, through zero-defects production. It is updated as «fitness for purposes», with focus on the users' requests. The last issue is ambiguous, unless the quality is measured through quantitative methods, within legal metrology frames, leading to "certified quality", established by third parties (as compared to sellers and buyers).

Anyway, the technology K comes as primary transformation factor, affecting the manufacture process throughput. The matter can be looked as well, saying that *quality* is technically specified at the design (conformance to specifications), or at the marketing (fitness to purposes) steps, if reference standards are defined. Then, the *quality* is widened to embrace the eco-consistency demands, and the "certified quality" has to be assessed with resort to **TYPUS** metrics, or equivalent standard. Quantitative performance functions are assessed, to connect the effectiveness up-grading and capital investments, while detailing impacts and returns of the (four) fixed assets. By that way, quality engineering becomes meaningful go-between for the company-wide information set-ups (assessed by formal standards).

Finally, to assess the actually deliveries value build-up, the used technology K ought to be accounted as primary production factor, not to be confused with the I-factor, as non-distinguishable acquisition, or in the L-factor, as operators inbuilt property. The complementary role of the K and the T factors is important feature, making the *extended* enterprise's operation especially useful.

The **KILT** models are opening steps in «knowledge» paradigms applied at the visibility of the «natural» capital exploitation. The subsequent steps are entirely dependent on the computer technologies and methods, through the resort to the PLM, PLM-SE, PLM-RL, etc. tools, and the widespread reference to networked facilities and functions, to operate lifestyle provisions. The **KILT** models and the **TYPUS** metrics are, both, preliminary suggestions, requiring further deepening.

3.3.6. Supply Chain Synergy and Net Concerns

The three *hyper-market*, *autarchy* and *altruism* scenarios come from coarse snapshots, bearing the many uncertainties of all political forecasts. The focus on the technologies deserves higher trust, notably when the attention remains on the existing «knowledge» paradigms and robot assessed capabilities. Besides, when one considers the *altruism* or the *bio-mimicry* desirableness, the reactions happen to diverge, assuming totally utopia, the acceptation to co-operate for sustainable growth, and imaginative, but doable, the «to re-materialise» prospects.

Actually, the law enforced *altruism* requests the personal involvement of all citizens, while the genetic engineering inventions come from individual scientists, not affecting the every-day life of the other people, unless, *a posteriori*, once the innovation is reliable reality. Therefore, it might be useful to look at a few more technical details that make the *post-global* approach reasonable, in view to make the *altruism* utopia, as well, a bit more realistic.

Among the many technical developments, we recall:

- the ambient intelligence, as the friendly interfacing of the people with their surroundings, to provide the safe overseeing of the current conditions;
- the robot autonomy and intelligent aid, as the effective means to carry out overseeing and control tasks, with reliable accuracy and effectiveness.

The technicalities of the above developments are example situations, in which the «knowledge» paradigms show the relevance of the *complexity* interchange, as compared with the *reductionism* route, up to «effects» separation. Both the topics illustrate situations, in which the dependence on the surroundings is leading truth, not to be separated, once discussing the technical properties of the applications. In both cases, the computer engineering plays the enabling role, according to views, not simply described by causal law, on the contrary, better explained, with resort to computer engineering techniques.

The ambient intelligence, AmI, is convergence of ubiquitous communication and computing, with duty driven interfaces adapting to the users, where:

- *ambient* embeds the ability of existing or being present on all sides;
- *ubiquitous* assumes that entity exists everywhere, at the same time and on a steady level;
- *natural perception* needs self-adaptation, to match the real-life conditions.

The ambient intelligence incorporates properties of:

- mobile (nomadic) computing, with resort to multiple-function devices and remote interaction capabilities;
- distributed interactivity, to support communication by invisible processing computer resources;
- self-adaptive multi-mode configuration, to interact with the people, while preserving the current human habits.

The AmI shows how computer tools become invisible witness of people life, so that social interaction and product functionality move foreground for further use. The AmI leads to co-operative settings, where entities communicate to each other. These entities can be humans, artificial agents, web/grid services, virtual-objects representing real things (not only human beings), descriptions of human knowledge (knowledge based systems), etc.; the whole layout leverages the full potentiality of net-centric facilities, for creativity improvement.

The ubiquitous computing is roughly the opposite of the virtual reality. If the latter puts people in computer-generated worlds, the former forces AmI to operate in the real world, automatically adapting with the people preferences. The option is important, when trying to understand the implication that AmI has on the world we live in, due to the ability of giving visibility of our current behaviour.

The option supports people to experience interaction with artificial agents, in their work place. By *ubiquitous computing*, the effective and unobtrusive presence of computing devices is supplied everywhere; by *ubiquitous communication*, the everywhere access to network and computing facilities is provided; by *intelligent user adaptive interfaces*, the perception of remaining in the real world is kept, as the interface automatically adapts to human (conscious or not) choices.

The AmI potentials are in the invisible, flexible and natural communication, providing input and perceiving feedback by uncaringly utilising all channels:

- the *ubiquitous computing* aims at invisibly and unobtrusively means, to free people from tedious routine jobs; in its final form, AmI leads to a computing device, which, when moving with the user, pace-wise builds dynamic models of the changing surroundings and configures its services accordingly; it is able, either, to remember the past patterns or proactively establish the new ones;
- the *ubiquitous communication* is major change, promoting data transfer, and allowing to integrate new blocks, like sensors or diagnosis modules, by natural procedures; distributed agents and wireless networking are enabling technologies, by sensing-driven modules, grouped into proper categories, say: visual-recognition and -output (e.g. 3D gesture/location); sound-recognition and -output (e.g. speech, melody); scent-recognition and -output; tactile-recognition and -output; other sensor technologies;
- by *profiling*, the ambient has the ability to personalise and automatically adapt itself,

to any particular user behavioural patterns; major importance is given to *natural-feeling* human and to multi-mode interfaces; humans speak, gesture, touch, sense and write in their interactions with other men and with the physical world.

The idea is that these *natural* actions are used as explicit or implicit input to AmI systems. The AmI services, for instance, may split on twofold goals:

- to support restoring and maintenance duties, for clients' satisfaction of the delivered *product-service*;
- to help data collection and vaulting, for tangibles consumption assessment and certification.

When looking at *product-service* monitoring on the lifelong span, one realises that AmI drastically changes the service engineering tasks. The *ubiquitous* mode unifies all processes, combining human and virtual entities, into product-oriented dynamic infrastructure. Benefits are:

- to concentrate information on the delivery;
- to create service-oriented bent, for the all lifecycle steps;
- to perform supply chain updates in real-time and by intelligent ways;
- to promote information sharing, by duty data accessing and managing;
- to enable products to carry and to process data, which affect their destiny;
- and the likes.

The robotics is second technical development deserving special attention, as knowledge society transversal and diffused option. What is more, the «robot age» moves towards entangled surveillance settings, where AmI plays the friendliness role. The *hyper-watch* abilities are common prospect:

- enjoyed by the *hyper-market*, under multi-national corporations influence;
- used by the *autarchic* set-ups, to control the secluded local surroundings;
- explored by the *altruism* ideas of the new citizens' behaviour lucidity.

The «privacy» protection, certainly, is challenge of tomorrow peoples, but this does not means to reject innovation technologies, rather to select proper visibility aids, consistent with the privacy protection.

The robot technology is highly interested in incorporating the AmI options; it differs, because the robot autonomy and intelligence are quickly taken in hand to behave as explicit synergic actors. The main technicalities are, again, summarised with explanatory examples. The robot-AmI co-operation is explored as the way to carry out overseeing and control tasks, joining enterprise competitiveness, privacy guard and eco-system protection, by ubiquitous communication and computing.

The robots, joined to collaborative net-concerns, are addressed, with focus on the consumable decay assessment (by appropriate metrics) on the all supply chain lifecycle horizon, disposal and reverse-logistics included. The potentialities of any innovation are not fully exploited, unless the enabling surroundings come in as the driving requirement supported by computer tools. Subsequently, the technology-driven options become standard requisites.

The robot-allied concepts are, here, assessed in relation with the net-concerns potentials.

Actually, the chiefly technical aspects, turning around the knowledge entrepreneurship (having the noteworthy issue in the *extended* enterprise), provide consistent example cases of technology-driven aids. Focusing with some relish to the many side issues, the robot-surroundings interaction permits acknowledging a series of behavioural habits, quickly generalised as collective features, shared by the ensemble, but not possessed by the individuals, when alone. These issues have been recalled at the national gathering level (see the paragraphs 1.2.1 and 2.3.6), and show the very relevant shifts in the lifestyle habits, in regard to western world trends. All together, the changes lead to the «robot age» innovation, in the present book acknowledged, not only as technology-enabled fact, but, as well, because of its cultural potential, in the eco-friendly *global village* socio-political contexts.

The robot technologies are widespread achievements, with standard outcomes in the current life, well known, and also exemplified by the case investigations of the next two chapters. Their cultural potential characterises, moreover, the added *technical* capital, to support the *cognitive* revolution, by means of the appropriate developments (see, notably, the paragraph 6.3.4). Here, the discussion of further specialised details is omitted, since, more than on technicalities, the focal point is on conditioning motivation, which promotes the cultural breakthrough.

The technologies are synergic help for the supply chain clearness, and support the *altruism* behaviour quite logically, with robotic aids express collaboration and ambient intelligence friendly efficiency, so that the two technical means are direct contribution to the «knowledge» paradigms deployment. The next two chapters collect sample project issues, to help introducing the «robot age» potentials, with the resort of already well assessed project outcomes, showing valuable trends that the changeful «knowledge» build-up is capable to offer. The commented sample projects are used to provide hints on the robotics technical developments, assuring suitable consistency to the *active post-global* approach, as it is over again outlined in the last chapter.

Chapter 4

4. ROBOT IN MANUFACTURE JOBS

The featuring condition technology of the «robot age» is vexed question, since starting point is, certainly, in taking off the irksome and annoying labour from the workers, as the Slavish root of the word reminds. However, the human operators are not easy to replace, when their *training* (or *programming*) makes (explicitly or implicitly) resort to self-explanatory series of actions and jobs, according to the *teach-by-doing* methods. The original «industry» paradigms are soon recognised as powerful means for the workers' instruction and for the robots' education, on condition to exploit the *reductionism* up to transforming every actually requested task, into sequences of independent elemental actions.

The controversial *reductionism* is critical innovation of the industrialism, and, as first instance, winning method to bring robotics in the productive surroundings. It is immediately apparent that the robot replacing function starts into very limited frames, when the changeover only concerns handling/manipulation abilities. At such level, the worker is machine, with educated dexterity and sophisticated five-fingers/two-hands/two-arms interplay, performed under sight monitoring. The manipulation robot, after effective *reductionism*, becomes quite simple (pick-and-place, etc.) device, leaving aside all the unnecessary jobs, out of the ones strictly requested by the considered duty cycle.

The robot anthropomorphism leads astray, because, at that level, the robot is duty-driven device, with little dependence on the fact that, previously, some man is trained to perform, with zeal and diligence, the devised machine's work-cycles. The *reductionism* means to separate the «decision», from the «execution» phases, so that the front-end actuation matches up some off-process established agendas. The dexterity-driven anthropomorphism shows the way to the *blue-collar* robots, where the on-process decision keeping is prohibited, because the *scientific* work-organisation cannot accept departing from the optimal pre-setting.

When looking at the industrial robotics, thereafter, vexed dilemma is:

- should the robot replace the *scientific* work organisation operators, having the corresponding handling/manipulation capabilities ?
- should the robot aims at the *intelligent* work organisation options, assuring the task-driven «decision» and «execution» functions ?

With the first approach, the anthropomorphism means to select the operation sequences according to the men's actuation capabilities, say, to limit the *artificial* world as if no technological alternatives could be devised. Subsequently, twofold restrictions might happen to follow:

- to remove the potentials of on-process task-adapting options, once the off-process

reductionism is used as mediation pattern;
- to have the *artificial* man as helpful manipulation substitute, with no focus on the education, perceptive, autonomic, etc. gifts.

The second restriction is, mainly, the starting reference in the instrumental (or service) robotics, to devise problem-solving apparatuses, moving from basically anthropomorphic views, namely, to design man-inspired contrivances. The views may be useful in the next chapter project developments, in alternative to the task-driven approach, there preferred.

The first restriction leads to the well prised fixed-automation solutions, widely exploited around the mid XX century, to optimise the productivity through scale-economy issues. This is compulsorily removed, if aiming at the *intelligent* work organisation. It is, perhaps, the industrial robotic most relevant feature, deserving the introductory section of the chapter. The central section, thereafter, takes into account the manufacturing surroundings, to give the task-reliance outlines. The third section, only, reviews example peculiarities of how the robot aids are made effective constituent of the manufacturing activity. The all is based on the special experience, gathered in past research projects.

The example topics gathered in the chapter are short overview of «robot age» premonition, outlined on its characteristic mark, with no technical details (given in the quoted literature), allowing explanatory hints, about why the «knowledge» paradigms are taken to start the divide in the wealth build-up. The concepts offer descriptive evidence of the interlacing between economics and technology, having on the backdrop, the role covered by the engineers' contributions. The «technical» capital is active input, with clearly independent functions.

4.1. WORK SURROUNDINGS STRUCTURES

The work surroundings structures bring about the «robot age» in the industrial contexts, as soon as the economy of scope replaces the economy of scale, and the *intelligent* work organisation permits to obtain effectiveness, with varying mix of products, manufactured according to customers' driven quality and environmental protection acts. The knowledge pervasiveness is fundamental requisite. Functional consistency and operation efficacy require views integration and lifestyle design. The latter implies information-intensive frameworks, to keep the task orientation centrality and to monitor investment return into value added. The former means the product-process-environment-enterprise method, to combine internalities and externalities, in view of the supply chain reliability and sustainability.

Robots become non-revocable innovation, being qualified issues of computer technology with, also, virtual-reality simulation as development means, to support expert decision for on-process govern and intelligent monitoring maintenance aid. The overview develops, including the following research lines:

- intelligent manufacturing processes, exploiting flexibility opportunities;
- collaborative facilities/functions management, with leanness efficacy;
- adaptive shop-floor logistics, allowed by distributed task-oriented robots;
- specialised computer aids, supporting the whole knowledge build-up.

A few hints are, here, recalled. The flexibility, before being technology issue, is economical request. The habits and living standards of industrialised countries prompt intensive marketing activities, with offer of equivalent goods, having price at several buyers range. The competing enterprises face clients with personalised whims, driven by changing fashions in the *world village*, quickly equalised by the communication means. The company cannot any more look only inside, aiming at best products according to core resources, skill and competencies; customisation requires offers extension and items become complex compositions, incorporating different parts and several technologies, so that the producer, by itself, possesses subsets of capabilities and can afford to accomplish only parts of the workflow, necessary for the productive chain. This early motivation of the economy of scope patterns are further stressed by the producer's lifecycle responsibility, requesting new skills, know-how and facilities, to support the product on-duty conformance and end-of-life take-back.

The ability to assess the business profitability is, then, a must for its survival. Quite happily, this is side effect of integrated information systems, characterising the *intelligent* factories, once every transformation transparency is established and the evaluation set-up is enabled and understood. The option is grounded on value chain monitoring, accomplished from manufacture, to lifelong. The concept of *value*, quite obvious in free market context, has, now, to be referred to all four the capital assets, notably to take into account the natural capital productivity over the lifecycle span. The engineers (and managers) need to deal with forecasts covering consistently wide sceneries, markedly, including the enacted eco-protection rules, as these are mandatory accomplishments to put in the market new offers.

4.1.1. The Intelligent Work Organisation

The manufacturing enterprises face huge changes to preserve competitiveness in the worldwide context, where information technologies supply the support for functions integration by means of knowledge intensive work-organisation. The changes aim at lifestyle supplies, to propose offers, which incorporate diversified technologies, making possible the properties mix, exactly matching the oriented requests of any particular buyer, and account of the eco-regulations. The one-of-a-kind products are challenging issue of the economy of scope, and involve relevant particularities, such as:

- investment concentration into strictly needed work-cycles;
- effectiveness improvement by versatility and adaptivity;
- resort to the facilities flexibility, to avoid equipment duplication;
- on-process management-and-control for bottom-up, just-in-time agendas;
- flow-shop break up, with decentralised production into autonomous units;
- team cooperativeness, by spread competencies and know-how addition;
- workers empowerment, with decisional proxy where problems arise;
- value added by entrepreneurial spirit and people self-motivation;
- net potentials incorporation, to assure information sharing and storing;
- achievements benchmarking, with recognition of individual success.

Effectiveness monitoring and evaluation has to be watchfully done (see, e.g., paragraphs 2.2.2 and 2.2.4), with proper facilities/functions integration (see, e.g., paragraphs 2.2.3 and 2.2.5), under the suited certifying body overseeing (see, e.g., paragraph 2.2.6). A plain approach is stated choosing qualitative standards, given (with the representational theory of measurements), building relative figures, for instance, through patterned interview method. The assessment, even if not without difficulties (due to snags, in evaluating intangibles), can be achieved by weighing each costs against the value added to the final offer. The recognition of the *value added* range is easily moved in linguistic figures (very low, low, high, very high), with the option of negative weighs for costs with no clear benefit.

In *intelligent* plans, the knowledge sharing, collaboration, empowerment, etc. technologies deserve special emphasis, in view of the enterprise leanness, which suppresses or minimises the unnecessary operations, suggesting to weigh out all material or human resource against the value, added over the company wide span. Example references are derived from small to medium size enterprises, operating in competitively special-technologies fittings, which shall be adapted to the final integrator requirements; high value added comes from co-design abilities, making possible quick response execution by co-operation achievements, jointly carried by suppliers and assemblers, into unified knowledge frames.

The on-process robotic equipment, quickly, widens, because: the product mixes ceaseless variability, matching buyers' satisfaction, can only be faced by high versatility manufacturing facilities; the changing delivery regulations and lead-times demand adaptive process and engineering rules, to react by quick-response and up-grading.

Lacking old methods, the industrialised countries are opting the *intelligent* manufacture, in lieu of productive break-up, with plants in low wages areas. The customer satisfaction is non-negligible issue, as for due time and certified quality, achieved locating the shops, close to the delivery stands. The net-concerns modify the earlier strictness, allowing freedom in the design housing; the remote steering and overseeing become common practice.

The knowledge intensity is firm business support, embedding the manufacture and the service phases, both established during the delivery design, indeed:

- the *intelligent* work-organisation enables effectiveness, with resort to the strategic, tactical and execution flexibility;
- the *lifestyle* purchasers' backing exploits PLM, PLM-SE, PLM-RL tools, for the lifestyle information and decision duties.

The *externalities* weigh, so far, cannot disregard that *intelligent* manufacture is the continuous enabler of every subsequent steps, mainly, when the integrated design activity is understood to be fundamental kernel, from which the knowledge build-up starts and around which the information flow spreads. The robots, in that context, exemplify the characterising *intelligent* actors, capable to join handling or processing, with monitoring or overseeing, and to accomplish on-process decision tasks. The all is the big departure from the earlier *scientific* manufacture, when the decision keeping is *optimally* set off-process, and the front end operators ought to be *passive* executors, whether diligent workers or automatic outfits. The change is at first accomplished, looking at the company *internalities*, or *product-process* 2**P** integration, and permits to investigate the work surroundings structures, keeping their inherent «complexity», with no resort to reductionism, to extract the element jobs. The next step, to *externalities*, or *product-process-environment-enterprise* 2**P2E** integration, is more recent achievement, where the «complexity» becomes standard reference, and reductionism is confined to be side help.

4.1.2. Collaborative Productive Flows

The «industry» standards aim at choosing *optimal* work setting structures, to enhance the process productivity. The most impressive achievements lead to the flow-shop organisations, with raw materials at the input, and final products at the output. The efficiency is maximised for given mass-products, on condition that no change occurs out of the pre-set structures. This entails right cares, *scientifically* established beforehand, leading to the planned throughputs, if the input requisites and scheduled resources combine, with no slips out of line. Such top-down plans are unrealistic, when the market does not take up the all forecast output, leading to the economy of scale end. This fact, as already well known, requires adopting the economy of scope, with buyers' quality offers mix, and «knowledge» paradigms, in lieu of the «industry» ones.

The change leads to bottom-up plans, to offer just what the customers desire, by flexible plants, typically arranged around robotic cells and machining centres, suitably enabling job-shop processing of the time-varying product mix. The time-to-market is competitive challenge, to allow proper reactivity face to the external drivers. Quite obviously, the manufacture process cannot anymore be beforehand cast; on the contrary, several decision-keeping operations need to be performed in real-time and on-process, interactively, when the product, itself, is conceived. The simultaneous engineering practice is winning scheme, leading the Taguchi/Toyota set-ups, to replace the Taylor/Ford ones.

The front-end workers ought to deal with time-varying schedules, enjoying, of course, the many supporting aids (from group technology, to robotic handling), still, being asked of individual initiative and collaboration. In the inter-personal relations, the harmonic character and mediation art surface, since the on-process activity cannot by fixed outside; on the contrary, the evolving situations need to be solved when and where they occur. The company successfulness is quite more entangled affair; the vertical hierarchic chains become critical, better replaced by horizontal nested meshes. The all-inclusive productive flow to final products is broken into intermediate chains, each leading to specialised components, later to be differently joined, so that the mix of final offers greatly widens.

The economy of scale is transferred at the specialised components level, so that the core business of several manufacture companies leads to midway items, not to have direct contact to the end-users. The business agility and leanness are important features, with the full visibility of the production quality, by company-wide certified systems, not to repeat the assessments by the purchasing partners. The information flow, parallel to the material flow, rises to highest function, and the cross-linking control flow needs ceaseless overseeing, with decision-keeping loops to modify the throughput according to the changing bottom-up drivers. The all can be accomplished by shop-distributed operators, filling quality charts and updating the process agendas and parameters. It can be fulfilled by robots, much more efficiently, and with objective repetitiveness. The industrial contexts «robot age» innovation brings in characterising details, mainly, because of:

- the driving surroundings: the previous «industry» paradigms, grounded on the *reductionism* rules, lead to highly structured patterns, and the reference set-ups provide coherently profiled decision lines;
- the information flow: the frameworks (input data, shaping up hypotheses, entrepreneurial goals, conditioning facts, etc.) are all acknowledged up to quantitative figures, with economic assessments.

The «knowledge» paradigms, then, have decision structures highly biased by sharp and well turned-out schemes, guiding the new *intelligent* work organisation along smart tracks. The extension of the «knowledge» paradigms, outside of the manufacture domains, is eventually to be performed, moving from sets of results, whose effectiveness is already verified under properly shaped surroundings.

The vexed question, opposing the robots' handling-to-reasoning capabilities is largely mitigated, because the *manipulation* is clear task-driven operation, even when accomplished by a man, and the *intelligence* is factually confined to work-cycle alternatives, requiring quite specialised know-how. The question is to make the man operate like a machine, and, at the same time, to push the robot to behave with standardised abilities. The other way, the *intelligent* work organisation is the keen method, to start exploiting the «complexity» into intrinsically well structured surroundings, exploring scattered collaborative solutions, keeping, on their back, the accumulated know-how, which assured the earlier «industry» profitability.

The «robot age», maybe, does not follow the economy of scope reorganising of the manufacturing business, since the innovation requests little effort to devise new «knowledge» paradigms; still, the issue is nice example to show the success of methodologies, alternative to the earlier *scientific reductionism*. Even if the full paradigms are not yet enabled, the robot technologies help establishing two facts:

- the front-end workers have their substitutes into *artificial* men, fulfilling, both, handling and reasoning operations;
- the design and development phase has, forcedly, to expand with *lifestyle* sceneries, tied by on-duty performance.

Both facts become standard prerequisites in the «robot age», and are object of introductory comments in the next two paragraphs, since the industrial robotics is very effective explanatory way to introduce the topics.

4.1.3. Handling and Decision Duties

The flow-shop planning optimises the process agendas, with right sequencing and tooling decisions. The basic parts or pallets travel the assigned lines, crossing the work-stands, where special purpose heads accomplish allotted jobs. The all is exalted by fixed automation, whether every detail is chosen off-process. If time-varying delivery is planned, the schedules address small batches of mixed items, up to one-of-kind products. The flow stops to be sequential; the items tracks have to include branching and crossing points, or to make use of free-travel equipment (AGV or the likes). The long path of the «industry» set-ups to optimal off-process work-cycle splitting and sequencing is destroyed, with no way to reach efficiency out of the frozen production policies.

The manufacture surroundings structures are, still, based on well established know-how, and provide the knowledge framework to, each time, select effective duty agendas, once the required provisions and the pertinent processing resources are available. The agendas foreshadow bottom-up «knowledge» set-ups, allowing nicely organised arrangements, where *flexibility* is dealt with at three ranges:

- organisation range: the entrepreneurial decision entails selecting the long-term policy, arranging the manufacture facilitie/function and ruling the marketing target;
- tactical range: the bottom-up sale requests involve choosing the production agendas, fixing up the item batches/mixes (by group technology, etc.) and optimal schedules;
- execution range: the on-line throughput overseeing engages, adapting the in-progress processes, to manage the sudden occurrences (failure, no input, new delivery, etc.).

The said «knowledge» set-ups exploit fairly structured surroundings, leaving little room to impromptu inventiveness. The «robot age» beginning has nice way to track, without big source of suspense. The oddity, if any, comes from replacing two centuries of *reductionism*, and to start looking at «complexity» as engineering problem-solving stipulation.

The job-shop planning is art, where skill and ability join to the availability of the sophisticated robot technology. The manufacture facility/function lay-out is made up by machining centres and/or multi-task cells, connected by time-varying dispatching means (programmable shuttles, automatic guided vehicles, etc.), and with interposed inspection and buffering stations. The parts handling is composite duty, combining path setting, manipulation task, posture control, and overseeing records. The multiple-function capabilities of the processing resources mean that each local work-stand is equipped with huge series of tools and fixtures, to make the re-fitting possible, each time. When highly varied and multifarious mixes are processed, the indispensable outfits exceed the cost of the process heads, and the dispatching and management of the tool/fixture kits rise to great intricacy levels. The job-shop planning, all in all, inherits the old structured surroundings, when dealing with the elemental machining work-cycles, but resorts to rather difficult decision-keeping choices in the task-allotment division and in the parts-and-tools path travelling.

The problem-solving contexts intrinsic variability has many origins, and time-scale horizons, because, for instance:

- the production agendas are bottom-up driven, to feed the market with the required delivering, permitting just-in-time plans, having proper tolerance, to maintain the due-date targets;
- the saturation of the manufacture facility/function resources is managed by the parallel processing of several product mixes, with erratic due-dates, to adjust the throughput lateness;
- the adaptable parts/tools dispatching courses permit to re-route on-process the related flows, in front of unexpected occurrences, assuring the delivery (full or partial) recovery;
- the economy of scope flexibility brings in smart production plans, making use of sorted out (by group technology, etc.) mixes, to optimally tackle the in-progress batches;
- the job-shop adaptivity supports leanness, leaving out gear redundancy, as, with proper dispatching, the suited work-cycles are re-arranged, due to the resources versatility.

The example features show that on-process decision-keeping is multiple-cause process, with *complex* conditioning drivers, which make *childish*, hunting *optimal* solutions, based on steady performance indices. The «becoming» is the permanent requirement, and the changing surroundings deeply affect the in-progress choices. The qualitative reasoning aims at plausible statements, based on empirical know-how and heuristic modes, rather than scientific laws and algorithmic blocks. This way, the front-end operators give *intelligent* answers, having comprehensive view and avoiding frozen bias of no more applicable details. Such capabilities have to be transferred to the on-process robots' handling and decision, with, conceivably, advantages of the procedural uniformity and fortuity absence.

4.1.4. The Simulation Frameworks

The robot conception and building bring into light the lifestyle design musts, because the equipment value depends on the *on-duty* performance, and rather not on the *point-of-sale* attributes. Of course, this is common to all instrumental goods and durables, and, today, the manufacturers' responsibility is like to be extended the delivery lifelong and even for the end-of-life take-back, notably, in terms of the eco-consistency. However, the lifestyle design methodologies started having critical relevance for the offers, which shall include the lifecycle technicalities as fundamental requisites, such as the robots. From that standpoint, too, the change is especial technological innovation.

The computer aids involve simulation abilities to replicate the robot behaviour and programmed tasks. In general, the data cover the facility lay-out, the fixtures and their process capabilities, the parts and tools inventory, the dispatching and labour allocation. The simulation enables the engineers to assess the technical specifications of the equipment, to balance the work loads, to accommodate the operation constraints (bottlenecks, etc.), to specify manning and ergonomics, etc.. The simultaneous engineering approach makes straightforward to jointly design the current product mix and the related manufacturing resources, using simulation for virtual reality testing.

During the operation life, the simulation packages, fitted with suited govern emulation blocks, generate forecasts and feasibility studies of fabrication agendas, work schedules, company strategies, delivery policies, etc., each time providing actual productivity figures. By intelligent automation, in fact, the effectiveness is reached by managing on-process the strategic (organisation range), the tactical (co-ordination range) and the execution (operation range) flexibility. The return on investment is, therefore, issued *ex post*, as the actual efficiency is the result of adaptively managing the available resources in view of market-driven requests.

Besides, the simultaneous design of the products, processes, robots, tools, etc. requires the up-dating of the govern information modules (sometimes based on realistic animated restitution), and profits by expert simulation, with knowledge-based architecture, to include decision aids grounded on experimental data and heuristic rules. The computer-based development aids, replicating the lifecycle behaviour, enable to design the robot structure and govern, in parallel showing the decision opportunities and acknowledging the actual performance.

The lifestyle design, on such premises, encompasses two accomplishments:

- the preparation of the reference knowledge-bases: this means detailing the relevant data, to extract the pertinent *Modelling* and *Simulation Features*, **M&SF**, through which the lifecycle assessments are fulfilled;
- the preparation of the computer aids: this support the programming of the causal models with *algorithmic* blocks, and decision modes with *heuristic* procedures, to show the physical/logic resources evolution.

The information frame represents a noteworthy change by respect of the up-now existing manufacture practice, oriented on the product-process *internalities* of simultaneous engineering. From now on, new *scopes* add, directly driven by the *externalities* of the on-duty eco-consistency and of the end-of-life call-back and recovery, according to the enacted sustainability regulations. So, if the robot lifestyle design provides technically sound frames, the *modelling* and *simulation features* **M&SF** become entangled data-bases, to explicitly deal with lifestyle aids and to provide the supply chain becoming outlook. Thus, the following quadruple features appear:

- robot equipment specification, providing the manipulation architecture, the govern abilities and sensing outfits, the programming decision logics, etc., to acknowledge the intrinsic **M&SF** data-base, in view of lifelong servicing and task-driven functional updating;
- development specification, detailing the manufacturing surroundings and resources, with varying production structures, to explain how to manage the on-duty **M&SF**

data-base, in view of granting the product-process joint adaptation to the users' requests;
- lifecycle specification, supplying the points-of-service (and -of-disposal) constraints, to rule the all delivery under unified responsibility, including the *M&SF* eco-protection rules, in view to implement the PLM, PLM-SE PLM-RL aids, each time, appropriate;
- corporate engagements specification, leading to assess the business policy for lifecycle responsibility of the supply chain performance, resorting to subsidiary enterprise *M&SF* data-base, to integrate the each time needed facility/function set/re-set and fit/re-fit.

Robots are evocative case, where the quadruple *product-process-environment-enterprise* data-bases are fundamental for the inherent technical constraints. Their conception and provision are, chiefly, the result of business rulers, integrating the robot equipment, with the application-driven outfits, running conformance-to-use and looking for supplementary aids, when the provision request the enhancement. The setting/fitting of a given intelligent automation plant comes out as «complex» business, and the sketched procedure is already clearly assessed issue. It becomes paradigmatic reference, when the lifestyle design has to be addressed as standard routine, in view of eco-consistency requests and lifelong producers' responsibility for all goods put in the market.

The «knowledge» paradigms in engineering involve, from now on, exploiting distinctive computer aids, permitting to assess the investigated object, preserving its inherent «complexity», with no resort to *reductionism*. In the lifestyle design, the knowledge set-up by digital mock-ups, because virtual reality tests, provides explicit assessment of the hypothesised bonds, without the reduction to elemental properties. The computer aids extrapolate the lifecycle, according to the chosen frames (*causal* orders and *behavioural* modes). By this way, the logic bonds are modelled on judgmental basis that factually affects the parallel selection of deep knowledge schemes. The functional dependence, in the models, indicates guesses between the entities without implying the closed-form causality; thus, the expert-modules provide qualitative progress, as compared to algorithmic deployment.

The resulting lifestyle design aids organise in nested multi-layer data-flow to combine structured information (algorithmic models), with empirical data, due to acknowledged expertise, vaulted into databases with the related heuristics modes. These aids refer to set of modelling concepts, presenting mixed dependence on the declarative and on the procedural knowledge, e.g.:

- type definition for entities, through complex conceptualisation «objects», *attribute-value-belief*;
- situation up-dating, done by off-process scheduled or on-process adapted occurrences, in view of the business project policy;
- situation-driven state transforms, for deep-knowledge structured models;
- context-driven decisional manifolds, for shallow-knowledge behavioural progressions;
- dynamic allocation and reallocation of entities, to accomplish on-process operation adaptivity;
- implicit de-allocation of non instantiated entities, to factually manage the process global properties;
- type hierarchies for relationships, through steering agendas that establish the tactical priority figures;

- message-passing co-operation and flag synchronisation, to enable the spot-adapted execution flexibility;
- function specification for the continuous interactions between concurrent processes;
- and other algorithmic/logic computer operations, to run the mixed-mode simulation checks.

The knowledge acquisition through simulation progresses by different tracks. Each behavioural property is the end result of the robot, according to the chosen model view. The link between pertinent views comes from the model-*federation*. The resulting *super-model*, once extended to investigate the whole lifecycle, is exploited: directly, to complete the given robot instance up to appropriate details; indirectly, to widen the (shared) reference database, for future developments. The spot modelling, comprised in the design computer aids, brings to modules for the virtual reality testing. Similar modules are filled in for real-time operation, to be later exploited, by service engineers, as monitoring maintenance diagnostics. The *super-model* architecture has nice properties, such as:

- varying-geometry boundaries, to expand over every view, full specifying the robotic facility operation functions;
- embedded simulation-emulation capabilities, to allow virtual behavioural experimentations, as off-process checks and on-process aids;
- co-operative infrastructure, to support problem-solving outcomes, open to multiple-domain experts, for the real-time facility management;
- automatic propagation of changes, by regular updating of the super-model data-frame, subject to hierarchically assigned priority set-ups;
- straightforward evaluation of alternatives, trends, risks, etc., grounded on reliable relational schemes, also, at the low priority levels;
- rapid producibility, affordability, etc. analyses, performed via intelligent decision supports, under the preset priority level allocation;
- fast and accurate exploration of lifecycle occurrences, to broaden concept-to-operation figures, and to get going the bottom-up process planning;
- efficient archival of globally accessible knowledge-bases, readily available in the current activity, of the collaborating partners;
- ubiquitous service through cross-connected facilities, enabling value added inter-operable aids, for the efficient management of the business project;
- any other advanced PLM opportunity that the today computer engineering provides.

The said series of computer-simulation virtues shows the prospects wideness, notably, as decision support to assess the on-duty and end-of-life attributes, and to investigate the setting/fitting alternatives of each implementation. The issues are obtained in real running conditions. Emphasis is focused on including *algorithmic* blocks, to duplicate the physical processes dynamics, and *heuristic* procedures, to emulate the choice procedures logics. The mixed-mode simulation is leading way for lifestyle design, satisfying «complexity» goals and requirements preservation, with provision of fully consistent behavioural details.

The opportunity, examined here for the robots, is deemed to become standard engineering practice, as soon as the *reductionism* is moved off, because unsuited as problem-solving methodology due the framework intricacy and variability. The robot development, thereafter, is by itself introductory case, where the off-process design has to be replaced by the lifestyle engagement: this is revolutionary, when compared with the old «industry» paradigms, and comes in as factual need.

4.2. THE SHOP-FLOOR LOGISTICS

When looking at the *intelligent* work organisation, the robot is man substitute, with closer anthropocentric, as compared with the earlier fixed automation rigs, as now changing handling and decision duties are allocated. Still, the robots are only front-end instrumental enablers, while the overall technical innovation comes out from the processing technological resources (machining, transportation, etc.), and methodologies (setting, fitting, etc.), whose resort makes possible the re-thinking of the «industry» paradigms, bringing back on-process the *flexibility*, with related decision keeping logics.

The section, on that premises, splits in two parts, to describe:

- the reference resources: the multi-purpose units and cells, the conveying, handling and buffering devices for parts and for tools;
- the reference methods: the strategic layout choices, the tactical/execution planning and running, the unmanned shop opportunity.

In the *intelligent* manufacture, amazing gap divides the (installed) *production capacity*, from the actually achieved *process efficiency* (or averaged productivity). The two data differ due to the manufacturing *coupling effects*, induced by the on-process adaptive planning, according to the economy of scope rules. With the old economy of scale, the *process efficiency* is kept close to the *production capacity*, assuring steady optimal mass-production, through resources redundancy, to never stop the flow-shop operations. With the economy of scope, leanness circumvents redundancies, and uses flexibility, for process recovery. The return on investment is obtained by the bottom-up schedules (the throughput corresponds to sold items, with just-in-time policy), and by the resources leanness.

The *intelligent* manufacturing, hence, is «complex» duty, where the *capacity* allocation does not corresponds to actual *productivity* figures. The process-plans (on available facilities/functions), the current agendas (of parts/fixtures), and the dispatching services (and annexed routing and buffers) are biasing surroundings, with coupling effects. The on-process decision logics adapt the current agendas, aiming at optimal schedules on the tactical spans, by best resources exploitation, and flows co-ordination, to approach the sought due-dates. The agendas, fixed at the *strategic* range, choosing sufficient *capacity* allocation, based on the company policy, are timely adapted, at the *tactical* range, driven by the market data. The unexpected occurrences (work-station failure, unfit dispatching, provisions want, etc.) are managed at the execution level, using re-routing and buffering choices.

The *flexibility* in structured surroundings has already been used to specify the paradigms in robot complexity, defining subsets of figures, for the simple framing of efficient issues, when suited organisational criteria exist. The few hints, given in the following, provide explanatory suggestions.

4.2.1. Flexible Manufacturing Facilities

The *intelligent* manufacturing, running with economy of scope, is established on multi-task *production capacity*, namely, machining centres, multiple-function heads, robotised cells, etc., connected by special purpose transportation shuttles and fed through suited manipulation and buffering units. The *capacity* allocation is basic step, to launch the company policy. It is less critical than with the earlier economy of scale, because the actual production agendas do not lead to necessary mass delivery, but allow manufacturing large mixes of items, driven by the actual market taking up. Anyway, the resources choice entails multifarious checks (see paragraph 3.1.1), looking at their actual on-duty potential, namely:

- the setting of the structural frames: components, facility-configuration and control, CFC, aiming at selecting the resources versatility, after identifying current product-and-process arrangements, planned because of the devised enterprise policy;
- the fitting of the information frames: monitoring, decision-manifolds and management, MDM, aiming at acknowledging the behavioural modes, in the actual surroundings, under visibly assessed operational situations and functional trends.

The on-process «reset» and «outfit» decision logics is design choice, and the study is done by computer simulation, based on the CFC structural models and the MDM behavioural modes. The functional scheme of the manufacturing plant is dressed, detailing the different machining areas, the transportation apparatuses, the buffering stations, the tool-room and annexed dispatching, the assembly areas, the inspection stations, the storing/shipment areas, and all linked fixture/service areas. The expert simulation uses causal (non-anticipatory) modules and heuristic (anticipatory) blocks. To assess the resources efficiency, conventional figures are:

- production capacity: nominal gross number of parts that can be produced with the existing resources and acknowledged work-cycles;
- manufacturing efficiency: averaged forecast, based on available machines, labour, on scrap ratios, set-up delays, etc., and just-in-time schedules;
- throughput level: averaged items inventory, embraced among the input and output stations, including set-apart and delayed components;
- utilisation ratio: active-time fraction of each manufacturing resource and annexed additional (loading, un-loading, etc.) settings;
- net production: *ex post* 'current inventory/mean flow time' ratio, the first given as shop global figure, the second computed at instant inventory.

The cross-coupling affects the production in such a way that the actual factory delivery can never be referred to the capacity and efficiency figures. The non-stop monitoring provides the utilisation ratio of each resource and the throughput level of the currently processed batches. The net production figure is made available as estimate, and can be split in disaggregated information for individual items of the planned mixes, in view of modifying the related tardiness, to reinstate the critical due dates. The flexible manufacturing facilities, in such an entangled framework, are chosen according to the following threefold scheme:

- setting flexibility, the lifestyle design data address: function figures (item mixes, batch sequencing, maintenance planning, etc.); programming goals (inventory/work-cycles/dispatching/etc. schedules, strategic horizon aims, etc.); and, in general, the enterprise policy forecast data;
- enabling flexibility, the simulation checks cover: monitored issues (current inventory, net production, utilisation ratio, etc.); management aids (guide, co-ordination, etc. tools, diagnostic signatures, etc.); and the likes;
- fitting flexibility, the decision manifold uses: knowledge frames (off- and on-line setting/fitting up-dates, etc.); control modes (adjust items schedule, adjust routing table, etc.); and similar other choice patterns.

The flexibility choice is made up by decision options in relation with: product mix batches, work-cycle schedules, production agenda plans, shop-floor logistics, capacity requirement re-setting, throughput level re-fitting, and so on. Each option offers several alternatives for how the manufacturing facility can be re-allotted, to meet requests, through available resources setting/fitting options. For up-grading, the decision loop begins by the flexibility managing issues on the faced (strategic, tactical, execution) horizon, following the underway logics (inventory-, planning-, sequencing-, dispatching-, recovery-, etc. -decision). The effects are related to the stored-up know-how. At the design and development phase, the flexibility scenery is important for the build-up of the pertinent conditioning knowledge, later to use during the on-duty and reset/refit phases.

4.2.2. Parts Dispatching and Feeding

The robot equipment is primary instrumental aid, to support solving the many manipulation and transportation requests of the *intelligent* manufacturing plants. The shop-floor, as said, is based on series of multi-task manufacturing facilities, and the parts need to travel and reach the appropriate stand, with *free* units, ready to accomplish the scheduled operations. The «industry» paradigms allow splitting the work-cycle files, into elemental sequences. With mass-production, these are optimally arranged, into steady agendas, along ordered lines. The flexibility aims at batches of diversified items, to yield personalised products, up to one-of-a-kind delivery. For the manufacturing efficiency, the production capacity has to be fully on-duty, and the diversified items ought to be simultaneously processed, each one according to pertinent diversified elemental sequences. The «industry» paradigms need to be revised, to aim at adaptive agendas, along time-varying lines.

A preliminary cataloguing, roughly sorting, the parts dispatching and feeding has to distinguish two facts:

- the processed individual items quantity: high, medium, low;
- the items inter-station transfer, either their local manipulation.

The first fact leads to distinguish alternative set-ups, grouped as it follows:

- the high volume productivity favours the flow-shop: work-cycle adapting is obtained by re-configurability and fixture adjust; items varying is dealt with by routing switch, through branching and crossing lines;
- the medium volume (especially) of little size items addresses the job-shop: large batches are reached with pallets assembling mix of items; routing is performed by steered shuttles or automated guided vehicles;
- the very low volume is tacked by robotised stands: single items are dealt, according to individual work-cycle and pertinent components feeding.

The return on investment is necessary demand. The last set-ups, of course, do not generally permit savings from inter-station transfer automation. The other two are typical deployments, distinguishing because:

- with the first set-ups, the effectiveness is chiefly managed on the strategic horizon, by resources re-setting;
- with the second set-ups, the efficiency is in effect governed on the tactical horizon, by resources re-fitting.

In the first case, the fixtures and outfits redundancy is fundamental requisite, permitting the facilities off-process re-setting, each time new production agendas have to be processed. The resources modularity is important, making cheap (and easy) the alternatives setting. The transfer-lines off-the-shelf modules allow wide changes; the rapid up-dating is managed by replacing the front-end fixtures, and re-routing the shop-flow main lines. The re-setting, even when quick and simple, requires stopping the production and ought to be sufficiently prolonged, to reach, in the average, proper manufacturing efficiency. Several example applications are known, notably, in the production of cheap objects in plastics, obtained by fixed-automation operation heads, readily interchanged and reset. The buffering can be kept at low ranks, and the throughput maintains fairly steady levels.

The second case is typical *intelligent* manufacture organisation. The resources redundancy is avoided, due to the facilities technological versatility. The layout is based on equivalent stands or cells, usually interconnected by AGVs. Locally, the parts dispatching is done by rotating tables, pick-and-place robots or closed-loop shuttles. Moreover, proper latch/unlatch rigs perform the front-end manipulation and positioning. The shop requires storing stands, where to put down the pieces if their work-cycle shall be blocked and delayed; furthermore, the stands and cells have buffering capability, where the in-progress parts are dropped/withdrawn by the transportation AGVs, and picked/placed by the local feeding equipment. The overall throughput is highly variable figure, and ought to be monitored (with the utilisation ratio of each individual transport/processing resources), to assess the manufacturing efficiency and to devise better alternative agendas.

4.2.3. Tools Dispatching and Feeding

The multi-purpose machining centres/cells cannot operate, unless the needed tools are present and assembled. When the parts mix is very large, the number of necessary tools expands, so that the investment in tools generally exceeds the cost of the matching work-head. Also, the tools ought to be monitored, re-sharpened and conditioned, to assure parts machining and finishing at steady high quality, granting zero-defect manufacturing standards. All in all, the tools management is primary accomplishment, much more intricate than with mass production, where the individual head work-load is fixed off-process, and the optimal planning is devised with full confidence. The *intelligent* manufacturing enjoy the advantage of the on-process information flow, and monitoring is helpful, when automation exists, to enable suited tools management flow.

The typical arrangement, actually, characterises the last recalled situation. The lack of tool-automation can only exploit hidden time between the plant re-setting, and the tools are part of the fixtures/outfits modification. The tool management, in any case, requires the on-process sophisticated tool selection/feeding equipment, especially, when programmed to work during man un-attended work-shifts. The efficiency increases by the off-process automatic tools conditioning. Thereafter, the standard arrangement leads to distinguish:

- the tool-room, namely, the specific area, where the exhausted, worn-out and unfit tools are collected and checked; the restoring and conditioning operation are accomplished; and the quality-class is allotted, for further dispatching to the machining centres;
- the transportation, between the tool-room and the different stands or cells, is, many times, accomplished by AGVs, taken from the regular parts fleet or specifically allotted; alternatively, proper dispatching paths, normally hanged to the ceiling, grant direct supplying, from the tool-room;
- the individual stand/head feeding is always automatic, being fast and risky job; the

local tool storages need to contains hundreds items, with indexed order and dynamic sorting (rotating drums or chains are in use); the pick-and-place arms and latching/unlatching rigs fulfil the changes.

The tool dispatching and feeding systems design is critical task, centred on simulation and beforehand cost/performance analyses and assessments. The tool-room deserves especial attention, due to the very high number and sophistication of the involved tools. With fixed automation, the *reductionism* looks at elemental jobs, mostly, to be accomplished by simple tools. The pieces, along the flow-shop lines, maintain steady attitude, and, once the set-up trimmed, the standard quality is permanent outcome. With flexible automation, the job background changes and affects the on-duty performance, unless the biasing off-sets and drifts are assessed and removed. This reflects on the tools, with combined issues:

- the visibility of the piece work-cycle progression has to be fixed into clear-cut targets, making possible the reference to process-embedded marks, to have the straightforward monitoring of the current quality;
- the visibility of the tool worn/conditioned status aims at managing in real-time the work-cycle agendas, by part batches and due-dates negotiation, to guarantee smooth company-wide quality deployment.

The each time on-duty tools (and fixtures) are dynamical allocation, to be up-dated, depending on the current production agenda. In the tactical flexibility pre-setting, *sufficiently* large product batches are pre-arranged, to limit the extra stops of the programmed re-tooling. This faces twofold limitations: operation leanness, not to widen the idle throughput into intermediate buffers; unwanted unexpected emergencies, not to re-route the batches flow, at critical situations. The suited tool management is example of on-duty complexity, whose solutions highly affect the manufacture processes.

We are not, here, to discuss into details all aspects in tools management. The software/hardware layout automation suggests standard remarks. In the tool-room, the automatic assessment, sharpening and conditioning processes are entrusted to special purpose measurement robots, leading to steady standards, out of manual operation reach. The information flow goes along with each tool, with encoded lifecycle updating, and data-sharing with the shop governing computer, for work-cycle agendas managing. The locally present tools, chiefly, distinguish between the ones in the indexed front-end storage, and the ones in the input/output buffers, waiting further handling. In the output buffer, not fully worn-out tools are placed, when their residual reliable life is beyond the agenda expected work-cycle. The central governor establishes the next steps: tool-room forwarding for recovery; either: front-end return for short work-cycles.

The tool transportation in and out the tool-room is accomplished with pallets, and only exceptionally, single items are individually handled. The all dispatching and feeding set-up in important part of the investment; with the whole tools series, the economical burden becomes central aspect of the intelligent manufacturing. The fact is, today, conveniently acknowledged, and the on-duty tool monitoring allows the explicit surfacing of the related costs, perhaps, too much missing in the shadow by the previous work establishments. Indeed, the tool/fixture dispatch and feeding is standard step of the *product-process* integration of the *old* simultaneous engineering, to deal with the enterprise *internalities*. The environmental acts open new economy of scope requirements, for tomorrow worldwide competition.

4.2.4. Resource Choices and Re-Fitting

The distributed knowledge processing system is employed for taking back on-process the decision manifold. Along the on-duty and re-set/re-fit conditions, the flexibility scenery is conditioning frame, to be dealt with as totality, accepting the cross-coupling effects, which are shown by the expert (joint algorithmic-heuristic) computer simulation. The goal *flexible automation* is coherently obtained merging the «patterns of action», into the «pattern of structure». The opportunity is created by equivalent lore building, through scattered knowledge processing. The *action-structure* interplay denotes the «complexity» of the surroundings reliance, when the background comes in foreground with on-process decision manifolds. As said, the surroundings, here, is well structured, and this helps managing the flexibility (see paragraph 3.1.3). A few remarks characterise the overall frame:

- the knowledge *structure*-patterns embrace resources (production capacity, facility/function layout) and entrepreneurial scopes, to specify the current policy and the forecast return on investment scenarios;
- the knowledge *action*-patterns include programming (strategic, tactical and execution) and scheduling (organisation, co-ordination and recovery), with resort to the surroundings conditions and monitored issues;
- the knowledge *structure*-pattern side accomplishes the setting/fitting up-grading, modifying the resources layout and the governing capabilities, to keep the business alignment at the market requests;
- the knowledge *action*-pattern side makes work adaptive governors, which transform the control according to current process data, on condition that the process structure consistency is perceived;
- the joint knowledge *action-structure* interplay assures consistency: *fitting* govern-for-flexibility options cannot be separated from *setting* intelligent manufacturing bottom-up ruled businesses.

The «complexity» management requires the suited closure of learning loops, to perform occurrence-driven reconfiguration of facility/function structures and of decision/govern actions. The interlacing of on- and off-process information loops is, perhaps, the most intriguing feature, showing the beginning of new ideas out of the *reductionism*, accepting *becoming* or *being*, *plausible* or *causal*, and the likes, to incorporate the self-conditioning knowledge.

The *intelligent* manufacturing profitability mixes technology-driven, market-focused, organisation-supported, knowledge-motivated, etc. frames, requiring the careful weigh of the many factors, in relation to the economy of scope issues. The computer engineering supplies the reference aids, enabling problem-solving ways, tailored to case-example situations, so that the «complexity» does not needs to be reduced to originating causes, and the effects interlacing can be considered natural outcome. The fact is significant departure from the original «industry» paradigms, and it comes out as gambit or ploy, to circumvent the saturation of the markets in the affluent society, too much tired of mass-products. The all is now reconsidered in the sustainable growth context, where the knowledge pervasiveness is deemed to move out of the manufacturing transformations, and to become standard means to accept «complexity» in the modelling and simulation features for the lifestyle design and supply chains eco-management.

As several times pointed out, the resource choice/fitting typically involves two levels, for the *internalities* either the *externalities* management. The first level has to tackle with the items' work-cycles, the shop-floor logistics (parts and tools), in connection with the market-required product mixes. The second ought to expand at *environment-enterprise* integration opportunities, looking at involving the apt facilities/functions, each time requested by the business project, with the resort to the new partners, ready to co-operate.

4.2.5. Adaptive Modular Setting/Fitting

Coming back to *flexible* manufacturing, the «knowledge» paradigms, as above pointed out, are exploited into (in advance) highly structured surroundings. It is, nevertheless, relevant that these innovative ideas, at first, were exploited in Japan, not in the western countries, where the *scientific* work organisation axioms enjoy long undisputed acknowledgement. The *intelligent* work organisation achieves its effectiveness by interlacing material to information flows; this entails knowledge distribution and decision decentralisation, leading to:

- *piece wise continuous betterment*: to yield the successful effort of adapting products to consumers' wishes (increasing quality and supply mix);
- *diagnostics and monitoring maintenance*: to aim at company-wide quality-control and at predictive/pro-active maintenance policies;
- *cooperative knowledge processing*: to enable reward ways, to single and to team creativity, for innovating products and processes;
- *leanness check-up*: to fully exploit the value chain and to remove all idle material or information, out of the enterprise profitability.

The adaptive *action-structure* interplay pace-wise embeds flexibility, keeping cost effectiveness by means of economy of scope. The iteration of the three steps: design/setting, testing/assessing, redesign/fitting, illustrates the interactive nature of decision-making, and shows how the behaviour of an alternative affects, which alternatives are identified for the next loop.

The application of decision cycle to intelligent manufacture is concerned with governing flexibility issues, so that varying market-driven requests are satisfied, whenever they emerge from bottom-up plans. The tricky fact is that the flexible plants are not used as they are, but after setting and fitting, according to the data collected by assessing the market-driven alternatives. The versatility requirements pre-setting at the development stage and the manufacturing flexibility on-process exploitation are preliminary guesses, based on:

- planned investment costs, with amortisation figures within the production programmes (business project definition);
- targeted productivity performance, to fulfil the technical specifications and trading goals (manufacture policy setting);
- delivery figures (time-to-market, etc.), according to prospects and buyers' expectation (production bottom-up planning);
- quality positioning, embedding conformance-to-specification and fitness-for-purpose (market assessing and managing).

The performance upgrading comes by exploiting the plant flexibility. The goal takes principal part in the product mix variability and critical role in avoiding idle resources. Return on investment arises from sets of rules, stating the objectives of «complexity» preservation, through *intelligent* task-assessment, such as:

- *functional integration* along the core manufacturing process, to support the synergetic co-operation of every factory resource;
- *total quality*, for conditioning the enterprise organisation to be purchasers' driven, assuring lifecycle fitness for purpose;
- *flexible specialisation*, to assure intensive facilities exploitation, expanding the offered mix, through technological integration;
- *lean engineering*, to avoid redundancy, in relation with the product-service delivery, over the enterprise planned policy.

The series of scopes are standard achievements, already repeated many times, with in mind to connect the scopes, into interlaced *action-structure* patterns. The all aims at step-wise tracking the customers' satisfaction, to widen the market by diversified mix of offers. The surroundings *action-structure* patterns become the reference conditions, which have intrinsic regularities, already acknowledged by the old «industry» paradigms of the *scientific* work organisation. The all leads to exploit such regularities, transformed into *action-structure* modularity. The role of the surroundings in the *flexible* manufacturing is well recognised, boosting the industrial robotic achievements. The modular *setting/fitting* operates on hardware and software, along well recognised tracks and aiming at granular objectives. This way, the «robot age» beginning in the said surroundings bears the special flavour of introducing the «complexity» management, within the (previously) simplified modularity of the *reductionism*.

4.2.6. The Full Un-Attended Factory

The unmanned factory is well discussed topic, which belongs to the common knowledge hints, deserving special emphasis in the «robot age». According to the remarks up now developed, the prospect is certainly technological challenge with relevant intrinsic value, but does not characterise the innermost innovation of the *intelligent* manufacturing, as compared with the front-end decision keeping and, perhaps, the just mentioned *action-structure* modularity. The unmanned solutions are, indeed, current layouts for, mainly, three reasons:

- the processing into hostile surroundings, making highly risky, either, not permitting the presence of human operators (nuclear plants, etc.);
- the production prosecution during un-attended periods (night shifts, etc.), to better exploit the investment, to grant proper due-dates, etc.;
- the manufacturing of items out of the man manipulation capabilities, such as with micro-electronics, nano-technologies, and the likes.

In the first case, master-slave technology is common solution, with, basically, the resort to on-process attendants, simply, replaced, in the perilous front-end, by the mechanical slave. With the third one, impressive new sceneries open in the micro- and nano-domains, and, again, the «robot age» supports the opportunities, without direct implication in the rethinking of the «industry» paradigms. The in-between case, on the contrary, is noteworthy example of the knowledge *action-structure* interplay, exploited to reach effectiveness by *setting/fitting* operations.

The full un-attended factory, out of demonstration pilot plants or stands, does not hold primary importance, having little economical soundness. Indeed, notably in Japan, great emphasis is placed on facilities, where the front-end operators are totally removed: the applications are widespread, from the automotive areas, with batch schedules of market-driven series of models, to the machine-tool domains, in the one-of-a-kind production is run, with very huge throughput (as compared to similar delivery by the EU producers). In these facilities, the man attendance is far from being removed, as series of engineers are present for monitoring, overseeing and clerical jobs. The company-wide quality control aims at economical profit by zero-defect manufacturing, with focus on computer-aided testing, CAT, and the expansion of the on-process checks, planned and supervised by technicians.

In that context, the un-attended factory, then, is the basic ending of countries, where unskilled operators are lacking, while trained, highly educated staffs give the potentials for sophisticated returns on investment. Depending on the allotted duties, the overseeing and clerical jobs might allow intermittent engagement, and the series of shifts are made to progress totally un-attended. If (exceptionally) the in-progress agendas show out of the standard trends, series of warning, alarms and emergencies require remote restoring, or call the technicians from other facilities, or definitely stop the planned schedules.

Hence, the un-attended factory, so forth, is technology-driven, more than real answer to factual requests. The remote diagnostics and maintenance are, perhaps, the most interesting fall-off, to be exploited in view of the lifecycle responsibility for the delivered products. The equipment presetting and related recovery outfits are developed quite similarly, and the company know-how is right away available to expand the business, with service engineering provisions.

4.2.7. The Mixed-Mode Processing

The whole removal of front-end operators, as said, can be obtained with resort to robots, when the return on investment reaches clear evidence. The manufacture of cheap but complicated objects, notably, in consumables domains such as shoes, textiles, garments, etc., cannot become profitable through the full automation. For instance, the clothes flow-chart production includes:

- the fabric rolls store, with sorting/dispatching fixtures;
- the laying-and-cutting cells, with selective picking;
- the sewing cells, granting appropriate number setting;
- the ironing and finishing cells, with annexed storehouse.

The automation level is quite different: pretty total for laying/cutting; suitably developed for rolls storing/sorting/dispatching; difficult for sewing; and tricky for finishing and packing. The productive break-up is well known issue, with manual sewing phases in low wages countries. The main drawbacks are:

- the shipping costs and delays, making impossible just-in-time processing, and problematic the control on the material flow;
- the delivery steady quality, difficult to guarantee, if the information flow control becomes entangled, out of unifying rules.

The quality dresses business is *intelligent* manufacture issue, utilising mixed-mode processing, with wide resort to front-line workers. The fabric store requests the accessibility of large inventories, with detailed records of quality data and of available quantities; the robotic handling joins effectiveness and reliability, with buffers distribution at the work-stands. The laying/cutting cells permit processing the scheduled batches and the one-of-a-kind orders of individual purchasers; these require laying-table free-positions for the just-in-time planning. The sewing cells immediately follow (at least, for the priority shipping), and set on quasi-automatic processing (robotic feeding and seam programming, under operators' assistance). The pressing and finishing cells, again, operates under direct workers' overseeing, with computer-aided monitoring checks, before packing and forwarding.

The example picture shows that the *intelligent* manufacture does not remove the on-the-process operators, at least, if the duty «complexity» is better tackled by the human expertise and versatility. The return on investment is, mostly, dubious, because the front-end robots should have to face rather delicate work-cycles. The handling of limp pieces is effective at the assistance level, but the man steering is dramatic complement, when skilful and accurate positioning is needed. The other side, the information flow is better acquired, managed, forwarded and vaulted by computer, and for such duties the robotic surroundings grant return on investment and higher performance and reliability.

The clothing is, today, quizzical domain. At the lowest choices, the affluent society habits look for disposables: the «use-and-dump» practices are consistent with cheap garments, not worthy to be reused and too expensive to be recycled. The waste and pollution highly increase, also because fabric and clothes are made of synthetic chemicals and treated with poorly degradable agents. The lifecycle responsibility and the end-of-life take-back are unrealistic hypotheses, and even the producers' eco-liability is request to come. It is hard to believe that the growth sustainability will leave unaltered that situation. Then, the «use-and-dump» way should be removed, and the more conscious behaviour ought to take back spread control, and proper supply chain visibility. The domains of disposables (not only cheap clothing), then, will be completely turned upside down, and, thereafter, the resort to low-wages work-forces could not be fitting, anymore. The mixed-mode processing, especially with due regard with the information flow effectiveness, is deemed to become relevant opportunity.

4.3. THE ROBOT AIDS INTEGRATION

The robot is conceived to replace to human worker, notably, if oppressive and burdensome duties are requested. In the instrumental robotics, the application jobs specify the operation architecture; the assigned functions and performance figures characterise the structural elements and control strategies. The dependence on the task orientation is basic constraint, possibly, for cost reduction, based on hardware and software modularity. As above noticed, the front-end workers, in the *scientific* organisation, receive *elemental* job-cycle allotments, purposely selected to grant *optimal* process scheduling, leaving no room to the personal ingenuity, in view to keep the full task orientation. This greatly simplifies the engineers' activity in the robot design, because the actual men behavioural abilities have little relevance, as compared with the work-cycle specifications.

The robot aids integration in the shop surroundings, thereafter, simply leads to straightforward accomplishments, because its functional bent already meets all the technical specifications of the task orientation. The prevailing design procedures organise along two targets:

- broad design goals (path planning, dynamics assessment, feedback choice, duty appropriateness, etc.), to assign the said task orientation;
- advanced topics (hybrid command, multi-robot co-operation, mobility or control redundancy, impact analysis, etc.), for special outcomes.

The rationale of any robot design process looks at keeping the task orientation centrality, while selecting the consistent manipulation architectures and command strategies, which lead to the suited accuracy, dexterity, efficiency and versatility. Computer aids are exploited, including simulation packages, to duplicate the robot dynamical behaviour, within real operation conditions. The solution effectiveness depends on empirical figures and previous results. The expertise and know-how are profitably stored into data-bases, with heuristically driven decision support, to help finding out plausible answers. The equipment instrumental bent pushes to the simultaneous design of products, processes, robots, tooling, etc.; the availability of expert simulation, backed by suitably tailored data-bases, helps to re-design the tasks and fixtures, anticipating the actual performance and acknowledging return on investment, when innovative and sophisticated options are devised.

The relevance of *industrial* robotics can be viewed on two prospects:

- the task orientation help showing the «complexity» dependence from the surroundings and to transform it into design specifications;
- the example project outcomes provide explanatory prototypal ideas, useful to devise equipment, suited for less-structured conditions.

The remarks of the present section move along the recalled prospects, while reviewing the typical design approaches and providing introductory comments on series of implementations of practical interest. The topics are well known, thus only bird's eye views are presented, to summarise established «robot age» facts.

4.3.1. Task-Driven Robot Architectures

The multi-axis manipulator is main equipment of the industrial robotics, doing specified tasks with effectors, mostly under computer govern, in acknowledged surroundings. Simple to sophisticated actions are accomplished, for manufacture and support (machining, handling, etc.). The computer advances expand the robot aptitude and effectiveness; the modularity allows cost cuts, making the *intelligent* automation affordable and reliable for wider number of tasks.

The application success vitally depends on acknowledging behaviour and task progression with continuity. The choice is repetitive matter, with trials and check sequences, which profits of computer simulation for the quantitative description of alternatives. The function features are concerned by set-up requirements, which face numbers of biasing effects, referred as 'outer' or 'inner', as compared to the equipment, namely:

- the surroundings itself can be cause of uncertainty (time-varying operation engagements, evolving driving occurrences, etc.); the outer problems may be solved exploiting sensor interfaces and information redundancy (such as position/force control);
- the manipulation non-linearity has noisy effects: *mapping off-sets*, work-space bias by respect to joint-space commands (compensation is built-in for repetitive tasks); and: *actuation bias*, robot members' inertial coupling (the feed-forward compensation permits balancing).

The surroundings affect the robot deeds, owing to the changeful conditioning knowledge, so that the *intelligence* gathering inserts 'external' disturbances. In the multiple-motility arm, the actuation of the distal member embeds all uncertainties (backlash, compliance, dynamic coupling, etc.) of the carrying parts, yielding to the 'internal' uncertainty. The kinematics discerns the robot architectures into:

- serial arms: the joint/link pairs are in sequence; an open kinematics chain establishes, each subsequent member having only one carrier;
- parallel arms: sets of joint/link pairs bind the motion of the carried part, so that the actuated joints need to follow constrained motion laws;
- mixed arms: parallel and serial architectures are addressed, to combine the two open and closed loop cinematic chains.

Another classification discerns the manipulation arms, based on the number of independent motors:

- under-actuated equipment: arms with less than six motilities, when the job does not requires reaching every work-space point with allotted attitude;
- iso-actuation equipment: arms with six motilities, assuring the one-to-one mapping of the motors motion into the work-space degrees-of-freedom;
- over-actuated equipment: arms with more than six powered joints, so that the same middle/final postures are obtained by different motion laws.

The customary development lines apply to iso-actuated equipment. The work-space six degrees-of-freedom have unique mapping into the joint-space (out of the possible *singular points*). The job-allotment is defined by the *forward-kinematics* transformation, through which the actuators accomplish the manipulation. The robot design requires defining the *backward-kinematics* transformation, when the input is the known handling duty, and the output the actuation law, to be planned. The multiple-actuation is basic key for path generation and motion stability.

The performance looks for high acceleration short work-cycles, and reflected dynamics brings in the non-linear coupling of axes. Today, the computer-based control permits cinematic and dynamic accuracy, once modelled the manipulation behaviour. Typically, twofold feedback closes: the overseeing loop, to secure the planned tasks progression, and the inner loop, to suppress the unwonted effects or disturbances. The controller reports to the monitoring module, and it records the accomplished functions. From the user point of view, the twin control supplies the following benefits:

- high operation performance: computer control makes compensation easy, to improve (cinematic and dynamic) motion stability and path accuracy; the job planning includes duty-tailored enabling actions (e.g., mixing and dispensing two-component adhesives, etc.);
- integration and interfacing options: multi-processors and multi-tasking layouts are (generally) proprietary hardware/software modules, but *open-system* concepts help to supply accessibility and to provide (off- and on-process) programming and testing opportunities.

Both, *forward* and *backward* kinematics transformations are nonlinear and the closure of the error loops requires the adaptive gain resetting. The computer-based control resorts to digital solutions of the local transformations (Jacobi's matrices), and the current feedback applies through the incremental up-dating of the nearby computed Jacobi's terms. The manipulation kinematics reduces to locally linear mapping, with time-varying coefficients, underway reckoned along the trajectory. The dynamical behaviour, as initial approximation, refers to series of rigid bodies, undergoing varying inertial coupling, owed to the transport, relative and Coriolis acceleration components each time impressed. The compensation cannot simply considers locally up-dated coefficients, and needs to include dynamics shaping decoupling, with feed-forward terms (at least, for the distal members of the high speed robots). Further corrections apply, to include the joints rubbing/stiction and the links compliance. The adjustment totalling is affected by the many 'internal' uncertainties, so that the feed-forward decoupling is further trimmed with resort to the sensors data, exploiting the twin control opportunity.

4.3.2. The Robot Nonlinear Control

The high performance robot is, perhaps, the nicest case of widespread digital control, applied to non-linear multi-variable systems. Most ideas started with the trajectory and attitude control of spacecrafts, in which the model-based computer-aids quickly became standard way to improve the earlier multi-variable control. It happened that many researchers, on both sides of the Atlantic, gladly turned to the industrial robotics, when the space-engineering activity cut down. The dynamics shaping gives dramatic advantage, having the ability to modulate the motion up to joining accuracy and dexterity, so that manipulation tasks (even out of the man's reach) are achieved. The instrumental robotics challenge is to exploit the process information, to compensate the systematic off-sets or drifts, which arise due to: transmission backlash and compliance; actuation nonlinearities; motilities inertial coupling; sensors uncertainties and bias; or the likes. The all comes by acquitting given demands:

- appropriate description of the «controlled» manipulation dynamics, there-through, the model-based computer-aids exploitation;
- coherent (direct or indirect) visibility on the process variables, combining system hypotheses and observer-driven sensors-data;
- convenient redundancy of control capabilities, making possible to reshape the robot dynamics by the context-driven input modulation.

The dynamics shaping is nice expedient. It aims at changing the manipulator controlled motion by modulated force-and-path feedback laws, chosen to match the task specifications. It presumes:

- command expansion, when feedback signals, driven by the manipulation dynamics, are exploited;
- state expansion, when the robot-to-surroundings relational structure is identified, and used for feed-forward compensation;
- combined command/state expansion, when hybrid feedback joins model-based and observation data up-grading.

The implied ideas are simple, and move from old studies, looking at control strategies, having two separate feedback loops for position, either force control, to be switched for driving the end-effectors along unconstrained, either constrained paths. The simpler (hybrid) arrangement supports impedance control, exploiting the force back-actions, mapped by the stiffness matrix from the current position data, once the constrained path starts. In general, the data redundancy can be used to provide task-orientation for position/force combined control.

The force modulation comes in, looking how control redundancy is exploited by (trained) living subjects to preserve stability of motion and to expand dexterity and versatility (e.g., the bike riding). Alternatively, the state can be expanded to include a set of sensors and feedback lines, so that the resulting system is forced to follow stable paths. The two approaches are, possibly, alternative views. In the first case, however, the dynamics has to be assessed by explicit models. In the second, the behaviour is measured and loops are locally closed, to preserve the posture, within bounded shifts aside; the underlying non-linearity remains hidden, provided that the measurement time lag allows offsetting the dynamics coupling.

The dynamics shaping, by model-based command strategies linking force and position controls, is primary goal, when reliable high performance is required for extended sets of tasks. Micro-manipulation, e.g., profits by graduating the forces and end-effectors displacements, with command redundancy. Due to dynamics non-linearity, every accuracy or efficiency figures change, further, requires task-driven adjustments, making necessary joint control- and path-planning, as soon as productivity brings operation rates above given thresholds. In fact, the new field of micro-mechanics develops as extension of instrumental robotics covering tasks out of man capabilities. To transform the control- into task-planning settings, two time-varying references are defined:

- the object frame is used to describe the trajectory by respect to the (mostly compliant) interface;
- the end-effectors frame is exploited for choosing the force either position control, with independence from the robot set-ups, and without limitations on the task characteristics.

Position and rate errors are controlled in the absolute object reference ('ideal' job-frame); feedbacks are closed for the force loops, transformed into consistent actuation signals. The command redundancy is important, when it is required to perform composite operation sequences, which include:

- the navigation phase (usually, a path-planned unconstrained motion);
- the engagement phase (say, a reflected-shock stage);
- the work phase (namely, a position and force steered constrained motion).

At high speed, the engagement phase leads to short impacts. The collision of the manipulator hinged members against rigid or compliant surfaces, mostly, does not keep central path, and requests 3D models, merging computer simulation and measurements. The standard approach, typically, organises to separately deal with the 'energy decay' and the local friction force/torque components. The velocity components along collision are weighed up by "bouncing coefficients", identified iterating experimental and simulation tests.

The dynamics shaping, chiefly exploited as model-based technique, can, also, manage the 'external' uncertainties, through the command redundancy, becoming suitable way to deal with the whole robot nonlinearities, after accurate data-fusion and state expansion.

4.3.3. Job-Integration of Industrial Robots

The robot on-duty behaviour and operation performance preview are relevant option, supplied once specified the modelling and simulation features, *M&SF*, and developed the pertinent virtual reality checks. The number of fundamental views and conditioning facts is very large, and federated mocks-up, based on functional modularity, are useful resort, to focus, each time, on single specific facets, and to work out the necessary knowledge build-up. The following list provides example topics, deserving especial attention:

- backward (and forward) kinematics, consistent with assigned work-space trajectories, to find out appropriate manipulation capabilities;
- open-loop dynamics, to give account of the reflected inertial effects and to compute the coupling terms arising at each actuation axis;
- control strategies with feedback/feed-forward compensation (at different levels) of the dynamics non-linearity and coupling inertial terms;
- animated restitution of tasks progression, to provide evidence of the robot attitude, to comply with the assigned duty cycles;
- command steering modulation, to obtain enhanced operation performance by hybrid position/force feedback loops;
- impact effects induced by stiff/compliant surroundings, to assess the robot behaviour along the engagement/work phases;
- functional redundancy by means of multiple-robotic equipment with co-operation, to enhance the operation flexibility;
- joint dynamics shaping and operation redundancy options, to achieve high accuracy and dexterity, for spot contact-jobs.

The listed modules provide case views: the first four, aiming at the design broad requirements; the subsequent four, concerned by more specific goals.

The job integration has, today, reached noticeably high standardisation level, and the different application segments possess the required performance, suitably adapted to the domain peculiarities. Without entering in more details, the hints on the following aspects show the technology extension and appropriateness.

- The robots are cost effective and reliable option for material handling and palletising with typically straight forward devices since early 1960s, up to thorny issues as case arises. Painting is proven achievement. Dispensing, such as applying adhesives, is obtaining improvements by microprocessor technology, enabling precise dosing and spreading.
- The manufacture fields cover the current durables (cars, home appliances, etc.). The tasks cover: transfer, machining, welding, assembly, painting, dispensing, inspection, measurement and any job based on manipulative skill. The extension towards the micro-mechanics and nano-technologies is challenge, today, classified between non-conventional applications.
- The robots are usual technique in gas-metal-arc-welding (GMAW), gas-tungsten-arc-welding (GTAW), plasma-arc-welding and -cutting and, at times, in submerged-arc welding. Resistance spot welding is mature job and GMAW concerns some 90% of the segment. Secondary machining deals with milling, grinding, de-burr and other metal-removal tasks with tolerances of about 0.05 - 0.1 mm.
- The assembly robots apply where manual labour does not satisfy accuracy and repeatability achievements or (in the electronic industry) cannot deal with small part size. Inspection and measurement robots are increasingly required for on-process quality assessments.
- Non-standard issues appear for food processing (e.g. the McDonald's ARCH, automated robotic crew helper), for remote-controlled jobs (in nuclear power-plants services, outer-space missions, etc.), etc.; robots opportunities in agriculture are, also, promising.

In the different areas, the job integration widely exploits the available know-how, and the extension to non-standard issues tries to replicate assessed schemes. The industrial robotics, when confined into conventional applications, is heavily task-driven and surroundings conditioned. The schemes replication, thereafter, is fruitful procedure, quickly orienting the architecture, with resort to design broad requirements (such as the four above listed), and adding the further special goals, each time needed. The idea behind the job integration through the functional bent is equivalent to a paradigm in robotic, stating: «as soon as the task is detailed into series of instructions, then suited equipment can be devised to do it». The idea is exploited to expand the *reductionism*, each time the ability to recognise operation models is assessed, and proper computer simulation allows virtual checks into real running conditions.

The instrumental robot design cycle, consequently, transforms into patterned sequences, notably, when knowledge-based aids are imagined and made available. The development of *new* robots, before assessing the *new* applications, becomes a non-sense, and the outcome effectiveness grows to be enabling feature. If a robot costs too much or its productivity cannot grant return on investment, it will fail in the marketplace. The development of the apposite robotic solution aims at fixing the operation capabilities and programming options, to grant the feasibility of the acknowledged tasks, regardless of «complexity» and with no reduction to element actions.

4.3.4. Knowledge-Based Design Aids

The task-driven robots conceptual design, regardless of complexity, requires suited computer aids. We might distinguish three knowledge reference stages:

- the data-bases, organising the available know-how, obtained from previous experiences on robotic equipment in actual running conditions;
- the specified modelling and simulation features, *M&SF*, and the outcomes of simulation campaigns, under the required operation conditions;
- the knowledge management aids, allowing information sorting and storing, in view of the shaping out of consistent task-driven arrangements.

The knowledge-based design aids, hence, do not aim at theoretic technology-driven constructs, rather at technical fit-for-purpose solutions. The design process, thereafter, follows simple steps, summarised into a series of accomplishments:

- path planning and architectural analysis: the end-effectors trajectory/force requirements permits mapping series of joint-space laws, and to establish typical robot-arm arrangements;
- control planning and performance assessment: the dynamics shaping rules give the feedback loop and observation schemas, for the tasks fulfilment at the requested effectiveness;
- engineering and operation details analysis: the simulation campaign gives behavioural overviews of different operation settings, including the critical occurrences (collisions, etc.);
- task extension and special opportunities: functional options (command and state redundancy, etc.), extra-motilities (out of the six-degrees-of-freedom, etc.), etc. are further studied.

Each execution requests series of steps. The architectural analysis recurrently resort to previous know-how, and *expert* inquiry blocks help the decision process, through the knowledge management aids, further accessed for the activity-modes recognition and dynamics-shaping assessment. The robot *intelligence* allocation is properly modulated, with the operation details analysis, choosing steering/govern strategies and observation/inspection interfaces.

The solution efficiency depends on the ability to exploit the device versatility. To reach task bent, the robots profit by number of peripheral devices and interface fixtures. These include end-effectors (at the arm tip) and external rigs (feeding, positioning, etc. equipment). The task control profits by monitoring devices, to assess the current state or to enable emergency signals that shut the process down, if thresholds are outran. In welding, e.g., the accessories include edge finding and seam tracking modules and, at times, welding parameters up-dating and weld-tip pressure control. Handling robots can be equipped with variety of tips (grippers, suction cups, magnets, etc.) and auxiliary devices (collision detectors, contact-force sensors, etc.); indexing signals (from feeding, positioning, etc. rigs) can be used for the start, progression and stop of any work-cycle. At times, the interfaced fixtures are themselves robots, rising at the level of manipulation equipment with co-operation, as the performance shall improve by the use of multi-robots, which operate by tasks parallelism.

The combined duty capabilities, typical of *intelligent* manufacturing, ought to link the duty specification frame, with the function execution craft. In terms of the decision schemes, the aspects to investigate include:

- job-driven functions, with definition of the advantages expected by mixed-operation, to improve the dexterity/versatility figures;
- execution concert, with specification of the tasks programming constraints, including expected accuracy/efficiency requirements;
- information set-up and govern environment, with indication of the selected observation and control architectures.

Computer-driven govern allows *information fusion*, gathering heterogeneous data from inner (controller conditions, etc.) and outer sources (measurement, etc.). The compensation of the reflected inertias, e.g., is model-based, and modulation counteracts the feedback coupling. When the friction or compliance effects add to the rigid model models, the biasing can only be avoided by observed information, and the control planning has to embed the extra data. Every surroundings change affects the robot efficiency and accuracy figures; it requires the operation details analysis, with account of the monitored drifts. The on-line measurement fusion is reliable way for the knowledge-based process tuning. The opportunity is dramatic help in the pro-active maintenance, which makes large use of on-line diagnostics, with procedural data-processing for the signatures detection, troubleshooting and resetting/recovery actions.

4.3.5. Control Redundancy Set-Ups

The control redundancy is witty way to improve the robot efficiency. Through digital governors, the option provides a set of alternatives, with no, or small, cost increase. In actual fact, the control redundancy is software enabled, programming the apposite commands, once acknowledged the in-progress changes. The options are numberless, and only introductory comments on common implementations are mentioned, relative to the, previously listed, dynamics shaping and position/force twin steering. Both the procedures show that, to make possible on-process control adaptation, two facts are needed: transparency of the dynamics effects, and ability of modifying the command strategies. The redundancy is to be enabled by models properly extended to cover (with the robot current behaviour) the interaction with the surroundings. The process visibility is assured either:

- by a state observer, generated according to the *a priori* process knowledge, supplying the reckoned behaviour: the control strategies exploit inner data, to uncouple the work-space requests and the feedback actions;
- by an observation scheme, based on self-sufficient sensor/processing units: the state co-ordinates are separately assessed in the joint- and in the work-space, with command (joint-space) up-dating and tip (work-space) setting.

The two options can be exploited in parallel, building combined resetting:

- the internal model for the real-time processing of the uncoupling feedback, for instance, in the feed-forward off-setting of the reflected inertia terms;
- the (work-space) sensors, to provide apt corrections, used to adapt, via the supervisory mode, the in-progress schedules and co-ordination demands.

To improve the current robot performance, the drifting surroundings and the reflected loads un-biasing shall, both, be 'accommodated', rather than 'resisted'. The idea is to balance the effects, by non-conflicting means. The thorough plans are recognised by the way that the dual behaviour gives visibility to:

- the position control of the unconstrained motion degrees-of-freedom;
- the force control of the constrained motion, during the engaged paths.

Examples discussion provides further hints. The explanatory value depends on the related practical interest, more than on the theoretical merit. The list covers:

- the feed-forward off-setting of given inertia coupling irregularities;
- the command planning, to join productivity and steady accuracy;
- the position/force control, to compensate coupled stiffness effects;
- the position/force control, for higher stability haptic manipulation.

The first example deals with common industrial robots, at first controlled with simple local Jacobi's transformations (without reflected inertial effects offsetting). The manipulation architecture is anthropomorphic, though, with out-of-axis trunk-joints. When used in very fast (painting) duties, a series of inconsistencies appear, related with specific work-cycles, but vanishing if exactly the given accelerations apply with mirror law. The robot manufacturer, looking at compliance problems, unsuccessfully modified the actuation stiffness. The irregularities fully vanished, with the feed-forward command modulation, done by model-based digital govern, accomplishing the combined kinematics and dynamics decoupling.

The second case considers a measurement robot, used for accurate tool-room assessment duties. The measurement setting, basically, comprises the carrier of an optical beam source (for the back-lighting of the selected tool), together, with the duly positioned camera. Spacing and alignment require accurate trimming, while the image processor establishes the detection. The collective govern plays on the carrier and on the borne instrumentation, taking in backlash bias and wiping out inertia coupling, at very high accelerations (less than 1 s duty-cycle), keeping the tools to be monitored into an indexed displacement chain.

The third case is standard option in high accuracy machining robots. The high speed entails engagement shocks, with poor finishing issues, if grinding/honing is spot task, to be repeated with several arm layouts, in many work-space locations. The solution considers switching from the trajectory, to force command, having proximity switches, with targeted thresholds. The variation makes use of the wrist force transducer, doing control switch at the tip engagement; the arrangement is suited for the widespread grinding of varying compliance surfaces, permitting to jointly adapt the work-pressure and the operation speed.

The forth case takes care of the (fairly academic) peg-in-the-hole duty-cycle. The peg insertion path can have off-sets in direction and in slope, leading to one-point or two-point impacts, even when proper chamfer helps the getting in. The control redundancy, in this case, takes advantage from an *expert* supervisor, able to gather information on the engagement line slant (by respect to the hole's axis), the front seat shape (peg/hole/chamfer tolerances) and the peg insertion features (stroke settling, once the wrist offsetting is within accepted ranges). The assembly task, thereafter, becomes occurrence-driven duty, steered by an expert supervisor, incorporating heuristic decision logics.

4.3.6. Parallel Kinematics Set-Ups

The parallel arms architectures provide a nice family of robots, with duty-bent in specific applications. Indeed, the two serial or parallel robot classes show dual characteristics, suitable for different jobs. The first class has gained all areas, with nicely remarkable performance, but intrinsic limitation in accuracy, stiffness and speed, only, overcome by quite costly tricks. The parallel family is conceived, re-inventing and actuating specially created mechanisms, to bring solutions, as case-oriented equipment. The parallel kinematics manipulators, PKM, chiefly, present structural solidity, to insure high precision machining; and compact moving mass, to allow high accelerations. The industrial parallel robots may be divided in two categories, antithetic in many points of view:

- high speed and accuracy pick-and-place devices;
- high stiffness machining/grinding work-centres.

The manipulators comparative study shall take into account sets of figures:

- the work-space smooth connectedness, say, the regions free of singularity, where the continuous path is feasible; the maximal connectedness is one of the most important figure in the machine tool performance evaluation;
- the tip stiffness, computed from the stiffness matrix of the all architecture, including the joints and links compliances; it is of primary importance, as it affects the force coupling, hence the accuracy and metal removal ability;
- the structural errors, estimated from the joints and the links tolerances; the PKM profits of the parallelism, with the main drawbacks at the assembly stage, at times, requiring re-work and trimming;
- the acceleration and inertial effects, with short duty-cycles and high speed, the biasing of the reflected inertias are not negligible, lower with the PKM due to the shorter cinematic chains.

The manipulation description needs to be accomplished with equivalent length mechanisms, using singular point and collision-devoid trajectories. The actuation is assumed to have comparable parameters (inertia, stiffness, backlash, etc), and the revolute joints, unlimited span.

The PKM task-driven design is, again, bounded to the chosen application area (for virtual checks into real running condition), and refers to a single performance index, suitably embodying the basic functions, such as the above detailed figures. The design methodology optimises the performance index, applying deterministic algorithms or heuristic rules. The size of the work-space smooth connected region is important feature, sometimes making critical the PKM solution. The ties among the PKM parameters and performance characteristics are entwined, and have to be deeply understood by accurate modelling. The computer aids, hence, are key help for new generation effective parallel kinematics equipment.

The said antithetic viewpoints, fast manipulation with many g acceleration *or* strong applied forces at comparably rigid stands, together, exemplify the dynamic loading properties and path accuracy preservation. Then, cinematic and dynamic models have to be carefully defined and studied. The cinematic analysis is used to specify continuous dexterous work-spaces, without singular points. The dynamic analysis permits selecting the suited control decoupling and compensation blocks. The forward and backward kinematics might be computer-generated through the local Jacobi's matrices; compact solutions are, at times, available, providing nice insight on the PKM nominal properties (dexterity ratios, force/velocity ellipsoids, work-space isotropy, outmost feed rate, etc.). The rigid body model is, here, more accurate, with, nevertheless, the caution that the nominal properties lead to critical points, plainly removed, when the friction/inertia forces or the compliance effects are taken into consideration.

With parallel mechanisms, the accurate specification and measurement of the spatial movements bear central importance, notably for the calibration purposes. The error modelling and calibration techniques cut by 60% or more the systematic errors in the platform pose, and are standard duty in precision machining. On the whole, the PKM solutions are, today, widely accepted opportunity, with markedly efficient implementation as cheap and reliable pick-and-place equipment, because they do not need sophisticated instrumentation for rightfully repetitive jobs. Also, the precision grinding is worthy application, due to the built-in stiffness, making useless the impedance control or more sophisticated govern processes.

4.3.7. Task Co-Operating Set-Ups

The redundancy is assessed option, to enhance the robot abilities. The issue can be obtain through control (mainly, by software additions) or through mobility (over-actuated equipment, with more than six motilities). The resort to robot with co-operation is further possibility. The duplication of robotic equipment, actually, leads to several issues, e.g., in terms of duty specification:

- *closed-duty*: the agendas are carried out managing the job parallelism; this leads to robots operating in the work-space, overseen by schedulers, with *passive* constraints (collision avoidance, etc.); the decisional schemes are moved off-process; the command logic is pre-set, depending on execution steps ruled by the scheduler;
- *sync-duty*: the agendas are carried exploiting proper function sequencing; the robots are governed by planners, with *active* constraints on the job to be fulfilled in parallel and/or in sequence; the action characterisation is detailed with *a priori* system hypotheses; the task co-ordination follows fixed logics, assigned by (static) procedural knowledge;
- *open-duty*: the agendas are built by procedural knowledge, shared by local units; the

job progression is ruled by controllers, with embedded decision aids that schedule the duty concurrence, based on actual state up-dating; the robot allotted functions are given, with the authorised tasks class; the co-ordination is adapted, following the on-process knowledge.

The co-operation task classification, also, ought to distinguish the linking tie of the allotted functions, namely:

- mandatory task co-operation, if two or more robots together fulfil the job, with compulsory constraints, such as: joint operated tasks, the robots are partially/fully doing the job, requiring coupled collaboration; simultaneous task, the operation requires more than one robot, e.g., one is programmed fixture for other equipment;
- concurrent task co-operation, if two or more robots carry out parts of the same job, having assigned work-space and independent charges, such as: the robots work together on different facets, reducing the total cycle time; the robot diverse abilities are exploited for specialised operations, e.g., singling, precision assembly, etc.;
- optional task co-operation, when any one of several robots can fulfil the all job, and only one is required, as the collaboration aims at interchangeable tasks, and the responsibility can dynamically be re-assigned to each robot, so that the job accomplishment is covered with full failure back-up.

The decision-and-govern mode of the robots with co-operation, what is more, can organise according to different philosophies, distinguishing:

- the structure of the decision logic: - *hierarchic information tree-structure*: the co-operation among the robots is assured by centralised control, under overtly established overseeing; - *parallel-distributed information network*: to co-operation exists in a multi-agent cluster of units (sharing common interest data), interfaced through an intelligent layer,
- the mode of the decision support: - to fulfil pre-scheduled steady jobs, by command decentralising; - to perform job planning, by programmed-mode conditions re-setting; - to recapture the real-time control of the multi-robot facility, at emergencies;
- the function of the govern module: - at *organisation* level, to manage the programmed tasks; - at *co-ordination* level, to achieve the collaboration between robots; - at *execution* level, to handle the underway duty-cycles and the up-keeping accomplishments.

The multi-robots with co-operation are sophisticated set-ups, and their design profits of simulation. The parallelism increases the productivity, as a robot shares portions of the job and perform larger varieties of tasks, as soon as requested. The task distribution needs unified govern, linking computer resource and input-output interface, via data-communication network. The multi-robot set-ups distinguish as for process and task properties, to supply added functions, to run parallel actions, and to interact in tangled duties. The simulation help to test «complexity», e.g.:

- the policy choosing activation, at upper management function, to select the resources with account of the market demands;
- the communicate-synchronise co-ordination, at middle planning function, to fix the sequence-conditioned run by thresholds consents;
- the logic timing, at low monitoring function, to comply with the execution schemes

of the programme-assigned tasks.

The govern modules enable hierarchic or decentralised policy. The decision logic aims at enhanced flexibility, being, as always:

- managed out-of-process: the policies profit to be prepared separately from execution; arm tasks and related paths (with motion-wait cycles) are fixed off-line and scheduled on-line;
- planned by a sequencer: the scheduler activates the tasks parallelism, once verified the programmed sequencing;
- governed by a supervisor: the controller adapts the actions to the ongoing job-progression.

The computer simulation permits: to specify the devised facilities function co-operation features; to assess the on-duty co-ordination features, after setting and fitting improvements; and to check the execution usefulness and reliability, under actual running modes. The value chain constancy is evaluated with the resort to emulation/simulation packages, along the lifecycle of the robotic facilities, with account of actually achievable goals. The all is standard selection of the lifestyle design methodologies. The robot, nonetheless, is striking instance, where the new approach enters in the engineering practice.

Chapter 5

5. ROBOT IN SERVICE APPLICATIONS

The «robot age», out of the industrial surroundings, is fairly more influential fact, being the result of multifaceted reasons, not simply related to the *intelligent* work organisation paradigms. A coarse classification leads to distinguish:

- the robots in the *service* applications, namely, *instrumental* devices having the capability to replace human operators, in fulfilling given task series;
- the robots in the *cognitive* domains, say, *artificial* men, permitting to copy the mental, learning, etc. operations, showing the human-like feeling.

With the latter class, the anthropomorphism is essential prerequisite, the main object of the developments being the individuals, in their understanding, thinking and conceptualisation abilities. Gigantic tracks have already been travelled in that domains, and will become driving support, once the *cognitive* revolution is turned into effect, with the *artificial life* as enabling process, building new resources, for the mankind survival and growth. The *self-generating* robot deserves remarkable interest, being the paradigmatic reference for the *artificial* life, leading to straight specimens' duplication by embedded properties.

Looking behind us, the former class is slickly linked to the industrial robots, as the duty driven aspects are dominant prerequisite, with, only, the big diversity of the operation surroundings, and the different weigh allotted, most of the times, to the effectiveness, as compared with new demands, such as friendliness, safety, accuracy, endurance, roughness, un-damageability, and so on, depending on the requested functions and application fields.

The chapter is entirely restricted to the *service* robot class, leaving totally out the *cognitive* one, requiring extremely different approaches. The idea is to look at the «robot age» from outside, when technology is instrumental opening, not at all enabling bedrock. Once the restriction accepted, the *service* robotics might split in many branches, according to the application domains, or divide in sub-classes, in view of the chosen criteria. Typical sub-classes could distinguish:

- the surroundings danger level, deferring to robots, the duties into hostile or noxious set-ups;
- the friendliness range, allocating to the robots, all risky, irksome, perilous, wearing or tiring jobs;
- the engagement peculiarities, entrusting the robots of tasks out of the man reach (micro-manipulation, etc.).

In fact, the *artificial* man does not suffer the many living men limits, and it is technically feasible to invent devices, which comply the example surroundings or operation or engagement requirements. The survey in the following moves along a different scheme. The initial section discusses the task-driven conditioning ideas that allow conceiving *service* robots, once their functional allotment is assessed. More general considerations are summarised in two domains, *ambient intelligence* and *networked organisation*, not directly involved by the example review, but of permanent relevance in current the «robot age» applications. The subsequent two sections provide explanatory examples of basically operator-driven manipulators, either, mostly autonomic service providers. The main stress is on the surroundings dependence, when the design deployment is tackled. The review aims at clear-cut implementations, having explanatory application potentials, to mark the extended changes that the «robot age» will bring in every day life, plainly, from the point of view of the actual robotic end-effectors.

When dealing with robots in *service* applications (or *instrumental robots*), the direct connection with the industrial robots is straightforward, keeping clear the original dependence from the slavish root of its naming, and its operation-driven specification. The robot design, thereafter, becomes conventional engineers' duty, readily accomplished through the described *integrated* approach, where novelty is in the lifestyle conditioning. The observation of the «robot age» from the related instrumental capabilities permits the smooth opening out of standard engineering methodologies, so that the robots in *cognitive* applications might be conceived, as well, by the same way, once the operation-driven domains are specified. The all is limited, today, as the genetic engineering and bio-mimicry deserve just pioneering interest. The approach shall be discovered, once the *artificial* life context is asked to play the fundamental role for the mankind survival. At any rate, the engineers' robot design activity will expectedly develop along the here sketched lines.

5.1. ACTION SURROUNDINGS CHOICES

Robots are dexterous devices, required to accomplish, with suited *autonomy*, given sequences of tasks, being able of exploiting their *intelligence* of the world, they are interacting with. They are conceived for replacing men each time this is worth; they are developed for performing tasks that are out of men potentialities, giving rise to the many fields of *service* robotics, depending on the conditioning applications, on the transferred amount of autonomy and intelligence and on the enabled functional figures (accuracy, dexterity, efficiency and versatility). The developments do not move from anthropocentric models, rather they acknowledge task-oriented solutions for the set-up of high effectiveness operation mode. The attainments are *knowledge*-pushed and *performance*-pulled, since the technologies availability most of the times exists, but the actual return has to be proved.

The technological appropriateness range is endlessly expanding, thus it might be said that effective solutions are quickly made available, each time new requests arise. In several fields, large opportunities are like to exist, once the competency level is achieved; this is expected to be the case of *micro-robotics*. The potentials are tackled to provide effective solutions, to the many manipulation tasks, which are out of man possibilities. Between the new fields of instrumental aids, the area of *micro-dynamical systems* deserves special notice. Example applications cover:

- the medical apparatus and services: prosthetic devices (with constraints on weigh, on articulation and function range, on mutual stability), etc.;
- the surgical equipments: catheter-based micro-manipulators for repairing blood vessels, automatic micro-lancets for remote operations, etc.;

- the scientific research devices: micro-actuators for force-balance sensors, micro-optical benches, micro-handling of biological materials, etc.;
- the manufacturing fixtures: components for aligning and assembly optical fibbers, active probes for micro-electronic applications, etc.;
- the clean room instrumentation: the down-sizing reduces costs of keeping free of particulate the drug and biochemical production facilities, etc.;
- the consumer products: printer mechanisms made tiny, micro-sized servo actuators with multiple force-path constraints, and so on.

These and similar other instrumental robot fields typify by the need to widely exploit the amount of information on the tasks given in charge, since the success critically depends on the ability of constantly acknowledging the task progression. Such issue is based on the availability of:

- appropriate models of the controlled manipulation dynamics;
- extended (direct or indirect) visibility on the process variables;
- convenient redundancy of the controlling capabilities.

Micro-robotics is, possibly, the area most severely asking refined model-based computer-simulation, observer-driven sensors-data fusion and dynamics-shaping control capabilities. These research efforts are, nonetheless, common traits in the instrumental robotics, to aim at specialised and effective devices with the low-cost properties for market profitability.

5.1.1. Task-Driven Operation Setting

The instrumental robots are intelligent automation basic objects. The addition, out of the industrial domains, to other areas (agriculture, construction, health-care, etc.) depends on the ability of establishing application-oriented solutions, and not of transferring technology-driven settings. The task dependence is ruling fact, and the operation course is obtained, purposely addressing the manipulation functions, and the annexed fixtures, control and sensors.

The multi-axis manipulator is central rig, acting under computer govern, with end-effectors fulfilling specific tasks in acknowledged surroundings. The service tasks (handling, modifying, etc.), are performed by simple to sophisticated cycles. The computer aids confer intelligence, skill and efficiency. Standardisation and cost reduction increase the robot-based domains, lowering risks, and making the intelligent automation affordable and reliable for wider number of tasks. The task orientation modifies the structural elements, in view of prescribed functions and performance figures. The service robot architectures are defined by the assigned jobs and application segments; there is strong dependence on the task orientation, and (for cost reduction) modularity is winning opportunity.

The manipulative architecture requires several fixtures to reach the task bent. These include end-effectors (connected with the arm) and interfaced rigs (feeding, positioning, etc. equipment). Especial attention deserves the sensors and control loops. Monitoring devices are often tied to alarms and emergency signals that shut down the process when thresholds are outran. The handling rigs can be equipped with special effectors (grippers, etc.) and back-up aids (collision detectors, etc.); indexing references are used for the duty start and stop. All opportunities, chiefly, duplicate the ones offered by the industrial robotics, with the fitting in at the task-driven ranges.

The robot control becomes standard routine. The devices, mostly, have arms with six (or more) motilities. Multiple-actuation input is needed, for path creation and motion stability. High efficiency requests short duty-cycles, and the reflected dynamics brings in the axes non-linear coupling. The computer-based controllers compensate for cinematic and dynamic accuracy, if the manipulation behaviour is properly modelled. Multiple feedback loops are common aid, using the inner loop to accomplish the (planned) tasks progression, and the outer loop to manage the undesired effects (suppressing disturbances, avoiding mismatches, etc.). Besides, the controller includes monitoring, so that all the accomplished tasks are recorded. The twin control loops supply the standard procedures, namely:

- operation performance: the compensation algorithms use computer aids to improve (cinematic and dynamic) motion stability and path accuracy; the job planning includes special actions (adaptive setting effectors, etc.);
- integration and interfacing: multi-tasking layouts, often through the *open-system* concepts, supply accessibility and support the (off- and on-process) programming and testing opportunities.

The equipment controllers take in quite a lot of options, from the standalone to networked set-ups, where data are dispatched to and shared among the connected actors. The information framework is ordinary complement, and is the integrated aid of every service robotic application.

The robot sensors, as well, are regular technology. The equipment is requested to autonomously fulfil sets of tasks, interacting with time-varying surroundings. They need to have consistent sensing tips, to close effectors feedback loops, and are progressively fit out, to accomplish process monitoring diagnostics, adaptive compensation or recovery operations. We distinguish:

- duty setting and regulation: several types of sensing devices apply (vision included), providing actuation axes (inner) feedback, and effectors tasks (outer) feedback, interfaced with (or belonging to) the work-space;
- surroundings surveying and assessment: data collection is used to build the out-side world model, for learning purposes (robot training, etc.) and for diagnostics goals aiming at safer and more reliable operation set-ups.

The robots monitoring auxiliaries look at redundancy, and explore data fusion, to remove inconsistencies from system hypotheses and in-progress measurements. In some cases, the surroundings watching and in progress duty monitoring are the service robot principal purposes, out of un-essential manipulation missions. These overseeing and guardianship devices are deemed to extend their operation ranges, notably, to assure *caution headway*, when the bio-engineering processes are used to provide sustainable growth, through «robot age» developments.

5.1.2. Data-Driven Information Tools

The task integration entails accomplishments at many phases, to start from the design and development and to cover the lifelong operation and management. The task orientation makes robots complex, joining manipulation, control and sensing devices; the computer aids increase the versatility and effectiveness, according to well known procedures:

- path planning (by algorithmic blocks for axes uncoupling, etc.);

- task performance checks (with model-referenced feed-backs, etc.);
- on-duty quality control (through on-process diagnoses, etc.).

These are technology-driven features, to be exploited as far as the application achievements provide return on investment.

The design and development phase, largely, exploits modelling and simulation features, inferred after throughout definition of expected tasks and surroundings. The computer tools include the robot behaviour and planned jobs simulation. The virtual test assessments accommodate the basic operation constraints, with proper account of warning/emergency episodes, so that the robot on-duty performance is acknowledged beforehand, with high reliability, even specifying the manning aids ergonomics, when suited interfacing applies.

The on-duty control exploits many computer helps (machine vision, ambient intelligence, etc.) for task planning, sensors setting, data-bases management, and so on. These general functions grant the robot govern with enhanced visibility at the operator interface for off- and on-process activities. The information set-up is highly concerned by the data-bases. Key data-bases are job-cycles, surroundings, and process resources. The basic outcomes are costing and performance records. The reference information includes: operation requirements; accuracy, dexterity, flexibility, leanness, etc. figures; tolerance and inspection data, job hierarchy and sequencing; fixtures and tools data, etc.; the secondary attributes deserve special interest for task trimming or recovery purposes.

Along the life running, the simulation aids, fitted by govern emulation blocks, generate forecasts and feasibility studies of behavioural agendas, duty plans, issue policies, performance achievements, etc., each time providing actual effectiveness figures. The intelligent automation, in fact, achieves efficacy by exploiting the on-process adaptivity by strategic (at organisation range), tactical (at co-ordination range) and execution (at operation range) flexibility. The return on investment is, thus, reached *ex post*, as the actual efficiency is the result of adaptively managing the available resources, in view of the market-driven requests.

As a matter of facts, the robots are powerful technical means, to be transferred in the practice on condition of leanness. When material or functional additions do not improve the reference operation agenda, they are nuisance, being directly (as for extra equipment) and indirectly (as for useless accomplishments) costs. Value-chain consistency can only be enabled and assessed through emulation/simulation tools along the robotic facility lifecycle, with account of the really achieved goals, as for performance and reliability.

The listed features repeat, to some extent, what already recalled for industrial robots. The leanness, now, is more critical achievement, because, in the service applications, the surroundings cannot, in most cases, provide entirely structured figures, and the designer ingenuity has constantly to deal with new traps.

The «robot age», once the information framework assessed, allows the notable *operation transparency*. The duties and surroundings *intelligence* is *implicit* claim in the behavioural agendas; it is, as well, *explicit* chance along the monitoring and vaulting actions. The service robotics is an expanding market, because:

- the number of worthy applications every year discovered widen, with the advantage of the robot users (and providers);
- the producers' lifecycle responsibility shall necessarily address robot-like supply chains, to assure certified quality.

Presently, the product-service market is opting the *intelligent* value chains, not the productive break-up or out-sourcing, with off-shore jobs in low wages places, because the lifestyle liability recommends the «robot age» ways. The customers' satisfaction is non-negligible feature, as for quick response and certified quality, achieved by locating the process design engineering near to the product-service production and delivery shop. The third parties protection and eco-sustainability are further motivation, to directly devise and run value added chains and process visibility, in agreement with the enacted requirements.

The advanced (technology-driven) robotics turns into market-oriented option, when intelligent automation eliminates any penalty due to higher labour rates, by joining delivery visibility, customer satisfaction and eco-protection. These issues are helping robots to do tasks better and at lower cost. The lifecycle responsibility may cause paradigms shifts in service provision and quality assurance, turning the cheap labour to become un-reliable and expensive alternative, once the computer-integration is fully enabled. The robots, initially conceived to relieve the operators from dirty, dangerous and demeaning jobs, are converting into basic technology, when process visibility is needed, or for tasks out of the man possibilities (hostile surroundings, etc.), preserving average high-quality tolerance.

The robot, as information flow enabler, opens new markets. The current trends look for systems integrators or brokers, to assure varying provisions, maintaining high quality and quick response, through net-concerns, assembling the each time needed functions/facilities. These service set-ups employ flexible arrangements to be competitive and meet evolving market requests. Example set-ups are given in the next 5.2.2 and 5.3.2 paragraphs. The «robot age» trading might cover rather dissimilar field. The change is obtained by knowledge integration and complexity ruling, specialising the robotic for each given application domains. The system integration is the winning option, making resort to distributed facilities, functions and competences, so that the core technology, capability and proficiency of every co-operating partner are, each time, fully exploited.

The knowledge pervasiveness is critical option. The standards and modularity adoption makes easier to integrate parts and technologies into readily operative set-ups. The environment and society compatibility is, also, source of paradigms, promoting given kinds of technologies (in view of safety requirements and growth sustainability); robotics is most like to offer valuable solutions, on condition to be aware of the requests, fully exploiting the data-flow transparency.

5.1.3. Ambient Intelligence Supports

The continuity between *industrial* and *service* robotic design is further evident in many side options, already discussed as typical contribution of the knowledge society entrepreneurship. Two opportunities are purposely mentioned: at the front interface integration, the *ambient intelligence*; at the ecology footprint clearness, the *net infrastructures*. Both have been already addressed as major technologies, and are shortly recalled, in the present and, respectively, in the next paragraph, due to their inherent relevance.

The «knowledge» paradigms, typically, lead to supply chains having merit in the embedded intangibles and disadvantage in their intrinsic entanglement. The *ambient intelligence*, AmI, is technical instrumental enabler and enhancement aid. The concept of AmI, as noted in paragraph 3.3.6, relies on provisioning:

- *ubiquitous computing*, i.e., useful, pleasant and unobtrusive presence of processing devices everywhere;
- *ubiquitous communication*, i.e., availability to interfacing and information-transmission facilities everywhere;

- *intelligent user-adaptive access*, i.e., perception of AmI outgoingness by people, who naturally interact with frames, which automatically adapts to their preferences.

In the «robot age», the AmI is part of people's activity invisible background that the social interactions and functions move to the foreground, as soon as the ensuing practices enhance everyday life. The AmI will spread out everywhere, in the human surroundings. The AmI vision leads to establish collaborative working set-ups, where entities communicate to each other. These entities can be: humans, artificial agents, web/grid services, actors standing for real things (not only human beings), human expertise descriptions (knowledge based systems), and the likes. All the entities interact together in the AmI surroundings; they leverage the full potentiality of network-centric layouts, improving creativity, boosting innovation, and enhancing productivity.

The resort to AmI aids requests putting up big, sophisticated, heterogeneous, distributed systems, built on platforms, capable to provide seamless networking, to support provision of value added services. The resulting layout, covering many interacting embedded components, needs to be ubiquitous, self-configuring, self-healing, self-protective and self-managed. The ubiquity is core concept of AmI. It involves the idea that something exists on constant level, everywhere at the same time; say, e.g., hundreds of sensors placed throughout the location, where agents, combined into network, monitor the operation of the on-duty equipments, on their whole lifespan. The AmI exploits ubiquitous computing and communication:

- ubiquitous computing means operating, no mind of the location; moving, it incrementally builds dynamic models of the surroundings and all jobs are accordingly re-configured; its services;
- ubiquitous communication means the resort to intelligent interfaces that create perceptive layers, coupled through visual, hearing, haptic, scent or taste information acquisition and processing.

With the users, the AmI easiness is implicit, through the ability to personalise and automatically adapt to the particular client behaviour patterns (profiling) and actual situations (context awareness), by means of *intelligent* interfaces. Whether the *virtual* reality puts people inside computer-generated worlds, the AmI forces the computer to live in the *real* world, with people. At times, the perceptive layers are joined (e.g., voice recognition, touch screen), to yield multi-mode settings.

Further of the hints of the paragraph 3.3.6, the impact of AmI on the lifecycle service engineering might be understood by looking at current already appreciated *ambient* technologies, e.g., mobile phone, satellite navigation, domotic managing, and the likes. The AmI vision building up ought to develop large, heterogeneous, distributed frames, giving ubiquitous spread of active computing/communication media. The infrastructures resort to multi-function platforms, capable of providing seamless networking, so to support the starting up of added value services or roles to the individuals, companies and administrations profit. In paragraphs 5.2.2 and 5.3.2, example developments are discussed.

The resulting frame, comprising several interacting bodies, with the embedded software components, shall be intelligent, self-configuring, self-protective, self-healing and self-managing. The new kind of service engineering will appear, the *ambient-intelligence/service-engineering*, AmI SE, assuring special flavour to the «robot age» behavioural habits. The type and number of value added functions are not yet acknowledged, but their indispensability will quickly appear, as it now is the case, e.g., with the mobile phone, unknown not so many years ago, and today necessity of every citizen life. The drawbacks are, as well, common snag, e.g., in terms of «privacy». With the cellular phone, each person is permanently tracked with high accuracy and, unbeknown of him, located. Similarly, trough AmI, the hyper-watch tasks will hugely expand.

Now, as before already pointed out, the SE is emerging new business, mainly, ruled by *voluntary agreements* along most supply chains lifespan. Similarly, the *reverse logistics*, RL, is becoming *compulsory accomplishment* at the end-of-life items disposal, with recovery enforced targets. These activities, according to the enacted bylaws, will request the impact monitoring and data vaulting, with third party certification. The above mentioned AmI SE, and the comparable *ambient-intelligence/reverse-logistics*, AmI RL, might develop into the standard computer engineering solutions of the «robot age». Important benefits are, already, clearly figured out. The maintenance tasks happen to be difficult and necessitate expert technicians. Maintenance working conditions are characterised by information overload (manuals, forms, real-time data, etc.), collaboration with suppliers and operators, integration of different data-sources (construction files, components, models, historical data, reparation activities, etc.). The remote *service* delivery is enhanced by the friendly integration into suited AmI SE envelopes. Similarly, the recovery (reuse, recycle) might combine into efficient AmI RL set-ups, granting the total transparency of the process eco-impacts.

5.1.4. Ecology Footprint Visibility

The *networked organisation* context is the second topic deserving comments, being implicit help in the next descriptive review. We all are aware of the changes induced by internet and world-wide-web; still, it is important to investigate some related options, at two levels:

- communication chance: the *net-concerns* provide valued benefits, allowing interactive contact, e.g., in the supply chain management, and the related computer aids, PLM, PLM-SE, PLM-RL, AmI, etc. are important example means on how networking modifies the lifecycle transactions;
- co-operation prospect: the *extended enterprises*, or tied *concurrent*, *agile*, *virtual*, etc. organisations, are options, for which the underlain networking is fundamental non eliminable technological perquisite, so that the set-ups do not exist, unless the below instruments are fully available.

The supply chain management, SCM, is old idea, but only now obtains high interest. Beforehand, the vertical flow integration is conceived by unilateral way, linking suppliers-producers-customers, as the manufacturers' logistic function, to be optimised (and frozen), with the associated information flow. With the arrival of *global* markets, the SCM evolves, to embrace productive break-up, work out-sourcing, etc., and to diversify the offer mix through component, technologies or functions out of the company strict core competences. The inter-firm alliance is option to handle the «*make-or-buy*» dilemma, by way of leanness and agility, in lieu of incorporating proprietary abilities. The old SCM *economical* opportunities turn in *ecological* mandatory accomplishments, as soon as the market knowledge transparency is forced to cover the «natural capital» provisions, and to ascertain balanced budgets through the supply chains bookkeeping.

When we look at the supply chain instances, the entrepreneurship happens to meet, basically, three ways:

- internal incorporation, where the management takes place including all the necessary facilities and functions;

- external provision, where the delivery integration is made possible through purchases and market transactions;
- business co-operation, where the delivery ruling is gained by means of the company-to-company agreements.

The resources incorporation permits enhanced efficiency through co-operative specialisation and synergic outcomes, on condition to fully exploit the investment with continuity. The transaction costs include search, contracting, overseeing and enforcement fees, with administrative and legal shares in the negotiation, drafting and defence of the company rights; loose contracts are cheaper, but the delivered items quality cannot enjoy guaranteed standards. The third way tries to get rid of the «*make-or-buy*» dilemma, inventing the collaboration solutions.

Today, several case collaboration prospects are sketched out, and all match up most of the «robot age» patterns. As rough overview, three ranges distinguish:

- market-driven motivations: the shorter lifespan, frequent demand changes, increased sophistication, rapid technological obsolescence, and the likes, require *agility*, minimising the internal permanent resources, but, at the same time, the «brand» value added is important trade mark, to be saved, and the structured collaboration is sought, to grant steady progression;
- purchasers satisfaction: the lifecycle conformance-to-use assessments are important competition feature, and the world-class firm is engaged to offer better releases than rival companies; the on-duty maintenance and service engineering provision is qualifying issue, to be reached involving partners and certifiers under *voluntary agreements* prearranged obligations;
- environmental protection: the pollution and waste are non tolerable fall-off of the manufacture market, and the governmental regulations impose new limits to put supplies in the market, including the take-back of end-of-life items, with recovery (reuse, recycle) *compulsory targets*, and the reverse logistics operators turn out to be the producers' necessity partners.

The *extended enterprise* emergence as organisational request is, today, wholly accepted prospect, with, however, open debates on how to imagine and govern the embedded facilities and functions. The incorporation shall face legal and technical difficulties, surmountable by alternatives, at the moment, not fully defined. From the already given definitions (see, notably, the paragraph 2.1.6), the technical side presents, chiefly, three prospects, depending on the corporation tie strength, from *concurrent*, to *virtual* and *extended* infrastructures.

The *concurrent* enterprise is standard company-to-company agreement issue, in which the partners co-operate, without loosing each identity. The networking is extra-advantage, decidedly enhancing the business reach (and profitability). The other two set-ups are typical *net-concerns* prospects. The dynamic resetting is the most relevant characteristic of the *virtual* enterprise. The temporary incorporated facilities and functions are only provisional property, and are incessantly changed, when the supply chain requirements modify. The layout is also known as *agile* (or *smart*) organisation, requiring little revision in the partners' attributes. Of course, the timely present resources allow the most effective current provision, and shall vary, if the requirements change.

The lifecycle producers' liability cannot be secured by temporary bonds. The commitment transfer shall be steady help obligation, to build accountable venture. The company-to-company agreements are expensive engagements, when drawn up within dynamically resetting organisations. Then, spot optimal arrangements are questionable benefit, at least, if the delivery lifecycle extra costs cannot be lumped into affordable fee and ascribed to a given reliable partner. This brings to the *extended* enterprise. In such case, the major manufacturer or project leader assures the lifelong and recovery responsibility of the supply chain, promotes the facility/function assemblies each time needed, and rules the whole business, under structured unified liability.

From the technical standpoint, the *extended* enterprise improves its efficiency by the lifestyle design, with resort to *design-for-maintenance*, *-for-recovery*, etc., practices, as already discussed. From the legal standpoint, the difference with the *virtual* set-up, granting lifecycle responsiveness, is important. In the «robot age», the alternative might become chief answer, enhancing the SMEs competitiveness, with lawful offers, readily established, combining complementary partners. At the back, the properly wide and reliable facility/function trading, with related running and networking structures, is fundamental prerequisite. In the example survey that follows, the focus is on explanatory robotic equipment, not involving big enough market domains. The entrepreneurial starting from small enterprises is quite easy opportunity, having at the back the just mentioned supporting infrastructures.

5.2. REMOTE OVERSEEN/CONTROLLED AIDS

The section provides introductory hints on sample projects, where the robot technology assists and co-operates to accomplish a series of duties. The examples are selected among the many investigations carried over in the recent years by the PMAR Lab [1] researchers, and might distinguish in three classes:

- the resort to the parallel data-flow to add effectiveness to current activities; the recalled projects are: postal automation and remote diagnostics;
- the robot backing-up for jobs in hostile surroundings: risky building tasks, and underwater or underground operations;
- the robot integration for expanded effectiveness: cheap and effective mine sweeping, and minimal invasive surgery.

In all situations, the human operators profit from the process information flow transparency, assured by the co-operating robotic equipment, but the job progress is, basically, governed through the overseeing instrumental aids. Certainly, in the hostile surroundings, the robot involvement is fundamental (cannot be removed), however, in all the quoted examples, the man intervention remains central, and in the actual accomplishments, the robot autonomy is reduced to minimal roles.

5.2.1. Auxiliary Interactive Robotics

The mixed-mode automation is nice way to face poor structure surroundings, when the changeful «knowledge» and/or the onerous equipment make difficult the return on investment. The clothing area is example case in the industrial fields, in which the *intelligent* organisation is essential for *quality* manufacture, but, notably in the sewing cells, the front-

---

[1] The acronym transliteration brings to IDREAM, *integrated design* in *robotics/expert-automation/measurements*; the core *robotics/expert-automation* is shared by many research groups; the *integrated design* and *measurements* are peculiarities of the laboratory, I promoted and ruled since several years.

end human operators play fundamental roles. Out of the industrial fields, the mixed-mode automation bears considerable larger weigh, because the *scientific* work *reductionism* background is weak, and the equipment expensiveness quickly outgrows all reasonable economical forecasts. The flexible automation to-day covers broader areas, as robotic equipment provides technically sound solutions to bring back decision on process. The return on investment, still, remains open point, and this is central question when competition is strong. The mail services worldwide face dramatic changes, due to new means (information and communication technology) and business options (government grant removal on utilities), with attention on the efficiency by lean work-organisation, say, the mixed settings, where human operators play chief roles in sophisticated automatic process lines. This is not without drawbacks, with upshots on the job ergonomics, while the all over productivity checks need to be run for actual operation lay-outs. The falls-off are extended in number and typology. Hereafter, the postal sorting and storing facilities are recalled, as illustrative example.

The communication technologies exponential growth has fully modified the traditional mailing services, which was consolidated as governmental branches or protected utilities. The worldwide-web supplies open space for information share and transmission, making unsound sheltered postal services, disjoined from duty-driven competitiveness, flexible specialisation and business integration. The full re-thinking of conventional mailing organisations goes through the identification of the main mission, by setting out the core business, up to developing processing machines and suited work-cycle plans, consistent with scope-oriented enterprises. These new horizons led to the *postal automation* special branch, with falls-off on facilities and techniques, to enhance the productivity, with inclusive check of the economical return.

Such challenge aims, on one side, at the *intelligent* automation (handwriting straight reading units, etc.), the other side, at the *man-machine* fair accountability (mixed human/robot-attended work-lines, etc.). The technology-driven aids, then, radically modify the postal set-ups, with requests on efficiency, since the national borders do not turn down the new outsiders, and the worldwide-web alternatives cover large extents of communication needs. The struggle to survive will follow, on condition to carefully balance technology innovation *vs.* investment leanness. The effectiveness choice needs lifecycle performance foresight, over actual work frames; the goal is achieved by digital mock-up and virtual reality testing.

The concept ideation, hence, establishes the function assessment starting from the facility/technology specification. Let consider, the regional mail delivery, e.g., *mixed-mode postal automation*, 2MPA, project, with the combined requirements:

- to fulfil the public service covering the territory, with neighbourhood nets and widespread organisation;
- to grant accredited due-time and reliability quality standards, balancing the universal service revenue and the value added dispatching.

The universal service requires facing large items mixes: letters, flats, packets, parcels, etc., further distinguished by types: surface, express, registered, insured. The inward flow has multiple origins: street boxes, post offices, large posters, sort centres. The outward one addresses the final distribution or other sort centres. The function assessment shows the inward flow peculiarities:

- from the street boxes, the letters and flats are mixed into sacks and require stamp cancelling;

- from the post offices, bundles of flats and letters are gathered (to help trays forming) and forwarded, already stamped, with packets and parcels;
- from the large posters, items arrive into sacks or containers, requiring the sample volume survey and franking checks;
- from other sort centres, the mail comes in, (suitably separated) into sets of homogeneous bunches or series of items.

In the first case, the mail feeding requires pre-processing, accomplished at the considered sort centre; or anticipated, on condition to propagate the pre-set choice during dispatching, to make ordered items and joined information travel together. At the outward flow, the processing aims at separating the items allotted for local distribution, from the ones forwarded to other district sort centres. The analysis can lead to several set-ups. Typically, a postal code specifies a district, and a sort centre feeds more than one set of districts; the sorting facility has a fixed number of parallel outputs, and the duty needs to iterate the processing to cover the full set of districts. The forwarding to downstream sort centres, also, requires proper trial, depending on the number of regional links to feed (say, the allocated outputs); the link up can be devised, distributing the items over properly wide neighbourhood, and transferring the all left-out mail to an assigned nearby sort centre; the loop has to cover the full regional mail delivery, possibly, after assignments updating, with account of the current work-on-progress.

The study progresses, listing the facilities involved by the specific sort centre duty cycle. The following sections are considered:

- unload docks;
- segregation and storage area;
- manual preparation and, for items from street boxes, culling, facing and cancelling machine;
- revenue protection, for items coming from large posters;
- buffering area and feeding/singling forwarding by manual in-feed or by automatic in-feed;
- optical coding recognition or video-coding support rig;
- carrier and sorter machine;
- amalgamation and loading area, for the outward flow fitting.

The last section represents the interface to ensuing steps and exploits storage and segregation areas, pick-and-place handlers and roll containers, to prepare the sacks, bundles or trays for the next duties. The second last section merits further specifications, to distinguish among the allotted items and duty phases:

- letter coding and pre-sorting or (at given outputs) letter sorting duty;
- compact flat sorting duty;
- bundles and bulky items sorting duty;
- parcels sorting duty;
- registered, insured and express items processing area;
- manual sorting area.

The special items follow reserved tracks to grant backtracking, security levels and allotted priorities. The manual tracks are planned for un-expected occurrences or at emergencies. The process bird-eye view shows that:

- the productivity is mostly affected by the carrier and sorter machine;
- the efficiency mainly depends on the inward and outward flows allocation;
- the effectiveness will ensue from the sort centre duty planning;
- the reliability comes out by servicing tasks and flexibility options.

Centrality is on the code recognition and automatic sorting. The latter grants autonomy to the material flow; the former creates value added, enabling the data handing, vaulting and communication options. The machine vision and characters acknowledgement have reached proper levels of accuracy and standardisation; the parallel track through the video-coding support rig is, thus, involved on rare cases, assuring manual machine-coding, with right consistency for subsequent data flow propagation. Around these two sections, the pre-processing and post-processing complements are devised and implemented, to balance the performance figures, with due account of the return on investment. These figures are only partial, when dealing with an individual sorting area, since the study shall consider the regional mail delivery lay-out, with composite linked up aids, having sets of sorting areas.

The analysis is not further pushed, but the bird-eye view of the set-up already provides explanatory hints on the problems. The flexible automation, with front-end operators, integrates attendants at the in-feed modules and at the mail sorting machines. The reference work-cycles shall necessarily combine with the machine-cycles, and the related ergonomic prescriptions ought to verify the outcomes of competing set-ups, in terms of productivity and workers' safeguard. The mixed mode automation, accordingly, propagates in time the drawbacks of the *scientific* work-allotment organisations, and the labour protection acts, issued into suited national rules, need to be fulfilled. The trade unions obligations, at social impact range, are only partially mitigated by the robotic equipment: the handling of high weights is removed, not the attention on the process and the promptness in filling the scheduled task. Actually, the robot is effective in clerical tasks (reading and processing postal codes, etc.), much less in handling ones (sorting and assembly of mail bulks, etc.), when the return on investment is essential. The universal mail service is good example of compromise, still open to future studies.

5.2.2. Remote Monitoring and Diagnostics

The tight interaction, for process-monitoring and decision-keeping, does not need necessarily front-end operators. The development brings, in this case, to set apart the information flow, including the signature detection and fault recognition, from the material flow, for keeping and refurbishing. The on-process diagnostics is current practice in nowadays lifestyle design, and *ambient intelligence* comes to be nice opportunity, to grant friendly interfaces, for assessing the conformance-to-use of on-duty machinery.

The idea is to develop non-intrusive surroundings, permitting the continuous monitoring, without affecting the ordinary operation states of generic mechanical devices, still acquiring all the useful information. The basic demands are:

- on-line operation, through non invasive observation schemes;
- real-time processing and restitution of the diagnostics signatures;
- resort to off-the-shelf instrumentation, with widespread acceptance;
- creation of friendly operation and interpretation set-up formats.

The answers can be reached by several ways. For example purpose, the *machinery monitoring-maintenance by vibro-acoustic signatures*, 3MD-VAS, project is now recalled. The acoustic emission is standard output of all mechanical systems, and wide knowledge exists, or can be collected, to recognise the normal working and the malfunctions, and to forecast the behavioural thresholds and trends. The full diagnostics frame can be made active, without interfering with the regular device operations, proceeding with cheap micro-phones or vibration pick-ups, processing data and signatures restitution on standard computers, and showing plain outputs, maybe, through colour code lights.

The resort to acoustic waves has long engineering tradition. The expertise is, perhaps, lowering, because more specific observation schemes are made available in many applications. In such cases, the 3MD-VAS device can operate in parallel, and integrate the additional information, with resort to the incorporated data-base manager. The diagnostics frame extracts, recognises and classifies the machinery behaviour, distinguishing in the observed waveform:

- the normal working features, or standard-modes taxonomy;
- the malfunctions, by respect to the conformance taxonomy;
- the anomalies, by respect to the standard-malfunctions (if any);
- the entity of the anomaly and/or the combination of anomalies.

The monitoring uses situation and trend detection, and resorts to a wide series of surveys, to discriminate: additive misdeeds, modulation glitches, spot hitches, burst irregularities, periodic wrongdoings, tone inconsistencies, timber variations, and further anomalies, corresponding to the characterising damage symptom. The features separation is done via apposite algorithmic modules, including especially developed *wavelet transforms* algorithms, to detect the local (timber, spot, burst, etc.) anomalies, when very small, compared with other ones (and the background noise). The 3MD-VAS device discriminates the anomalies and uses, via patterned opposition, taxonomic scales, to track their time evolution. The seriousness level is evaluated this way, with sets of thresholds, and, whether useful, the request of additional checks.

The on-duty diagnostics demand links to the conformance-to-use certificates, lifelong requested to guarantee the machinery fitness (for environment and people safeguard). The demand quickly expands the application domains, requiring, most of the times, to involve third parties (out of the suppliers and the users), to certify the machinery suitability, with resort to legal metrology standards. The technical certification business will, accordingly, become not negligible activity, with falls-off in the research and development of adequate instrumental means and methods, and in the management of the related information flow. The availability of totally non-invasive observation schemes deserves noteworthy value, together with the autonomic remote operation capability. The «knowledge» patterns in the domain have the mentioned twofold purpose:

- the socio-political mandate, to safeguard environment and third people;
- the clients' satisfaction, to grant lifecycle delivered machinery suitability.

The commitment is entangled, because of the sophistication of the equipment, affecting the everyday life of users, not inevitably expert, but trusting the delivery intrinsic reliability. The trend is towards critical issues, with risky or embarrassing situations, for the at home and at work damages, the road and railway accidents, etc. that, even whether not catastrophic, affect the current routines of ceaselessly more exacting societies. The availability of observation and processing appliances makes possible the overseeing and warning tasks, adding the proper transparency as standard kit of every «robot age» supply chains.

The 3MD-VAS device is just example outcome, where the selection, detection, restitution and interpretation of the monitored signatures operate with autonomic consistency, giving direct benefits (the conformance-to-use visibility) and related falls-out (the up-keeping and recondition measures). Among its applications, the railway monitoring is quoted, to provide explanatory hints. In the past, in the EU countries, the rail service was basically operated by single national companies, in monopolistic regime. The opening to competition means to have separate partners for the rail network and for the (passenger/merchandise) traffic. Several trains, in competition, travel the same tracks, paying for the use of the infrastructures. The correct fees depend on several facts, including the rail/car fitness.

The wheel-flat is typical car wrongdoing, generally induced by the protracted skid of braked wheels, over (harder-steel) rails. More generally, (fatigue) damage and (service) comfort are greatly affected by the wheel/rail contact, so that the matching outlines deserve accurate assessments. The 3MD-VAS device consents the early detection of polished facets, with resort to wavelet transforms, tuned to extract the pertinent quite tiny residuals, out of very noisy signals. The finding is possible with stands, close to the rail track, or with on-board instruments. The on-duty monitoring, moreover, can be tuned to detect different other anomalies, both of the rails either of the wheels, providing the basic data, to show the network or the users responsibility of the extra-wear, as compared with nominal values. The monitoring can be done by each partner, duplicating devices and attendants; can be accomplished by a third-party certification body, transmitting with continuity the data to the involved companies. The records are important, as well, in case of accidents, providing the up-dated description of the involved materials fitness, in view to assess possible negligence and/or culpability.

The example rail service provides a nice picture of «robot age» potentials. In the earlier time, the operation visibility is not relevant, with the all-comprehensive operators, directly backed by the national governments. Presently, the monitoring maintenance becomes a must, to assure fair competitiveness, together with clearly assessed reliability. The *fitness* databases build up can become standard benefit, at low cost and pervasive spreading.

5.2.3. Risky Tasks Backing-Up Robotics

The hostile surroundings are robotics old acquaintance. If no man is allowed to operate, the labour shall necessarily be accomplished by the robot. The nuclear power plants are typical example, where remote manipulation is the standard, and the master/slave arrangement, the basic technology. The approach refers to a pair of identical rigs: the master is handled by the operator to carry out the manoeuvre, and the slave repeats exactly the same sequence, in the hostile surroundings. This way, not previously programmed actions are performed, facing any particular new situations, emergencies included, without front-end human attendants.

The master/slave set-up entails worthless redundancy (the master is of no use), and is worthy proposal, when the slave space geometry is equal or scaled account of the one at hands, especially, with the force/haptic feedback replica. The *virtual reality* concept is obvious evolution, replacing the master set-up by the computer-generated surroundings, assuring the faithful duplication of all situations faced by the slave, while performing the requested tasks. The computer aids allow:

- to re-centre and re-scale the work-space, to keep the forestage *best* vision, highly important, today, in the micro- and nano-manipulation tasks;
- to add autonomic modes, through macro-instructions, encoding standard sequences,

putting the slave in *planned* (instead of *pace-wise*) mode.

The master/slave concept, in recent years, evolves into more flexible set-ups, in which remain the remote robot backing-up, while the operators are located in safe stands. The task progression is basically un-manned, under, nonetheless, the constant overseeing of attendants, charged to control the correct execution, and to face the possible emergencies. The rocky wall firming-up is example application, tackled through the ROBOCLIMBER project.

The firming-up of rocky walls is, today, done by specialised workers, which operate manually, exploiting fully bridling rods and total or local scaffolds. The consolidation is done by deep piling (15-20 m depth), with beams distributed few meters afar. The resort to mobile cranes is possible only nearby roads. The main drawbacks are:

- the intrinsic risking and fatiguing engagements;
- the preparation costs and laying times;
- the environmental damages.

The ROBOCLIMBER project looks at the safe and sound alternative, through the build-up prototype, based on:

- the ROBOCLIMBER hanging vehicle, having active rigs (legs, braces, etc.) to move across the sloping surface of the wall to be treated;
- the on-board fixtures: drilling head, drill-rods (tips and extensions) buffer, pick-and-place arm, and (grouting mortar) injection pump;
- the on-board overseeing unit, with sensors, camera and controller;
- the remote governing stand.

The hanging vehicle is held by two main ropes, through four cables, allowing the horizontal displacements, impressed by the four legs. The (hydraulic) power is fed by umbilical, from the wall top or bottom. The drilling head is fed with rods in sequence by the arm along the forward duty, with recouping in the backward path. The overseeing unit monitors the task progression, enabled by the suited macros, unless warning facts appear, switching to pace-wise mode (with attendant remote control), or to emergency stops. The monitoring covers all operation data (thrust, speed, etc.) and geologic observations (stratification, etc.), so that at the duty end the site morphology mapping is automatically vaulted.

At the remote stand, the friendly interface shows the cameras views and all the current data. Suited displays provide the special messages (warning, etc.), and the on-board overseeing unit assures the non-stop data collection and processing, to deliver the expert technical advice to the remote attendants. The basic advantages of the robot backing up are:

- lower operation costs, avoiding the overall scaffolding;
- higher efficiency, due to the robot piling-work continuity;
- operators' safeguard, with no workers on the scaffolds.

The ROBOCLIMBER idea exploits known techniques, such as basket slinging or balanced four-legs gait, permitting to plan out the piling map, and to carry out the overall rocky wall firming up after quite simple yard up-dating, by displacing the two main ropes. At each yard location, the ROBOCLIMBER is made to operate at the fixed heights, each time horizontally sweeping the assigned wall path. The hanged vehicle carries a considerable weigh (drilling head, pick-and-place arm, drill-rods set, etc.), but the structural frame can be optimised, at no operators' risk, because these are located at the remote governing stand.

5.2.4. Undersea Maintenance Robotics

The undersea zones are, surely, not friendly for terrestrial living beings, but, in our overcrowded world, they need to be inescapably more and more accessed, for the everyday men life. The underwater robotics, thereafter, has a long and fruitful history, with many outstanding achievements. The paragraph simply focuses on a few rigs currently used in the submarine oil fields. The, so named, *diving tools* are standard option for controlling and up-keeping duties on the submerged plants, to replace scuba divers and frogmen in most routine jobs. They characterise by sets of properties, such as:

- small weight and high structural sturdiness;
- material selection based on marine corrosion endurance;
- special shape to avoid energy losses during way down and up motion;
- pressure watertight conformance;
- typical work area constraints;
- operation autonomy extension;
- effectors interchange modularity;
- safe docking and un-docking cycles.

The first four points are general; the subsequent four provide the basic hints towards modularity and specialisation. The maintenance of the off-shore oil plants is properly fulfilled by series of *diving tools*, governed and fed through umbilical, from the surface tender. Three typical interventions are planned:

- to run operation monitoring and conformance checks of individual items;
- to enable alternative set-ups, by on/off switching the proper equipment;
- to accomplish recovery/resetting tasks, replacing local modules.

The precision docking, with related accurate handling of sensors, levers and modules, is common feature, and parallel kinematics set-ups are especially prised, having the actuation in the internal base (simplifying watertight safety), and high tip stiffness (no mind if the workspace is limited). The project has, today, issued prototypal items; the market, however, is expected to widen, allowing return from apt series of standard *diving tools*.

In the submarine oil fields, the sea-beds decommissioning at the exploitation end is another relevant business, notably, in the North Sea area, where strict rules are enacted. Reclamation is global duty, with focus on contamination prevention and on disposal activities side effects, including materials and refusals discharge: e.g., the objective of off-shore Regulations (Geneva Convention, art. 5) includes the *non-interference* of the removal activities, with the marine habitat resources, and the resort to *clean technologies*, with minimal pollution risks. Then, the sea-rehabilitation demand pushed the European Countries to regulate the procedures, aiming at 1 to 5 m below seabed soil, the removal depth for the sub-sea structures (*jacket* legs and piles, wellheads, etc.) and fixing strict rules for the handling and scattering of the removed materials, with particular focus on the oil-contaminated residuals.

The robot backing pushes to the jacket *sub-bottom cutting*, SBC, project, with the design of the special purpose robot apparel, carried and let down by the tender ship, nearby the oil-platform leg to cut. The main parts are:

- the sub-sea *remotely operated robot* apparel, with twin boring heads, and control stand on board the vessel barge;
- the *dig-and-saw effectors* (after the preparatory drill to the correct depth), with diamond cutting string, for pollution safe operation;
- the *process monitoring* and *condition maintenance refit*, to join the system risk and the operation risk planning.

The apparel accommodates actuators and sensors, to keep the task evolution always in hands. At the govern stand, the operator resorts to the camera displays and command console. The controller operates modifying the task progress:

- positioning and anchoring, acting on the hanging ropes and on the below grippers;
- stand-by and rack-tilting, to select the boring heads slope compared with the sea bed;
- digging and cutting, up to the complete pile sawing;
- warning, when anomalies occur, and the job progression needs to switch in the pace-wise mode.

The cutting rig is core part. In the standard mode, the monitored data include: the inlet and return pressure and speed of the driving hydraulic motor; the string stretch and position (for diamond wire diagnostics); the idle pulley velocity; the process noise and vibration; and so on. The set-up allows to select and to code the macros for the work steady progression. The warning suggests more care, and the speed and/or pressure resetting, to recover the effectors convenient balance.

The hostile surroundings and marine habitat suggest peculiar caution series, to turn the prospected solution, into highly effective technique. The key advantages of the approach, also in comparison with other ones, are:

- the resort to quiet and clean processes, not interfering with the equilibrium of the marine habitat;
- the noteworthy efficiency of the *dig-and-saw* process, permitting to lower the overall costs.
- the unmanned operation, ruled by the *intelligent* remote govern and drive at the on surface station;
- the guarantee of the completion of the cutting task, provided by the remote monitoring;
- the integrated design of mechanics, hydraulics, control and the underwater functional components;
- the optimisation in terms of environmental impact, overall efficiency and underwater system reliability, and low energy consumption in relation to the total power applied (250/300 kW);
- the *unaltered* overall efficiency of the removed structures and materials involved in the cutting process, thus allowing the re-use for the same or different work-scopes.

The SBC technique joins conservativeness, through impact and removed parts reusability, and efficiency. It replaces old techniques (explosions, flame cutting, etc.), with proper operators' safety. The innovation (careful drilling, wire cutting, etc.) lowers the pollution falls-off (also, as compared with water-abrasive jets), up to easy site remediation. The North

Sea decommissioning programme profits of the robot apparel, as standard technology.

5.2.5. Underground Remedial Robotics

The protection against environment contamination requests active actions for land remediation, with removal of polluting and pathogenic agents. The waste landfills are great dangers, with the percolate infecting the water beds. The fact needs worry, and the EU enacts decommissioning bylaws. The today techniques transfer and encapsulate the landfills, by expensive and humans unsafe methods, without cancelling out the leachate risk. The MIDRA project overcomes all limits, through robotic technique, capable of creating controlled draining manifold, with resort to remotely operated fixtures.

The scheme combines micro-tunnelling technology, to create fitted collecting lines under the landfill, with implanting, in the waste bulk, the SIDRA® draining elements [2], to obtain safe and cost effective landfill remediation. The innovation is based on the robotic equipment, capable to move and position in the small micro-tunnels, and to accomplish the drilling from the bottom, up to 20-30 m afar in the waste bulk with the piping system implantation. No worker is allowed (nor would accept) to drill below the percolate pools.

The MIDRA idea leads to the robotic equipment:

- able to move in the micro-tunnel and to stop at the draining pipes insertion locations (inner micro-tunnel diameter 1.6 m to 1.8 m);
- capable of orienting the drilling head, from horizontal, to vertical boring lines (permitting, also, to bore the micro-tunnel concrete wall);
- fitted out with SIDRA® elements feeder, carrying a buffer, for the laying down of 30-40 m long draining branches;
- equipped with remote operation monitoring, to oversee the micro-tunnel autonomic task progress.

The MIDRA project requires solving sets of engineering problems, covering: the multiple-task boring/drilling head, with speed gears and (high) thrust control; the pick-and-place arm, with form-closure pliers (to not damage the slotted rods), and positioning aid at the mast unlatching; the on-board buffer, with special tools, for wall boring (each time, recovered) and for landfill drilling (left, to form the all draining pipe); and the likes. The overseeing talent is special feature, providing:

- resort to the robot backing-up, for the all duties in hazardous zones;
- ability of the remote setting/resetting and work-cycle autonomy;
- continuous overseeing, providing human perception on current jobs;
- on-process diagnostics, with autonomous management of faults;
- high reliability levels, with visibility on the decommissioning issues.

The diagnostic frame and overseeing functions move from acknowledging the on-going equipment states, with annexed hierarchy. For instance:

- *mobile robot positions*: the set include: the before chosen locations along the micro-tunnel, where the draining pipes shall be placed; (each time) the boring/drilling

---

[2] The elements SIDRA® consist in hollow rods, with draining slots covered by water soluble plastic caps, to avoid debris occlusion during drilling (patent 6 375 389 B1, Apr. 23, 2002, by Tecnigest Srl).

location firming up, by hydraulic jacks, after levelling and attitude trimming;
- *drilling head positions*: standby position, when "out-of-duty" (e.g., robot moves along the tunnel, buffer refills, etc.); duty position, when the mast is turned to the angular slope of the new duty boring/drilling cycle;
- *rod feeder/buffer positions*: generic angular locations of the standard rods; special location, to draw/bring back the concrete wall boring tool; special angular location from where to take the drilling tool of the new drain train; special angular location for rod with sealing collar, to end the drain train.

To ascertain the warning logics, the duty-cycles are addressed, disjointedly listing the *nominal* operation set-points, for example:

- the robot moves along the micro-tunnel, to reach the pre-planned location;
- the robot is levelled and firmly fastened, to fulfil the drilling operation;
- the robot rotate the mast, to orient the head in the required radial direction;
- the boring tool is handled, and the reinforced-concrete wall is perforated;
- the drilling tool is loaded, and the lay-down of the drain piping is started;
- the rod series is loaded, and the landfill drilling done, leaving out the rods;
- the last rod is loaded, and the drain piping is ended, with bottom sealing;
- the levelling rod is handled, pushing the piping, to not jut out of the wall;
- the robot keeps the standby position, and the buffer refilling is done.

The sequence includes multiple-actions (e.g., to bring forth/back the boring tool and the levelling rod, to repeat the standard rods loading and insertion, etc.). In the controller, the *nominal* operations are viewed at two levels: as «macro», for the steady progression; as elemental steps, when the warning switches on. The all is standard option. The robot fitting out benefits of the *interchangeable virtual instrument*, IVI, approach, to explore, during the off-process development phase, competitive lay-outs, and to enable friendly on-process restitution, displaying to the operator, what happens on the remote machine. The duty chaining is deferred to low-qualified attendants, and the *human machine interface*, HMI, needs to help distinguishing the implementation and debugging tasks (performed by a high level language through the IVI approach), from the steering tasks (fulfilled, according to the standard sequential drilling routines).

The robot controller is located on-board, doing the sensors data fusion, and extracting the diagnostic signatures, so that the warning occurrences are directly assessed (without involving the overseeing attendants). The client-server lay-out grants autonomic ability in front of power or communications black outs, with the local processor, fully operating at the emergencies, according to pre-set modes, if the remote master is missing. The basic trick in developing the remote overseeing and command, is to allow only sequential actions, having just two exceptions: the drilling and the screwing commands need two simultaneous actuations (the head rotation/torque and the head advancement/thrust). The sequential jobs simplify the setting of, both, duty-planning and emergency-ruling. The planning addresses two levels: at the lower, the attendant consent is requested, to initialise any subsequent action; at the upper, the duty cycle is fulfilled automatically. The emergencies are, similarly, managed according to multi-level schedules:

- high level emergency: - *work state*, the mast fulfils standard drilling duty, either the process sensors switch to "*failure state*" or the communication breaks off; - *emergency actions*: all parts stop moving; the on-progress rod is cut at the tunnel

inner face; the holding jacks fully withdraw; the refill dispatcher (if inside) comes out of the tunnel.
- mid level emergency - *work state*: the robot holds in the tunnel with active firming up jacks, either the process sensors switch to "*failure state*" or the communication breaks off; - *emergency actions*: all parts stop moving; the in-progress rod is cut at the tunnel inner face; the holding jacks withdraw; the refill dispatcher (if inside) comes out of the tunnel; the drilling robot is moved out of the tunnel by the tow rope (or equivalent fetching means).
- low level emergency - *work state*: the drilling robot is tracked along the tunnel, or fixed to the dispatcher for rod refilling, or the head is rotating to required angular attitude, and the communication breaks off; - *emergency actions*: all parts stop moving; the refill dispatcher (if inside) comes out of the tunnel; the drilling robot is moved out of the tunnel by the tow rope (or equivalent fetching means).
- warning alarm - *work state*: any; - *special actions*: all parts stop moving; the on-going duty sequences abort and reset; the drilling robot control switches to the manual (pace-wise) operation mode.

The example list explains the philosophy of the mixed-mode decision process. The remote attendant direct commitment is required for exceptional steering; the current operations are deferred to autonomic chaining, including the basic safety cautions. At the low level planning, the attendant consent for any further action is supported by the current diagnoses, delivered by the local processor. In the low level mode, the computed signatures are transformed into intuitive formats, and displayed at the HMI, with colour codes. At higher level, or for the robot fitting out, the actually computed signatures are provided, with the related back-tracking to look at the originating mismatches. The IVI approach bears the architecture of conventional instruments, with software-implemented hardware parts, namely, the duplicated acquisition/actuation physical rigs, with intelligent signal conditioning blocks for gauging, compensation and trimming operations. This way, the robot development is carried on-line and off-line, permitting to analyse alternative set-ups and/or recovery procedures, making possible to quickly up-grade the all.

The client/server lay-out considerably improves the on-process reliability and friendliness. The local processor updates twice per second the main parameters displayed by the HMI. The graphic interface, supplied by the local web-cameras, gives the visual perception of the on-going operations, at the remote location. The MiDRA project is completed, with the prototypal implementations operative on-the-field, exploited, also, for draining the ground water tables of the landslips, and preventing the risky accidents widening.

5.2.6. Robotic Humanitarian Demining

The landmines are weapons created to be disseminated on the ground, or close beneath the soil, and to explode because of contact or proximity with a person or vehicle. The AP *anti-personnel* and AT *anti-tank* landmines distinguish on their disruptive potential (and cost). Both were developed for military purpose to create forbidden zones, where the enemy could not enter. Actually, the low-cost of the AP mines transforms them into munitions, to be spread in the country, making highly risking any further use for civil (agrarian, etc,) exploitation. On these facts, the Ottawa Convention prohibits in all circumstances their use, manufacture, trade and stockpile and requests their destruction, because the armed forces objectives cannot justify the falls-off in humanitarian disruption.

In regions with suspected mine presence, the reclamation discerns:

- he military mine-sweeping: to allow reliable logistics, specific tracks shall be made safe, by adequate technologies and efficient work-organisation;
- the humanitarian demining: to achieve safe re-appropriation of the ground (for farming, etc.), the all land needs to be recovered, at acceptable costs.

The study considers the Sri Lanka situation, where the in-progress guerrilla is endemic threat. We shall recognise that:

- the landmine munitions typify the terrorist-driven warfare theatres, where historical socio-political motivations exist; the simple demining shall not work, unless also the implications are removed or neutralised;
- the humanitarian demining characterises as necessary activity to assure the fit-for-use land operation; the goal has to build up as routine duty, having the community involvement at the mutual decentralised level.

The two aspects suggest series of guesses. First, properly aimed measures are dealt with, rooted on the local routine habits. Second, the behaviours usefulness is deferred to results, and these are consequence of the achieved benefits. Third, the routine business involvement shall utterly neutralise exogenous or illegal biasing damages. Fourth, the techniques and workflow schedules should amalgamate and organise on the shared know-how.

Both Sri Lanka parties judge to benefit from mine giving out, and are certain that, under the current situation, the damages produced to the opposed party are cruel, with effects on the spoilt populations social and economic growth, due to the terrorism component. The endemic warfare cannot be contrasted by legal way, as little is left to deontological schemes, since the ethical motivation of the parties belongs to the sphere of entrenched patriotism, on average, coloured by ethnic and religious spurs.

On the outlined facts, the engineering approach to the mine clearing reduces to devise the instrumental process and work organisation, to be enabled as routine counter-measure. The operators are enrolled on place. The technologies exploit the local know-how, with resort to bare agricultural equipment. Process safety and reliability are fit for the required duties. Should this be reached, the mine terrorist spreading ceases to be winning operation, during the temporary occupation of the enemy lands, as the counter-measures minimise threats and widely avoid injuring upshots. The analysis is not pushed further, only retaining that:

- the engaged technologies need to use special purpose outfits, having duty-driven consistency, and to employ operators adapted uniformity;
- the work-flow pre-setting ought to detail work-cycles and standard targets, and to specify the on-going failure protection rules;
- the operators' instruction and training aim at off-process optimised work-flows, notably, to circumvent the emergency of risky engagements;
- the effectiveness comes from organised routine jobs, fulfilled by the work-force diligent activity, in entire conformity to the allotted tasks;
- the local Civil Service is entitled of the authority to accomplish the mine clearing operations, and the involved community is solidly concerned.

In short, structured work surroundings are planned, and robot effectors are set up. Indeed, the resort to the harrow/brush rake technique is current reference. If carried out by trained and careful workers, it assures fair outcomes, having costs within acceptable ranges, directly allowed to local people (which receive wages, independent from the actual land productivity). The lack of mechanisation hinders from structured work and safe productivity. The today state of the art shows two robot-like techniques:

- heavy armoured vehicles, fit to withstand the mine blast, eliminating the danger, because the mine handling is never required;
- sophisticated robotic demining, by the three step cycle: localised detection; picking/removal; neutralisation/reclamation.

Both have the barrier of the special-purpose outfits cost. Alternatively:

- robotic mine sweeping should be privileged, enhancing the productivity by front-end automation, but at low cost and avoiding sophistication;
- manned contribution should be included, exploiting on-process decision-making, to widen the engaged resources adaptivity and flexibility.

The mixed-mode processing capability, embedding mechanised end-effectors and back-drop man intelligence, is the alternative, in view of technologies transfer aptness and of the socio-economic impact as for the Civil Service appropriation. The RAMS, *robot aided mine sweeping*, project moves from such facts, exploring a bottom-up design procedure:

- to plump for the on-the-field equipment, depending on the mine-sweeping strategies to be enabled, satisfying the constraints: low-cost, leanness and technological appropriateness;
- to programme the workflow agenda, adapted to the land conditions, mine spreading, effectors type, etc., and to encourage the on-process operators training for the optimal productivity;
- to help facing un-expected occurrences (break-down, dead-stops, blasts, etc.) by autonomic alarm management and assisted diagnostics/recovery decision-making.

The on-the-field resource selection requires looking at agricultural equipment, widely available in Sri Lanka, to incorporate:

- the self-powered carriers, say: given types of tractors, power-tillers, etc., to bestow the proper mobility;
- the task-adapted demining outfits, to support appropriate, reliable and safe effectors for robotic execution.

We are not here to give the RAMS project details. The single-axel power-tiller is retained as basic carrier, with added pneumatic actuators for remote command, stabilised by the front effectors axle. Two demining outfits are conceived:

- ground stripe lifting, and mines singling out by gentle sieve descent of the shifted earth;
- land sweeping by (forward displaced) striking flaps and flounders, causing the mine blast.

The former, modified potato-digger, applies on recently laboured pastures or soft sandy lands; the latter is used on tough meadows or compact dry areas. The work organisation requires series of actions, such as the following:

- to modify the picked up farming machine, assigning the looked-for work-cycles and specifying the duty-driven fixtures, to achieve demining;
- to design and implement the rigs, preserving technological consistency, so that the outfits modify, run and maintain by existing know-how;
- to define and programme the basic work-cycles (targets, thresholds and timings), to assess the strategic, tactical and execution productivity;
- to devise and check the performance figures (safety, efficiency, reliability, etc.) for averaged agendas and anomalies (incongruities, failures, etc.).

The all is standard engineering activity, and, if fulfilled, provides the detailed agendas, with basic behavioural assessments, covering:

- duty-steered functioning on the strategic horizons, to prove the technology suitableness (becoming the terrorist counter-measure) and appropriateness (getting the shared habits acceptation);
- occurrence-driven performance on the tactical horizons, to help selecting optimal off-process plans, in view to achieve the suited high productivity, during the steady running conditions;
- anomaly-coerced evolution on the execution horizons, to show the benefits of adaptivity and flexibility, enabled through on-process decision patterns, assigned to the operators.

The RAMS idea successfulness, hence, requires exploring low-cost robotic aids, incorporating:

- the *intelligent* task-driven paradigms of flexible automation, with mixed-mode steering schemes that make use of on-process operators to adapt the on-going work schedules;
- the available agricultural machines to devise implementing mobile *robotic helps*, capable to incorporate the (set of) task-adapted specially developed demining fixtures;
- the ideation and adjustment of simple front-end effectors, heading to the safe and reliable land reclamation, with great productivity and protected man involvement;
- the (purposely conceived) *remote-control apparatus*, conferring direction-steering and navigation to the mobile robotic outfits, governed by the on-line operators;
- the *adaptive process-planning* agendas, in order to rule and manage the strategic/tactical/execution flexibility opportunities, to achieve enhanced operation performance.

Once the demining resources and management policies are detailed, the duty planning is obtained, distinguishing the three flexibility horizons:

- the strategic horizon deals with off-process versatility; e.g., the checks aims at maximising the process effectiveness comparing the series of the mobility providers and front-end effectors, actually, implemented; the strategies considers the remote control accomplished by the on-the-field operator, which has the direct visibility of the governed robot, through the connected instrumental data (course, speed, thrust, etc.);
- the tactical horizon deals with on-process adaptivity; e.g., the demining, by the

power-tiller endowed by the ground stripe lifting, is detailed, defining competing agendas; the pertinent decision aids are developed, with issues shown to the operator, who can switch between the agendas, to start again the enabled duty-sequence;
- the execution horizon deals with sudden occurrences (equipment failure, course stop dead, mine deflagration, etc.); e.g., the software/hardware fit-out is studied on multiple-level (warning/emergency) alarms, depending on the relative risk and frequency, each time showing restoring/healing tracks, and requiring the operator consent for the subsequent steps.

The careful understanding of the currently enabled functional settings shows that the main constraints, for the project success, distinguish a series of objectives:

- the set-up of effective management structure, with the bottom-up spur of the local society, to promote, steer and check the endowed Civil Service;
- the choice of standard work-flow charts, to reach organisation productivity in mine clearing, with assigned work-in-progress targets and due times;
- the conception of consistent task-oriented outfits, derived by shared know-how and resources, to warrantee high on-process helpfulness;
- the truthful assessment of process-agendas, with resort to strategic, tactical and execution flexibility, for the bottom-up mine-clearing planning.

The focus on robotics abilities is deemed winning weapon, making possible to involve the local people in the *intelligent* work organisation, and offering the spur towards the old "industry" meaning, say, «*diligence and assiduous activity at any work*». The damaged farmers' commitment opens the way to the recent "industry" meaning, say, «*structured organisation, or systematic work or labour*». The blend leads to modify, in the trend to innovation, the role of ideas behind technologies (chiefly, the ability to prearrange structured surroundings and assiduous labour, to assures the economic growth), because the diligence of front-end operators (out of the mastery of individual craftsman or scientist), is winning enabler.

5.2.7. Co-Operating Robotic Surgery

Mini-invasive surgery merits great attention, to lower the post-operative stay in hospital and to lessen the end complications. This leads to robotics new trends, to provide help for safe and accurate remote manipulation, as chance of integrated computer-aided implements. Out of front-end haptic effectors, the back support is turning to inclusive on-duty functions, e.g., surgical planners, operation assistants, etc., making possible the rethinking of protocols, today, based on anthropocentric dimensions, to gradually embed the micro- and nano-technologies innovation.

Actually, the robot options are technology-driven aids, with twofold fall-out:

- information tools: data acquisition, processing, handing, diffusion, storing, validation, etc. are continuously expanding options supported by the ICT; new computer tools ceaselessly appear to support remote supervision and control; tele-medicine is recognized technology, and remote-surgery has experienced noteworthy accomplishments;
- execution effectors: specialised tools and fixtures are challenging chance, possibly, too much tied up with human handling scale; the future surgeons shall continue to

deal with standard sizes; the inner-body equipment will timely evolve towards micro- and nano-apparatuses, as soon as effective new solutions are conceived and made available.

The twofold fall-out leads to the paired outcomes: «computer integration» and «duty-driven instrumentation»; together, they lead to «computer-aided surgery».

The joint request of local micro-manipulation and operation wide-handling is big challenge. The difficulty is solved by master/slave, to grant neat handling by miniature slaves, scaled into natural size motion at the master controller. Such setting, however, needs special purpose information tools, with focus on the direct and indirect potential of the effectors, and right mind on the computer integration aids. These nice issues are the enablers to expand the robotics to plan and to carry out more accurate and less invasive surgical interventions, confining the effectors into narrow spaces. We expect to see the emergence of four matching tracks:

- surgical planners, to integrate accurate patient exact models, surgical job optimisation and a variety of execution protocols, permitting the plans to be fulfilled accurately, safely and with minimal invasiveness;
- surgical effectors, to provide progressively new operation tools, conceived as special-purpose task-driven devices (out of the anthropic scale limits), evolving with the emerging micro- and nano-technologies;
- surgical ambient-intelligence, to support the man/robot interfaces, which assure the transfer of the surgeon expertise, know-how and proficiency, towards the local, minimal invasive, intervention theatres;
- surgical assistants, to work co-operatively with the surgeons, carrying out precise, highly effective and fully reliable front-end interventions, always incorporating the full continuous monitoring and supervision.

Over time, the tracks merge in the family of aids, assuring multi-disciplinary growth to the interventional medicine. The «computer integration» for surgery is growing domain, with proved achievements. It combines images and other data on the individual patient, with "atlas" picture about human anatomy, to help the clinicians to plan out the treatments. In the operating room, the patient-specific plan and model are updated using images and other real-time data. The option has a variety of means, including servo-helpers and "augmented reality" displays, to assist the surgeon in carrying out the procedure safely and accurately. The similar technology will be used to assist in subsequent patient follow-up, and in enabling quality control, to help improving the overall healing efficacy and security. The computer-aided surgery inherently involves four synergistic areas:

- modelling and analysis of the patients and surgical sequences, in order to support more effective planning, execution assistance and follow-up of the intervention procedures;
- end-effectors technology, with centre in combining the specific treatment requirements, to the less destructive intervention agendas, by proper resort to the micro- and nano-devices apparels;
- interface technology, including robots and sensors, connecting the virtual reality (computer models and surgical plans), to the ambient intelligence in the operating room (patient and surgical team);
- integration aids for ensuring the equipment/protocol safety and reliability, for characterising the expected performance in the presence of uncertainty, for analysing

how subsystems and components will interact, and for the overall intervention performance validation.

The issues in the four areas address the knowledge conditions, distinguishing:

- the background knowledge: in the domain (surgery expertise) and in the ancillary aids (actuators, sensors and information handling tools) ;
- the foreground knowledge: pre-operative (models, simulation and plans) and intra-operative (process overseeing and execution assistance).

The few remarks wittingly addresses «robot age» aids, but the understanding of the whole is vital, as the synergies among the areas require continuous checks, involving the background and foreground knowledge: all up-grading steps result in combined issues. The fact characterises computer-integrated surgery, which is forced to tackle with subtle opportunities, ranging at two layers:

- forefront features, e.g., artificial fingers haptic abilities, straight steered by the remote man hands, co-operating with other instruments, steered by the auxiliary remote man steering (by feet, speech, etc.);
- backdrop aids, e.g., on-process adapting the end-effectors location by co-robotic carriers, with real-time up- or down-scaling the monitored images, by ambient intelligence interfaces.

The *split-duty* approach directly affects the second layer, but major up-grading is expected by synergy of the two. Thereafter, the *co-robotic handling appliance* (CRHA project) development is especially relevant. The CRHA concept aims at discerning the front effectors (*robot executor*), from the back carriers (*co-robotic* support). By the *split-duty* approach, the robot setting/fitting separately addresses the carriers and the effectors. Several competing lay-outs can be considered, but, in general:

- the *carrier* is regular manipulator, with distal platform held, controlled in position and attitude, near by the patient's body; it provides the co-robotic handling appliance, CRHA, of the architecture;
- the *effectors* are (three or four) arms for the in-body intervention, possibly, articulated to assure convenient insertion and re-shaping; they consist of task-driven rigs, with end tips, quickly adapted to the current duties.

A hierarchic controller manages the all: the carrier position and attitude are set at the beginning, then tuned, depending on plan progression. The *split-duty* aims at opportunistic optimality, taking as first instance the effectors' requests (based on the sensors data and visual checks fusion), and further hints in the arm make-up. The all redundancy is turned to satisfy separate constraints, driven by the task assignment. The shape up-dating for better engagement, the friction compensation of trocars and effectors, the force feedback from the operation interface, etc., are added assistance, offered to the surgeons as case arises, while the routine headway will deploy with continuity, upon their consent, following the programmed rules.

The *split-duty* requires motility *redundancy*, managed, with sensors data, by the control, to grant *recovery* in operation, actuation and intelligence. Many robot architectures and functional strategies might be devised, in the *split-duty* lay-out. The current mode addresses:

- the handling appliance, granting the correct positioning and feeding of the overall set-up, as co-robot (implicit) surgery assistance;
- the front-end effectors, performing, after the in-body access of camera and tools, the intervention, under explicit surgeon's control.

Each sub-system is hierarchically governed by the remote location, following pre-set protocols; the interfaced operators receive the sub-set of information at the selected *degree* of *immersion*, with suggestion of trustworthy (*vs.* not advisable) actions (under priority restriction). The robot augmentation traits fully incorporate mechanical architectures and operation strategies, as the background knowledge (surgery expertise and ancillary engineering aids) operates. The *split-duty* vision, simply, provides redundancy in manipulation, for higher reliability in task-doing, without implying the definition of specific surgical interventions, or the choice of given technological tools.

For further insight, the robot action procedure is sketched. The lay-out permits to separately address the CRHA and the effectors. The pre-operative planning is established according to the patient's anatomy and the given operation, followed by the patient registration. Then, the arms entry points, the robot optimal positions and all other preparations for the intervention are chosen. Finally, with the patient stabilised and monitored, the operation begins with the trocars insertion, through which the surgical tools and the endoscope are moved to the intervention theatre. The CRHA locates itself close to the patient (near to heart, head, or upper/lower limbs according to the protocol), holding the pre-established position, in order to allow the end-effectors to reach the points of the patient's body, as required by the intervention. During this phase, the arms keep neutral positions, with no activity.

The CRHA firmly holds the fixed configuration, and the arms are actuated to move and to position themselves inside the respective trocars. Then, the surgeon starts the operation from his console and controls the surgical tools endoscope, to fulfil the intervention. Each time the surgeon needs different tools, he applies for assistance, and the special-purpose fixture STAC, *surgical tool automatic change* operates: the arm at issue, selected by the surgeon, removes itself from the patient and reaches the tool carrier; this stores the before used tool, supplies the new one and performs the change. During the phase, the CRHA holds and grants stability to the arms and special-purpose fixture, by closing the pertinent position-control loops, to keep the requested accuracy and stiffness.

By computer integration and *split-duty* idea, the cycle suggests revising many assessed features. Today, the effectors exploited in robot surgery are special tools of self-sufficient conception. They aim at assuring the functional efficiency of the device, with account of the requested tip mobility. This means redundant arms, to join location and slope invariance of the arm member at the trocar entry, to whole manipulability and dexterity at the intervention theatre, obtained by task planning, through the combined motility of seven actuators. The resulting device becomes multi-task equipment, with sophistication corresponding to the required functional effectiveness. The resort to the co-robotic positioning appliance, holding the front arms, makes likely to lower the command complexity: on one side, the motility to reach the base position assures the reference conditions up-dating, without need of sleeping kinematics redundancy; the other side, the carried arms always act within optimal force-transmission effectiveness.

The *split-duty* idea, in line with the above remarks, matches up robot surgery with computer integrated opportunities. When the CRHA is used to manage the base positioning of assessed-technology manipulation arms [3], the integration steps are quite simple, with little or

---

[3] Such, e.g., as the KineMedic® arms, developed by the DLR in Munich, Germany.

no parallel actuation of co-robot and effectors. The technicalities, such as the optimal robot base positioning, are tackled by the pre-operative planner, whereas the surgeon concentrates on working with the special-purpose effectors in optimal performance configuration, automatically managed by the main processor. In the near future, the effectors are deemed to develop in mini-probes, incorporating mini- and nano-technologies [4]. The CRHA help in the manipulation needs to be continuous, providing stable co-operation actions. The pre-operative planner shall include the task parallelism, as the co-robotic support directly co-operates in the actual force transmission. The mini probe is thought to reach the intervention theatre with *optimal* attitude of the front tip, along *optimal* trajectories, duly avoiding every sensible organ. The rank *optimal* is established at pre-operative planning, but, certainly, corresponds to time-varying settings along with the intervention, even if fully self-exhaustive patients models and surgical sequences are established. With the *split-duty* approach, the operation mode has to follow 'open-duty', specification, requiring on-process variation. The off-process task-planning prospects cannot, anymore, bring to simple information flows, as the local situations are brought forth with interlaced chains.

The STAC idea, then, follows, to enhance the surgical process. The design of such fixture is standard engineering robotics, although with caution, as the tools sterility has to be maintained. In long term view, the entire circulation of sterilised tools has to be made automatic, enabling autonomic handling, logistics, treatment and delivery, leaving to the surgeon (and/or to the assistant) the basic charge of overseeing and controlling the insertions, after proper pre-operative planning.

With today computer-aided surgery, the preparation of the activity modes is disjoint from their execution. The tasks given in charge to the co-robotic device, and the job fulfilment agendas (the motion/wait conditions) are programmed off-line (planner level); only the synchronisation is enabled on-duty. The co-operating robots govern requires an on-process communication structure between the units, and, using the robot with co-operation abilities, this leads to (see paragraph 4.3.7):

- monitored *closed-duty* agendas, at the scheduler level;
- sequenced *sync-duty* agendas, at the planner level;
- coordinated *open-duty* agendas, at the controller level.

The scheduler allows task parallelism, once checked the job consistency. The controller, into varying surroundings, requires the tasks progression full visibility, to exploit the up-dated knowledge, to modify the plans, depending on the changes and, eventually, to adapt the robot behaviour. The context brings to the hierarchic knowledge reference frame, to distinguish *external*, from *internal* conditions and to prepare fully new procedures. The planner, in-between, is especially addressed by the *split-duty* approach, at least at the present state of the arts, when the CRHA assures the outlined task-planning prospects, with embedded function sequencing. The small thrust/torque capability of future micro-effectors might require turning to more sophisticated CRHA agendas. The controller level could represent better solution, with the co-operation enabled by the procedural knowledge, in common between distributed processing blocks.

The task adaptivity on the intervention theatre is basic requisite, to expand the surgical robot effectiveness and reliability. The goal can be tackled from different prospects. The systemic approach moves from the (above mentioned) knowledge prerequisites, to not disregard the foreground one, when planning the instrumental proposal. Besides, the background knowledge assumes distinct areas, depository of the medical, either, of

---

[4] The prototypal mini-probes developed by the LRP, in Paris, France, are existing implementation.

engineering sciences. Then, computer-assisted surgery expands as technology-driven option, to encompass:

- passive aids: navigation and aiming devices, enhanced restitution displays, advanced surgery effectors, etc.;
- backing aids: action guided interventions, based on planned duty-cycles, finally enabled by the surgeon;
- autonomic aids: task sequences performed by the robot, under the watchful eye of the surgeon.

The information infrastructure supplies precision aids and co-operates to make assisted or autonomous actions. The effectors replace the human actors' front-end actions, with benefit in size, accuracy, daintiness, dexterity, efficiency, flexibility, reliability, safety, etc., once the effectors are optimised in terms of the duties to be fulfilled (and anthropocentric limits are overrun). Robot surgery, according to the (initially recalled) four tracks (surgical-planners, effectors, ambient-intelligence and job-assistants), offers great support for minimal invasiveness, by two means:

- the inherent handling effectiveness of the instrumental front-ends, joined to the remote control abilities;
- the functional extensions of the integrated interventions planned and ruled through ambient intelligence.

The *split-duty* approach mixes the two, boosting the basic surgeon's expertise, with higher handling accuracy, versatility, reliability and sensitivity. The issues complementarity requires that: on one side, the regular rules in the *instrumental* robots designing ought to be followed, with duty-bent account; the other side, the understanding of the intervention peculiarities has to develop, as the task-reliant surgeon's obligation. The surgeons' activity asks for the highest trustworthiness and reliability, and only fully tested and safe devices are deemed to be taken into account. The *split-duty* idea, even so, addresses the handling set-up of the robotic surgical planners, aiming at widening the duty versatility by background potential, while leaving the pace-wise technological up-turning to the foreground additions, each time, if the proper confidence is attained. Clearly, the human decisions will never be replaced by robots, and this principle exists in surgery. The outlined idea follows the approach of the "best-of-skill" co-operation, combining the strengths of the human mastery, with the robot reliance and the ambient intelligence. The surgeon is in control of the whole procedure in all times, with non critical tasks entrusted to autonomic robot abilities and to complete knowledge-managing aids.

The paradigm shift in the *split-duty* idea leads to manage function redundancy by intervention fusion methods and tools. When dealing with future prospects, these methods and tools are crucial challenge, and need to be warily accounted for to devise the robotic surgery to come. In the present survey, the mini-probe and CRHA association is mentioned, to show exacting prospects. For instance, the probe compliance and looseness make difficult the firm deployment of current modules; proper guiding contrivances shall add at the trocar entry; the all, very soon, becomes 'open-duty' agenda, calling decision-and-govern protocols. The study of these protocols shall develop in parallel with the mini-probe deployment. The concept design of the *co-robotic handling appliance*, thus, permits to show the problem-solving potential of technology driven equipment. The focus on the *split-duty* idea moves accordingly; it brings insight on innovative prospects, by out-lining their methodological support. The question is not to invent a technical answer for an, up-now, unsolved problem;

rather, to devise alternative devices and protocols that could represent more effective ways out. The decision about the alternatives is totally left to the surgeon, who bears the whole responsibility of the intervention and shall feel fully confident on technologies and methodologies.

5.3. AUTONOMIC TASK-DRIVEN EQUIPMENT

The example developments, grouped in the section, characterise by autonomic behaviour, namely, by the un-manned accomplishment of the allotted tasks. The series of applications is unlimited, at least, on merely technical point of view. The actual implementations are pretty limited, because of poor return on investment, or of shortage of interested end-users, or insufficient producers' commitment. The survey in the following recalls example studies carried by the mentioned PMAR Lab researchers, driven by specific research inputs, and, mostly, in co-operation with interested partners. For classification, we might distinguish:

- developments leading to sample issues to solve demands, whose potential is large, with today poor return; the survey addresses agriculture, technical service and domestic chores implements;
- developments concerning special demands, with little expressed market request, in the current contexts; the car fuelling and the dispatch/watch functions are recalled examples;
- developments finding, now, latent producers' little favour, but, plausibly, destined to play significant future roles; the quoted examples consider the submarine prospecting and the reverse logistics business.

The chosen projects are sketched only by starting remarks, with helpful hints to show the wide mix of situations, where the robot technologies grant technically sound answers. The return problems might motivate, in the first case, the lack of widespread advanced robotic developments, but the market evolution is expected to change, as it is shown by the many contrivances, today, classified at pre-robotic level, already existing. The second case examples lead to niche robotics, and only specialised producers can find their interest in the domains. The last example area gathers potentials to come, basically emerging to diversify and expand the sources of wealth, in our closed-system earth; the marine habitat and the recovery (reuse, recycle) domains are like to be more than acknowledged prospects.

5.3.1. Robotics in Agriculture

The agriculture is key man activity, providing the subsistence means through, mostly, renewable resources. From the olden man-intensive farming and breeding settings, the present approaches turn into highly mechanised establishments, with, basically, mixed-mode automation, leaving to the attendants the decision keeping duties. This keeps robotics in the shadow, with separate information and material flows, and curtailed on-process monitoring with remote overseeing. In that overall situation, nevertheless, spot attainments exist, notably, to look for fairly dedicated equipment.

The fruit harvesting is typical duty, dispatching the robots along appropriately arranged fruit trees, to detect the quality and ripening level and to gently collect the items, into distinct containers. The citrus harvest is example condition, where the robotic set-ups reached the prototypal implementation stage, with successful on-the field campaigns. The oranges are carried at the tip of tiny peduncles; their colour sharply distinguishes from the foliage, allowing the ripening appraisal. In the developed project, the robot sorts out into:

- the base vehicle, moving between the orange tree arrays, guided by the pre-set beacon arrangement;
- the instrumented arm, to detect the items, assess the quality, pick, handle and (softly) place each orange in the proper container;
- the specially conceived effectors, to accomplish the singling and breaking off operation, with no damage to fruit and tree;
- the on-board instrumentation for the autonomous navigation and process control, and for communication of the in-progress situation.

The effectors need further remarks. The following options are retained:

- tip path-planning to approach the item from below; the orange is engaged in a cup, and: three pneumatically driven fingers clasp it, before twisting the stem away; a suction bellow holds the orange, and the stem is turn off or cut;
- tip path-planning for lateral approach, and driving the stem into a guiding slit; the fruit is detached, either, by: taking away the fruit by direct rotation and talk twisting; tearing the stalk by combined tip departure-and-rotation; cutting the stalk by pulling it against a sharp edge with a bill-hook.

In the first case, the fixture is simple, but needs the suction rig. In the second one, a series of tests provide proper information on the all effectiveness, notably, of the combined departure-and-rotation tear.

The development was done within the special purpose project, to bring fresh fruits, from the field to the end-consumers, keeping original genuineness. Today, the oranges are treated; the conservation requires actions before and after harvest. With the strict automatic handling, the fruit conservation is enhanced, by utterly avoiding all contamination risks.

The greenhouse automation is another case of successful robot aid. The area is relevant business, as it makes possible to supply the market all over the year of produce, selected for customers' satisfaction. The tillage follows highly intensive schedules, confining into crowded ground crops, with optimal surroundings as for air humidity/temperature and for soil manure and fertilisation. The grown species botanical heterogeneity, the co-existing varieties mix, the crops sequencing with overlapping periods, the quick changing farming goals, etc. make the pathologies detection and protection a very challenging duty. The glasshouses, besides, being characterised by continuously recycled supporting supplies, have critical epidemic ratios, enhanced by the moisture and lighting continuance (for farming outcomes), favouring the spread of unsafe mosses and moulds and other biotic agents.

The today answer depends on heavy chemicals, spread at given time intervals, making the greenhouse inside space highly toxic. The secure management of the protection against diseases requests the effective diagnoses, to fully eradicate the pathogens as soon as detected, jointly with the habitat prophylaxis and control, to keep the plants growth optimal conditions. The robot solution aims at:

- monitoring the basic operation conditions within the greenhouse;
- overseeing the current plantation grow and health situation;
- performing the spraying operations, at the acknowledged timings;
- assessing the re-entry safe time, after each spraying operation;

- estimating the chemical residuals in produce and surroundings.

The farmer robot shall be the *agronomist*, provided with rigs for monitoring, overseeing, analysing, spraying and recognising. The symptoms detection is done by visual sensors and image processing; data fusion follows, with the help of data sorting and storing means. The administration of drugs is accomplished, aiming at goal-driven operations, by eradication or extirpation duties, without interfering in the uncontaminated growth of the surrounding healthy plants. In the following, the GRA, *greenhouse robotic automation*, study is shortly reviewed, with further hints on the agronomist business demands.

The glasshouses are closed spaces, with proper natural lighting and protected air and soil, typically 20 m or more wide, and 100 m or more long. The individual parcels are (rectangular) areas, 1 or 2 m large and the same length as the building, with interposed paths as narrow as 40 cm. The floor-travelling robot shows space and stability troubles for handling or inoculation tasks; when on duty, it obstructs the circulation. The studied solution exploits appended arms, moving along rails, used as auxiliary beams of the covering roof. The robot becomes built-in feature, of the husbandry shops.

The project checks a set of fixtures. The basic is a device, on circulatory rail, for the transfer from one track to the next. Twin arms can, at once, operate on the opposite strips, possibly, with superposed independent camera carriers, to make, in parallel, diagnoses and treatments. The set of effectors is purveyed, depending on the tasks to be accomplished. The surroundings monitoring is left to humidity and temperature sensors, positioned to cover the all space. Several work-tools are available for doing spraying, inoculation, injection, eradication, removal, log, etc. duties. For the overseeing, the robot carries the video-camera, for on-line features detection and patterns recognition. The data are recorded for further processing, to work out diagnoses and to adapt the growing up plans. As soon as the misfits are recognised, the restoring actions are undertaken, and purpose-driven analyses are started to maintain or to recover the fit-with-norms conditions. The fulfilments are related to agronomic requests and phyto-pathology assessments.

Out of glasshouses, in agriculture, the robotics finds widespread usefulness in disinfestations with chemicals and drugs, to avoid or cut the losses induced by the parasites and related diseases. The medicines administration is not without risk, as the poisons are handled, with toxic falls-off, due to the chemicals concentration, if scattered as preventive aid, to avoid the nearby plants contagion. The environment induced effects and the consequences on future life safety cannot be disregarded any more. This suggests looking at more conservative approaches. In the coming agriculture, the robotics may help to accomplish selective missions, to lower the poisonous agents' amounts, just limiting the treatment to the single vegetables and plants requiring it, after the discriminating monitoring and prognoses.

The new methods make necessary to better understand the phyto-pathology, to support reliable diagnostics and agronomic provisions. Actually, the happening of an ailment or sickness is detected by means of symptoms arousing by respect with healthy behaviour or by identifying the occurred pathogenic elements. The phyto-pathology, typically, takes into consideration:

- biotic factors: insects, viruses, fungi, bacteria, etc. or illness emergences in terms of bugs, spongers or other phyto-pathogens;
- a-biotic factors: environmental causes generating un-healthiness: in the air (acid parts, etc.), in the soil (toxic concentration of magnesium, etc.).

The fight against diseases, today, usually applies to the whole crop population, rather than selectively addressing individuals. The actions undertaken aim at:

- prophylaxis, by enabling farming conditions which reduce illness risks;
- resort to (genetically modified) plants, resisting to pathogenic agents;
- pathogens extirpation, by destroying the seeds of potential misfits;
- blocking treatments, with drugs to kill/neutralise parasites/a-biotic factors;
- recovery treatments, to reduce/heal the established diseases outcomes.

The switch from population to individual treating is out of actual possibilities, unless highly work-intensive set-ups are established for disease diagnostics and for care taking with purpose-driven acts. The other hand, the widespread scatter of toxic stuff is increasingly contaminating the environment, with pollution effects on the (present and future) life quality. The in-progress decay consciousness starts to rise. Many national authorities are issuing acts, requiring the complete revision of the current trends. Then, automation will impose, giving return on investment, at different ranges, to enable selective watch and diagnoses, and to fulfil oriented interventions and treatments. The hothouse tilling, most likely, will have, the first, to comply the new prescriptions. Happily enough, the surroundings can approach structured conditions, where robotics properly assures effective technologies.

5.3.2. Lifecycle Service-Engineering

The lifecycle service-engineering is duty area, extending the entrepreneurial businesses out of the earlier shop-floor traditional domains. The manufacturers' responsibility is moved, from the point-of-sale, to the on-duty conformance-to-specifications and the end-of-life take-back, fostered by environmental acts, due because of demands obliging the consumers vis-à-vis of the society (third people and future generations). The fact, obviously, brings to regulate the trading on the market of the new material goods, through series of accomplishments outside the conventional manufacturing. The efficient answer, as already pointed out, leads to the *extended* enterprises, which encompass the supply chain *externalities*, in the main business, expanding in scopes the earlier *internalities*-minded set-ups. The situation is far from understood by most (old-fashioned) manufacturers. With the environmental acts escalation, service-engineering might develop as independent business. In fact, the users *demand* «functions and results» (rather than products); the same *demand* may modify, when different contexts are acknowledged, so that specialised «service» providers might carry out self-sufficient activities, achieving the purchasers satisfaction, out of the usual manufacturers' *internalities*.

By now, it is not easy to say, if the «service-engineering» providers might be permanent business, or shall, inevitably, merge into *extended* enterprises (or fit in *virtual* enterprises). Anyway, even if the core service is incorporated, the subset of certification activities ought to remain independent, with lifecycle obligation, as noticed, vis-à-vis of the society. The robotic tools are recognised aid, basically, on their ability of providing information visibility, in parallel to the enjoyment of the material goods.

The remote monitoring, cited in paragraph 5.2.2, assumes man co-operation in decision-keeping and troubleshooting. The service-engineering provision ought to make a step ahead, having the distributed intelligence joined to self-sufficient on-situ equipment, able to keep in charge all the operations: monitoring, diagnostics, up-keeping, repossession, certification and duty levying. The servicing applies to machinery, plants, transport and communication means, shopping centres, bank and trade appliances, entertainment and education stands, etc., and the business is progressively expanding, as the eco-protection acts ask to operate at full coverage. The service-engineering provision shall, thereafter, warrantee:

- the lifelong knowledge acquisition and management supply;
- the self-consistent troubleshooting, mending and healing supply.

The former duty requires: to capture knowledge in many form, to process and store it; to support reuse and retrieve methods, to vault and maintain the data. The diagnoses trail the standard reasoning techniques: causal and temporal chaining; deductive and probabilistic algorisms, frame-based union, case-based association, rule-driven heuristics, etc., where knowledge build-up and maintenance are vital options. The latter duty entails front-end resources, system-embedded and/or exo-supplied, to fulfil the suited regulation and reinstatement interventions.

The knowledge and recovery management can be performed by small/medium enterprises, with expertise in the service-engineering domains. The objective aims at the provision of *functions* (in lieu of *products*), with high AmI features, seeing the crucial competitive edge of the new opportunity. The AmI involvement allows enhancing the friendliness, lowering the down times and cutting the maintenance costs. The knowledge management assures, on one side, the remote supervising, with efforts/costs reduction of the servicing; on the other side, the full visibility of all the caused eco-impacts. In the following *AMIR, Ambient Intelligence Ruling*, project, the whole intelligence lay-out comprises the following modules:

- the shared knowledge-base, central repository of the problem-solving data, with the system characteristics and hypothesised/acquired know-how;
- the customer support, personalised data-base for the specific service, with the delivery instructions and the problems-avoiding advices;
- the AmI interface, to process the monitored data (diagnostic aid input), to interact with the main/local data-bases, and to help the users;
- the diagnostic module, to enable the programmed reasoning methods, fast showing the problem-solving suggestions, with risk-assessments;
- the overseeing module, to organise the data, doing the communication, up-keeping, remediation, etc. jobs, and storing the whole data-frames.

The knowledge acquisition and management modules are designed and tested in collaboration with interested (SME) service providers. The *AMIR* project cases cover example situations, including:

- continuous processes watching: the everyday life is compelled to exploit increasing entangled machinery and equipment, requiring number of built-in checks and refits, to maintain or restore the correct performance; the use of knowledge-based analysis of the ambient data enables adjustments to be continuously made, when required (with no need to send out, each time, the field engineers);
- intelligent home-support: the networked system deals with multi-tenancy buildings, including a large number of professional/company offices; the AmI, through proper sensor lay-outs (video surveillance, temperature and smoke detectors, IR motion inspection, etc.), build-up the clients specific profile, and supply the expected security figures (with no people on-situ);
- remote after-sale monitors up-keeping: the real-time business disturbances require immediate corrective actions, whose critical lag-time depends on the allocated duty (bank terminal, tele-medicine post, etc.); the effective intervention, via AmI aids, is of major relevance, to achieve the customer satisfaction and to increase his trust;

- remote managing of technical installations: the combined monitoring and actuation of specific facilities/fixtures (shopping centres, sport plants, etc.) is built-in option, needing, however, proper expertise of the attendants; the service-engineering provision by on-process AmI, allows real-time control or the immediate alert of responsible operators.

The link between service-engineering and ambient-intelligence is deemed to reach full success in the «robot age». The resort to knowledge-based instruments, such as the above recalled ones, has practical relevance, with falls-off in:

- the automation of the refitting control, by the resort to turn-key solutions, with throughout reliance on the on-process intelligence, acknowledged by the off-process confirmation;
- the security provision, with the users trust, grounded on measurements and observations incessantly acquired and analysed, to grant safe continuation to the local activities;
- the safe preservation of the real-time function-suppliers, in front of sudden disturbances, so that the programmed supply does not cut short, avoiding any kind of operation deficiency;
- the remote management of the pro-active restoration of fixtures, notably, to be enjoyed by wide public, without time delay (and cost) of dispatching specialised technical staff.

The sample implementations provide clarifying hints. The number of actual cases is deemed to continuously increase, and the service-engineering field, even if now at its beginning, is expected to be growing business of the world to come, in the spirit of robot-driven falls-out.

5.3.3. Domestic Chores Automation

The domestic robots are widely appreciated aids, with valuable kits in use, to accomplish: food preparation, etc. in kitchens; cleaning tasks, etc. in flats; watch and overseeing duties, etc. of homes and gardens. All these devices belong to the well known domain of «domotics», entering in the current habits, each day, with new proposals. In the average, all apparatuses are notably simple and cheap, easy to integrate in the every day life, with minimal efforts and expertise. The all field is conveniently classified at the pre-robotic level, even if, at times, sophisticated contrivances are made available. The «domotics» is assumed to play larger roles, in the near future, each time specific necessities are singled out, and appropriate solutions are offered, at competitive performance/price ratios.

Besides, the sophisticated contrivances happen to be developed as prototypal specimens, suggesting more intricate problem-solving hypotheses, with falls-off, perhaps, in side achievements. In the paragraph, the GECKO (and GECKO-COLLIE extension) project is recalled, as example advanced suggestion. The study moves from the request of a multinational home-appliances manufacturers, compelled to face hygiene regulations in the big, professional cooking plants, where cleaning and sanitising need to be accomplished and certified, through visible and assessed patterns, leaving recorded evidence.

In general, the inhabited premises professional cleaning is huge business, with millions employees in the only EU, also, engaged in the parallel information flow, to assure the outcomes transparency. The robot alternative can become attractive, on condition to offer integrated *char-robots*, joining leanness with sturdiness, and transparency with automation. The choice of instrumental robotics, to replace the man in unpleasant duties and wearing assignments, is smart solution, explaining why, in the household appliances advertising, the domestic chores are entrusted to robots, being repetitive and harassing jobs. The reality lags behind fantasy, due to the factual issue of the humans' competitiveness, when unsatisfactorily structured surroundings are tackled.

The tidy up-keeping, of the food and beverage processing rooms, requires big concern, as the Health Authorities expand regulations, with meticulous rules. The human operators replacing by robots is likely to reach economical return, to grant tiding up standards and achievement repetitiveness margins. Looking a little more into technicalities, the task analysis shows inherent conflicts between the cleaning of vertical walls and areas needing full sanitising, as compared to collecting piled up garbage, and scraping and sweeping the floor surfaces. The resort to combined fixtures enhances the task setting, as it separately distinguishes *climbing* (GECKO), and *tender* (COLLIE) functions, each one with proper work-cycles.

The *char-robot* GECKO project develops the climbing vehicle, using a vacuum gear, based on four suction cups, having pumps, to assure sticking stability during operation (all cups enabled), and gait chaining (two cups, acting each time). The duty planning avails of the valve-and-manifold device, for the suction sequencing. The cleaning is accomplished by steam. The duty requires:

- to generate steam, from water supplied by the carried tank;
- to sprinkle on the surface each time covered by GECKO;
- to recapture the condensed water, into a separate tank portion.

The boiler, nozzles and tank are nicely conceived, to grant proper autonomy. The all is electrically powered, with umbilical linked to an outer source (mobile stand, where the cable rewinding block is located). The cleaning up, on the whole, follows vertical tracks, and horizontal dislocation is used, to start the new track. The GECKO project requires fixing the work-cycles and the governing logics; the main targets cover:

- path planning, to establish effective *char-robot*'s tiding up policies, and GECKO's dispatching strategies;
- assessment of robots-at-work engagements, to show their soundness in front of unwanted occurrences;
- analysis of the GECKO's gait, for vertical (and lateral) trajectories;
- on-process checks of the steam spraying and dirty liquid recapturing;
- lay-out investigation, with component's choice, to rise the effectiveness and lower the cost;
- consistency examination of the structural details, to inspect stability and reliability.

The GECKO device is designed to work on vertical walls. It can act on all flat surfaces, and is, really, very effective to sanitise the horizontal tops (tables, etc.). In principle, it can be made to collaborate with attendants, in charge to enable the programmed job sequencing. The collaboration requirements lead, quite easily, to the GECKO-COLLIE extension. The tender COLLIE is a mobile platform, carrying a six degrees-of-freedom arm. In the project, the mobility is obtained by an idle ball and two powered disks, assuring nice path tracking, with sharp turns. The bottom frame carries two long-hair brushes, a trash chopper, two short-hair brooms, the cleanser supply and means for the cleansing recovery. The list shows that COLLIE is conceived to accomplish three tasks:

- removal of solid refusal and floor washing, with resort to the platform said special fixtures;
- cleaning of cupboards, ovens, refrigerators, etc., using proper effectors at the arm tip;
- carrying and positioning GECKO, which is, thereafter, charged of cleaning vertical walls and surfaces, deserving particular tidiness demands.

The design of COLLIE avails of standard technologies:

- mobile devices are extensively used for floor washing and garbage picking and example implementations are offered for autonomous running;
- conventional arms could be added, with tools for special tidying (shelves, cabinets, ovens, etc.) and for the GECKO handling, as the case arises.

The electrical power comes from fixed stands; the link uses cable rewinding blocks. Together, the GECKO-COLLIE pair acts according to supervised autonomy: COLLIE starts the allotted duties, following structured contexts. Routine paths are tracked, with proximity sensors, to keep out the barriers. The dirty kind and level are recognised by optical sensors, to repeat the cleaning sequences to the desired quality. The collaboration is requested by GECKO, when fulfilled the given work-cycle. The arm has box-shaped limbs with shoulder's driver, special wrist for high dexterity (in view of tiding up constrained spaces), and *cast* attitude (for operation safety) conferred by irreversible remote gearing. The automatic tip replacement is scheduled, with two series of effectors:

- brushes and wiping tools to clean internal spaces, with sharp inner corners;
- coupling plug for GECKO handing and unlatching at the required locations.

Several duty-cycles are considered. Basically, COLLIE starts by choosing the surface to be washed by GECKO, and move to clean up in parallel given cabinets; typically GECKO is charged of longer tasks, and COLLIE is get free to accomplish the tidy up of the floor. Unmanned sequences might be planned; if not, supervised duty-cycles are done, so that, at unexpected occurrences, the worker is alerted and enabled to modify the ongoing work.

The *char-robot* is apposite suggestion for service automation. The health and safety protection acts requires certified levels of cleanness and full contamination freedom, to be achieved with transparency. This pushes toward robotic handling, when food processing or related duties are accomplished. The exclusion of front end human operators grants the task schedules and issues standardisation. The last remark is, possibly, basic aspect for the return on investment, aiming at the fixture leanness, strictly within duty-cycles stated by the health protection authorities.

Indeed, the tidy up-keeping and cleaning of the food processing spaces belong to domains where robotics should expand. The community kitchens case is judged with interest by specialised manufacturers, with offers that take automation into account. The concept design characterises by modularity and step-wise inception. The (tender) COLLIE is conventional (autonomous) floor-washing machine, with an added arm for special purpose cleaning and, possibly, for GECKO handling; it might be used as stand-alone device, at different completion levels and at tailored task planning abilities. The task distinction into classes (floor sweep-out, cabinets tidy up, surfaces steam cleaning) simplifies modularity, each time bounding the device functions at the proper sophistication levels.

5.3.4. Robot Car Fuelling and Servicing

The car fuelling faces risks for operators' health and environment protection. The robots are being considered, to avoid nuisances in handling the today fuels, providing, also, hints for more critical supplies (liquid hydrogen, etc.) filling up. The regulation fosters a few points, such as:

- fuel composition, listing the not allowed components, additives, etc.;
- no pollutant emission during handling, by suited outfit and vapour suction;
- no assistant at the fuelling stands (self-service stations).

The sucked vapours are withdrawn to the underground tank, removed during the station refilling and transferred back to the central storage plant. The resort to robots is generally investigated for basically two reasons:

- the exclusion of frond-end assistants;
- the monitored performance/safety guarantee.

The technical problems to face are, above all, the following:

- the steady positioning of the fuel input: vehicles with different shapes and sizes are on the road, and their exact stops depends on the specific driver;
- the fuel input aesthetic protection and inlet closure cap: characteristics of carmakers (and models), typically, requiring screw/unscrew actions;
- the protected handling of the piping layout, to avoid the risky falls-off in dealing with inflammable and poisonous fluids.

The studied set-ups require proper sophistication. The client stops at the stand, on the station path left side; two laser scanners (on the path right side) assess the car position and monitor the evolution; the data are compared with a transponder message, from the vehicle platform, to identify: car model, filler port location and opening procedure, fuel type and total delivery, filling rate and job sequence. The robot, following the acknowledged path, approaches and opens the closure (with a sucking device); three IR optical fibre sensors trim the effectors location, and two rope-driven actuators push the pipe into the (properly modified) port. Such stand prototype has already been built, but only as exhibition sample.

Factual progresses are expected by modular standardisation of the fuel input assemblies, removing aesthetic peculiarities, replaced by unified self-acting plugs. The modularity can be achieved by companies' fleets, and robotic fuelling stands are on duty, notably, for coaches and trucks of known types. A hanged Cartesian robot fuels vehicles, with left or right plugs, tracking sensor-driven paths, thank to transponder data and vision trimming. The on-duty flexibility mixes low and high level control, say, robot joints steering and fuel delivery, with event-driven plans and decision up-dating. Besides, the limitation to low level control needs previous restrictions on the vehicle types and feeding shapes/caps.

The sketched considerations bring to the **SARA**, *systems for automotive refuel automation*, project, suggesting the modular approach to offer sets of solutions at different sophistication. The attention is on detailing the duty-cycles, so that the consistency is inherent property. Typical issues are:

- the task surroundings setting: the vehicle approaching path shall avoid any collision

risk, and bring to stops at acknowledged locations;
- the duty-cycle overseen safety: the pipe coupling shall grant zero-emission conditions, and the robot stand-by state ought not to encumber the vehicle entry or leave path;
- the operation execution: the fuelling needs to be fulfilled without blocking the passengers exit, and allowing inlet port oscillations when people steps in or out.

The stand organises into islands, with lanes (between footpaths) separate for cars and trucks, to take into account:

- the driver window and fuel inlet heights;
- the side (rear view) mirrors protrusion;
- the vehicle width and cap distance from mid front window;
- the (standard or modified) fuel cap type and side location;
- the further useful particularities of the acknowledged vehicle.

The client interface is provided with a keyboard to forward the order, and a slot to pay by credit card. A display gives the on-progress operation basic data. The lane width compels the vehicle natural positioning, limiting the driver odd behaviours. The distance between the user interface and the refuel site depends on the vehicle model, while a fixed relation establishes at the keyboard (accessed by the driver's hand), and the fuel inlet (accessed by the robot effectors). Modularity permits diversified solutions along two ways:

- the island size, diversifying two or more car/truck classes;
- the pattern recognition type and robot fixture adaptation means.

The **SARA** project develops assuming that proper outfits exist, say:

- every vehicle possesses a radio-tag, where the car-type is coded;
- the fuel-inlet cover has automatic opening, directly actuated by the driver;
- the special fitted fuel-inlet devices are standardised, with spring-actuated opening cap, easily operated by the arm tip.

The standard operation applies to the current sedans. A camera, on the island top, gets the car co-ordinate frame, as compared with the stand frame, using fast image scanning. The robot motion is identified in five steps:

- recognised the car model, the reference outline is taken from the archive;
- the matching 'outline-camera image' is used to define the body centre and to locate the car-frame in the stand-frame;
- the horizontal position of the fuel cap is referred to stand-frame, through proper co-ordinate transformation;
- the robot-base is approached, to have the fuel inlet within the arm reach;
- a second camera, carried by the robot, help locating the cap vertical site, following the said matching.

The robot consists of a mobile carrier, for the horizontal displacements (along the side footpaths), and a three degrees-of-freedom arm, with spherical wrist and multi-function effectors. The hollow arm bears the internal fuel pipe, surrounded by enveloping sheaths, to bring the sucked vapour back. The roll-pitch-roll wrist, having 45° slant axes, with shared spherical symmetry, avoids local singularities, including, both, inner pipe and outer sheath. The wrist supplies roll full rotation and pitch-and-yaw 270° angular spans. The arm and wrist make use of pneumatic actuation, and spring clamped brakes, for proper safety. The all technology refers to proved (registered) criteria, not to be here repeated.

The filler and connecting inlet cap augment the arm safe duty. The fuel inlet includes two elements: one solid with the modified tank cap, one carried with the effectors. The opening of the cap cover is made by the driver, as soon as the light gives the consent, because the tip-to-cap docking is accomplished. The inlet opens by means of a compliant flap, giving access to a nozzle (equipped with non-return valve), into which the filler inserts. This is telescopic, with torus shape: the inner pipe is moved in the nozzle; the outer sheath combines with the external cap. The effectors end with a compliant hull, so that refilling is done, allowing limited car oscillations. The overflow is avoided by the usual pressure mechanisms.

In case of sudden vehicle departure, the robot stops delivery and moves back. The radio tag originates a sharp peak. The compliance of the all assures that no damage occurs on the arm or the car. The incorrect cap-to-filler connection does not allow the process to start. The SARA project is just an example solution. The modularity is exploited to devise scaled lay-outs, permitting simple vehicle-island engagements, while preserving the basic ideas. The fuelling regulation, should the eco-protection be enacted, will include the suited car outfits, such as above listed. The robot aided set-up becomes easier, whether the vehicle identification is done through radio labels, and the refilling task can exploit unified linkage devices. The other benefits, similarly, follows from standardisation: driver window height and size, fuel-inlet shape and location, etc.; the strict unification, maybe, contrasts the today clients' penchants, but considerably simplifies the implementation, because only low-level control is required, without the preliminary vehicle identification and coupling adaptive trimming.

5.3.5. Robo-Ranger Dispatch/Watch

The automatic guided vehicles cover broad class of services, meriting especial attention, on condition to be easily adapted for ample series of tasks. The common features are:

- short/medium journeys into poorly structured surroundings, requiring high manoeuvrability, at moderate/low speed;
- extended autonomy in covering the allotted missions, with fully embedded intelligence and communication talent;
- structure and function modularity, to permit reconfiguring the vehicle, the govern abilities and the job aptitudes.

The ROBO-RANGER project moves according the said objectives, with expected job areas allowing very broad alternatives, such as:

- city roamers, for eco-compatible personal service into the downtown areas with traffic limitations;
- district dispatchers, for home delivering of mail, parcels, dairy, drugs and, in general, daily wares, to elder people and the likes;
- autonomous guard and caretaker, for territory overseeing and watching, and remote reporting;
- yard shuttles, for protected leisure-time resorts, fairs/expositions grounds, junk and

salvage areas, golf courses, etc.;
- people ramblers, for clients mobility into airports, harbours, train stations, parking garages, shopping centres and malls, etc.;
- personal carriers, for visitors transport aids into hospitals, nursing homes, industrial shops, museums, commercial complexes, etc.;
- similar other travellers/transporters, used according to individual needs or whims, within the chosen autonomy ranges.

The ROBO-RANGER idea looks at conceiving sets of modules, easily assembled into the specially-aimed specimens, for the requested application. The modularity is well prised engineering practice, with nice falls-off. The module is a group of standard and interchangeable parts; it allocates a function to the all, which can be changed and replaced, whenever useful, widening the final assembly variety. We might conveniently distinguish four sets of modules:

- mobility or locomotion modules, permitting guided displacements;
- control and govern modules, assuring actuation/duty supervision;
- navigation modules, conferring trajectory programming capabilities;
- job-specialised modules, namely, the function-characteristic payload.

The survey, in the following, addresses the first two sets, which represent the core of the ROBO-RANGER project. The subsequent two sets are more application-oriented: the payload (people mover, instrument watcher, ..) typifies the entrusted ROBO-RANGER mission; the navigation options are heavily dependent on the needs each time to satisfy. For both, modularity, basically, means to add the necessary functions, which make the ROBO-RANGER fit for the planned mission.

The mobility modules address the wheel locomotion: the ROBO-RANGER is assumed to operate into urban contexts or chiefly anthropic territories, and wheel is fine solution in terms of simplicity and efficiency. The basic modules embrace:

- locomotion blocks, with one powered/idle, fixed/steered wheel;
- frame blocks, with batteries and/or auxiliary equipment.

All blocks have standard size, and connect through vertical grooves, allowing the limited settling of the locomotion blocks, through spring/damper suspension. This way, platforms with three, four or six wheels are possible option, permitting to have all the wheels on the ground, even on uneven surfaces. The frame blocks accomplish: the structural link between locomotion blocks, up the desired vehicle size, the payload support, with vibrations dropping off; the housing of the power supply and control/navigation modules.

The full-inclusive locomotion block includes a powered wheel, directly driven by an electrical motor, carried, through a fork, by an actuated axel. The mobility is obtained, quite regularly, assembling two idle wheel blocks with one powered-and-steered wheel block. If two powered-and-steered wheel blocks are used, the controller needs to direct speed-and-slant of the (independently powered) wheels. Generally, such control does not create problems, face to quiet manoeuvres, over smooth grounds. Anyway, the control redundancy is important feature, always to be taken into account, if two or more independently powered-and-steered wheels are in use. Now, the multiple-powering might be essential for many missions, and the standard treatment of the control redundancy is important practice.

For example purpose, the vehicle, with four powered *fixed* wheels, is recalled. Several inconsistencies appear, when the friction (road reaction) is not constantly mastered. The actuation redundancy makes the model more complex, but, at the same time, opens new opportunities, if proper information is exploited. In fact, at low speeds, the controlled wheel slippage permits fine manoeuvring, with no steer explicit actuation. Actually, the ROBO-RANGER shall be able to follow given paths and slopes, with set performance (payload, velocity, acceleration, autonomy, etc.). The controller interface, as usual, includes the steering, power modulation (torque setting) and brake commands. Further options (neutral position, emergency brake, gear switch, etc.) depend on the application. All commands repeat, in natural way, even when the function (e.g., steering axel) is implicit request. This simplifies the vehicle path monitoring, to oversee or backtrack the mission, and the job revision by the remote up-dating of the current manoeuvres.

The ROBO-RANGER control and govern module, in the quoted example, avails of the vehicle dynamics (centre instantaneous trajectory and oscillation around it), with compliant tyres and suspensions, when four motors operate, on varying soil surfaces. The govern sets the four actuation torques, and combines the data with the individual wheels speed, to reckon the traction force, required at the tyre-soil interface, once the adherence and creeping conditions are assessed. The vehicle is made to follow the desired trajectory, controlling the local slip (with no turn angle of any wheel), and, even, to accomplish in situ rotations (by forced slippage).

The suspensions and tyres modify the (six degrees-of-freedom) body motion, by the action of (four) swinging appendages. The tyre-soil coupling is evaluated, comparing the model-generated motion, with the measured displacements, given by observing the (fixed) surroundings. The ROBO-RANGER governor has always up-dated friction estimates, to modify the torque requested by each (four) wheels. The set-up follows the standard redundant-control arrangements:

- an inner loop oversees the correct angular velocity setting of each wheel, based on simple cinematic steering constraints;
- two outer loops modify the torque (and the power) output of the individual motor, depending on the driver commands: one to fix the vehicle turn, one to set the cruise speed.

The overseeing governor enables the two timely loops. The technique, mostly, applies also with six powered wheels. The model, this time, has three appendages on each side, and averaged figures are estimated, making the velocity setting and the torque command recurrent processes. Indeed, when several powered wheels operate, the replicated tyre-soil interactions, in general, do not exactly match the cinematic conditions at the contact areas; the resultant force might yield yaw turns of the platform, and transversal slippage of the tyre thread, unless torque adapts to the (unpredictable) local adherence conditions. At low speed, when the Coriolis effects are negligible, the control of the wheel rotation velocity provides fine hints on the vehicle manoeuvrability, with proper compensation not exceeding the tyre creep thresholds, once the vehicle motion law is acknowledged.

The ROBO-RANGER idea has been devised for different jobs. The surveillance mobile robot, used in a regional airport, resorts to a rectangular shape (1280 x 960 mm), with four wheels and six frame blocks (320 x 320 x 390 mm). The top plate carries the surveillance kits, with cameras and infrared sensors, at 2.5 m from the soil. The watching is performed with continuity, with randomised path planning. The data are transmitted to the remote stand, where the attendant can modify the robot mission. A slightly modified vehicle was considered as district dispatched, to be exploited in the Middle Age city centre of Genova, to grant the home service of elder people. The narrow lanes request agile aids to carry the shopping and the navigation is pre-planned, to assure the delivering at know timing. A somewhat bigger platform (1800 x 960), with six wheel and six frame blocks, was studied as personal carrier, to grant smart help to low-mobility people. The work highlights how the ROBO-RANGER modular concept gives remarkable advantages both to the manufacturer (allowing customisation with minor design changes) and to the end-user (supporting new layout settling and reduced time-to-repair).

5.3.6. Underwater Autonomous Navigator

The growth axioms (of the industrialism) fight the risk of decline, and, face to bigger world population, ceaseless natural resources consumption, and to ruinous environment pollution augment. It becomes imperative to find new raw materials reserves and to have access to not exploited areas. The marine habitat will enter in the business, with the hostile surroundings drawback and unreceptive work areas shortcoming. The under-water robotics is factual help. The availability of efficient autonomous unmanned vehicles, AUV, with high performance/cost figures, starts being prised with interest, and oriented research lines exist, such as:

- to develop the reference set-ups consistent with the sea-culture paradigms, which enhance of the marine surroundings productivity, by (artificial) man actions (in lieu of the, today, opportunistic economy);
- to figure out effective operation means/methods that permit accomplishing the un-manned and remotely supervised tasks, according to the established (artificially) enhanced organisation set-ups;
- to acknowledge the off-shores technologies and information frames of the marine cultivation yards, with throughout assessments of productivity and eco-impact figures, to assess the business return.

These are example issues, though, showing that the marine surroundings yield is possible origin of benefits comparable by the ones given by the old agricultural revolution, if the specialised cultural means/methods are made available, with due account of the pertinent technologies and know-how. The above lines request:

- to conceive and to develop proper sea-culture paradigms reference rules, extending already existing practices (e.g., in fish-breeding) or developing totally new techniques aiming at the (artificial) deployment of productivity factors in the marine biological processes;
- to make available the multiple-functional robot facilities, say, task-driven autonomous diving tools, un-manned monitoring patrollers, etc. and other auxiliary equipment and procedures for the on-process direction and watch and/or the remote overseeing and govern;
- to build the marine-cultivation knowledge, integrating the previous issues, into highly effective frameworks, consistent with broadening the materials and processes domains, which guarantee the production of artefacts and of commodities, to be traded and enjoyed.

Together, the sea-culture reference rules, efficient robot instrumental tools and pertinent combined know-how bring to un-explored prospects, along the *spirit of adventure* of pioneering tracks, opening creative and risky out-comes. The picture splits in suited actions: to establish sea-culture knowledge and marine cultivation know-how, which define the operation theatre; to devise and perform the means and methods, which make possible unmanned task progression; to conceive the robotic kits, with all functional requisites (autonomy, task-driven abilities, etc.); to check the coastal survey, spot, etc. missions, assessing the related path tracking and/or docking effectiveness.

The job is reached by the SWAN, *sub-sea wobble-free autonomous navigator*, project, summarised hereafter. The focus is on developing a low-cost AUV, with highest on-duty performance/cost figures. This leads to:

- a vectored-thrust symmetrical body, assembled in low cost materials, with ballasting through suited mass distribution;
- original wobble compensation, by means of the three degrees-of-freedom actuation of the propeller and counter-rotating duct.

The existing underwater vehicles are, mainly, multi-function rigs, having high cost and complexity. They, initially, appeared for military purposes, or, at times, for enacting and severe duties (e.g., benthic prospecting), in chiefly *unstructured* contexts. The new outlined domain requires low cost rigs, widely spread over the properly *structured* operation space, to accomplish specified tasks. This suggests twofold specialisation: functional bend, leading to sets of specialised tools, one for every singled-out task; the all-inclusive motility actuation, to standardise the diving and manoeuvring requirements.

The function bend, recalled in paragraph 5.2.4., is typical of the instrumental robotics, even if, at the moment, the whole set of tools is unknown, depending on the new knowledge, to support sea-breeding and marine farming. The underwater mobility distinguishes from the surface mobility, since four (rather than two only) degrees-of-freedom shall be actuated. The SWAN idea gives an efficient solution, as the wobble-compensated vectored-thrust aims at:

- providing the forward speed, by modulating the power of propeller shaft;
- selecting the veering angle, by modifying the thrust yaw deflection;
- selecting the heaving angle, by modifying the thrust pitch inclination;
- assuring steady attitude, by acting on the (roll) motion of the tail duct.

The SWAN technology resorts to the (symmetric) parallel kinematics device, with three motors at the base-frame, actuating the yaw, pitch and roll motion of the tail duct. This carries the co-axial (jointed) propeller shaft, driven by a fourth power motor. The wobble compensation around the AUV axis is assured, either:

- by varying tilt fins: the duct-linked hydrofoils pivot by respect to the main body, and the twist compensation depends on speed and slant angle;
- by counter-rotating screw: the continuous rotary duct carries a set of vanes that generate the twist compensation by opposite screw effects.

The SWAN idea considerably simplifies the navigation setting, through the power motor and the three wrist actuators. The mobility planning requires multi-variable control, to enable manoeuvring capabilities, in view that:

- accurate path tracking and spot positioning without towing and umbilical;
- no sea-worthy vessel, out of special recovery duties;
- standard heaving or veering turns, with radii of 10 m or less;
- accurate hovering and docking by continuous attitude angles trimming;

- widespread missions through navigation autonomy;
- dangerous, exacting or un-safe missions through un-manned running.

The navigation avails of the all-inclusive vectored thrust command rig. The combined actuation simultaneously acts on: the propeller power, for thrust setting; the duct rotation, for twist balance; the yaw angle, for veering; the pitch angle, for heaving. The thrust, applied by the propeller, balances the hydrodynamic forces, on the body and hull surface. The motion deals with equilibrium conditions, and buoyancy pole not always coincident with the centre of mass. In still water: for straight paths, the command operates for the thrust setting and twist balance; for horizontal veering, the propeller generates forward and transverse components, and the drift off-setting shall be compensated; when the vertical hauling is sought, the hydrostatic term appears. The model needs to be derived at several levels of sophistication, with account of all effects: inertial terms, added transport masses; centripetal and Coriolis terms; field-driven actions; viscous coupling (lift and drag included); surroundings effects (whirls, streams, etc.); every time reckoning the lumped-data *equivalent* estimates. Several example navigation cases are studied, showing the SWAN abilities and performance, in on-duty conditions.

The prototypal SWAN patroller is devised, with inherent abilities for accurate tracking of the path, keeping the requested attitude, and for the safe hovering and docking at previously specified locations and orientations. Proper control/govern capabilities are needed, as the navigation is accomplished in sea conditions, where streams and whirls vary in time, possibly transporting and bending the vehicle by non-constant quantities. With the SWAN technology, four independent feedbacks provide the continuous control on the vehicle location and attitude. Presently, the conceptual design addresses a small and properly shaped outer casing, with length around 2 m and 30-60 kg gross weigh. The body, in mission trim, is easily carried by a couple of men and hauled down from a small boat. No expensive support is required, and every mission is easily planned, with very little outside impacts and apparent marks. Proper outfits might up-grade the individual operation trim, e.g.: in shallow water navigation, SWAN could exploit neutral buoyancy, along paths adapted to the sea bottom line; for deep water tasks, SWAN would take in ballast, for powerless descents, and progressively jettison it for surfacing. The final fitting up will depend on the required mission. Up now, shallow water tasks are retained, but paths, down to 500 m depth and 12 hours autonomy, are regular prospect.

The SWAN issues, keeping the sketched design lines, begin from specifying the operation surroundings. Quasi-structured conditions should apply, as the wild world is moved into cultivated spaces. Proper divisions will establish, splitting up the marine habitat, into focused fields, more or less following the way used by men to turn wild lands, into fertile farms. The sea-farming, then, leads to exploit given fixtures, duly located in the tilled spaces. The AUV moves inside the fitted-out zones, where the auxiliary supports (fishing beacons, energy supply stations, etc.) are distributed.

The lack of effective communication channels hinders the under-water tasks. The umbilical drastically limits the work area. The sonar band, time-lag and decay do not permit proper data exchange. The patroller autonomy is, thus, vital request, forcedly limiting the data transfer to overseeing and check functions. The SWAN project exploits the *tele-presence* approach, based on deep-knowledge *modelling & simulation features*, to reckon the current dynamical behaviour through the on-board processor and the remote supervisory computer. The communication stream performs the timely up-dating of threshold parameters; as an all, the *tele-presence* creates the on-duty SWAN behaviour, with marginal resort to data exchange. The outlined approach, joint with the all-inclusive multi-variable propulsion versatility suggests incorporating low-cost micro-devices, into processing blocks, to exploit:

- at the inner level, the real-time controller, to guarantee the AUV steering, according to the task-driven local demands, directly exploiting the robotic actuation peculiarities;
- at the mid level, the on-process governor, to fulfil the fusion of measured and simulated data (after the time-shifting synchronisation and redundancy compression), for the dynamics identification;
- at the outer level, the remote supervisor, to provide off-process monitoring and steering, based on the AUV dynamics simulation/emulation reckoning and the *tele-presence* decision aids.

This way, the sea-cultivation avails of *tele-presence* mode steering, allowing the operator to fulfil remote task steering by closed-loop mode, without being on the AUV. For efficiency, the operator must feel the accurate situations, receiving inputs duplicating the real engagement, and performing actions equivalent to on-process commands. The big hindrance, in the sea habitat, is the time-lag set up by the sonar channel, joined to the drastic restriction in the information content. The real-time scenario up-dating cannot be done, and it is crucial to generate the twin processes at regular time steps, in the local governor and in the remote supervisor, so that the parallel evolution keeps on with the synchronisation, to cancel out the transmission lags. Actually, the observations come in with the varying time-delay, and the commands need suited advance time-shift, to be consistent with the duty frame; besides, the communication channel exiguousness requires decimating the up-dating parameters: most of the data on the twin processes evolution are locally generated, limiting the data exchange to characterising checks.

The *tele-presence* is technique, having wide application ranges. As compared with earlier jobs, the sub-marine distinguishes due to the communication channels criticality, bounded to use low-frequency acoustic-waves. The technique can, e.g., instance, exploit:

- the *autonomic* mode, to accomplish the manoeuvres with the (local) on-process governor, which has resort to the simulation/emulation tools to select the up-dating of the control strategies, as soon as the AUV sensors data are acknowledged;
- the *supervised* mode, to interact with the (local) governor for the mission up-dating, combining the remotely run process, with the additional data provided to the human operator, to accomplish specific actions, without breaking off the current mission.

Other options are devised. The *tele-presence* mode was originally interested to enhance virtual details friendliness (by 3D cameras, by hectic feedback, etc.), in view to enhance the exchanged information quality. In the underwater case, most information is confined locally, and the twin processes evolve in parallel, using decentralised knowledge for refreshing and up-dating purposed, while the shared additions, at the most, concern spot checks and/or mission up-dating. The option is here maximally grounded on the *modelling and simulation features* availability and quality.

The technicalities of the SWAN idea make easy to realize potential sceneries: undoubtedly, appropriate devices and fixtures happen to be built, as soon as the demands are recognised. The un-manned solutions develop, with performance and reliability at the needed levels. Return on investments, directly, corresponds to the achieved productivity; meanwhile process monitoring and eco-impact recording will transparently assess the safeguard figures, for humans and surroundings. The list of SWAN example outcomes covers:

- the development of the patroller, aimed at marine tasks, provides access to sea-culture policies, to expand the foodstuffs and commodities availability based on

resources, with resort to un-explored processes;
- the project will stimulate the eco-conservative transformation economy, by ways not dissimilar from the agricultural revolution, on condition to bring robotics for the un-manned operation in the underwater surroundings;
- the autonomous navigator will contribute to the mankind safety, expanding the farming/breeding areas, into common use, with the productivity in sea-foods and other marine natural yields;
- the fish-breeding and sea-culture techniques will open occupations in the primary areas of commodities provisioning, and in the assistance robotics technologies;
- the low-cost, duty-driven, robotic, submarine equipment will contribute to the industrial countries with competitive advantage in sustainable growth, joining new businesses deployment and eco-impact monitoring.

The sea-culture involvement to sustainable development is emerging outcome, requiring coordinated researches in a wide science and technology spectrum. The comments on the instrumental robotics aids provide ample feeling on the project viability, on condition to work integrating multi-disciplinary efforts and qualified expertise. The idea falls in the natural (*hard*) sciences area, with focus on marine biology and on multifarious opportunities that the cultivation of the ocean huge spaces could open to the future of mankind. The sceneries, here briefly outlines, aim at integrated researches according to the quoted three lines:

- the marine biology investigation, with proper invention of innovative and effective modes for fish breading and for sea-space farming;
- the robotic equipment ideation and build up, leading to cheap un-manned exploitation, based on a set of duty-driven diving tools;
- the study of off-shore modifications, assuring the sea-culture options by joined enhanced productivity and eco-conservativeness.

The innovation, mostly, has to bring to wide series of *diving-tools* (see 5.2.4) with task-driven characteristics, in addition to the low cost AUV, whose success resides in the self-contained path guiding and attitude trimming arrangement, due to the robotic wrist, carrying the vectored thrust. Main SWAN benefits include:

- deep-sea, straight-line missions, with low consumption and sound speed;
- enhanced versatility and research rate, through navigation autonomy;
- surroundings protection (e.g., mid-water column biology, etc.) due to the independence from towed fixtures.

The expansion of the productive domains seems to be valuable goal, free from the unrealistic goal of getting totally rid of tangibles, and from the threatening risk to play with artificial life. The trend has been experienced by the humanity in the old times, before even the industrial revolution, when staples were in farm's yield; then, to expand victuals, new lands ought to have been turned to cultivation, or to more intensive exploitation. The switching, from wild surroundings, to cattle- or poultry-breeding and agriculture, is acknowledged as *revolution*, permitting the *agricultural* economy, as the nomadic populations *opportunity* economy was no more consistent with the *natural* restoring abilities of the habitat they could reach. Now, once accepted that the *agricultural* economy of the *thrifty* society is unlike to preserve the desired quality of life, and the *affluent* society *industrial* economy leads, by itself, to non-sustainable development, one could try more conventional ways, based on:

- expanding the cultivated areas, looking after fish-breeding and sea-culture techniques, not dissimilar to the old *agricultural* revolution ideas;
- protecting the bio-consistency, by monitoring the eco-impact and assuring the conservativeness of the enabled breeding/farming balances.

The oceans exploitation is, today, at the stage of *opportunity* economy, with fishing done, more or less, with nomadic spirit, looking after *natural* restoration of the original situations. Moreover, the under-water and off-shore surroundings are felt high hostility habitats to work in; thereafter, the marine spaces cultivation could only be accomplished by un-manned means, with resort to robotic aids and remotely steered procedures. The project mentions these topics, based on the fact that already available tools and technologies supply all the basic means as for the task autonomy and the impact monitoring.

5.3.7. Car Robotic Dismantling/Recovery

The «robot age» is turned into legend on its abilities of decreasing the decline of the industrialism, expanding the breeding/farming areas (the marine lands) and carrying the reverse logistics recovery, because of the *intelligence* means brought on-process, thus, enabling the «knowledge» paradigms of the *cognitive* revolution to come, to sustain the «to re-materialise» prospects. Out of the myth, the survey of acknowledged robotic issues ends with a few remarks on the recovery (reuse, recycle) business. The argument should appear within the industrial robotics, as the manufacturers' responsibility includes the end-of-life (free) take-back of the items put in the market, making the reverse logistics just a phase of the company business. However, up today, the concepts, such as *extended* (*virtual*) enterprise, are, themselves, myths, and most manufacturers, so far, do not recognize that the «industry» paradigms of the past cannot be exploited any more.

Along this sketch, the *reverse logistics* is as the *service engineering* instance ending each supply chain. The considerations of the paragraph 5.3.2., then, fully apply, with the addition of the law mandatory recovery (reuse, recycle) targets, at least, for some durables (the EU directives on the ELV and WEEE). The premise explains why, in the next outlook, the focus is on the information framework, not on the instrumental appliances, and the web-based emulation/simulation tools are specially mentioned (further than the ambient intelligence tools). The compulsory regulations are fundamental motivation, with biasing effects on the costs build-up, unless the transparency provides sufficient information to fight squanders and role profit (or stealing).

As general rule, the *reverse logistics* is recent entry, which distinguishes from the earlier waste treatment and reclamation, because of:

- pollution remediation, lumping the costs, into affluence side-effects;
- resource revival, to lower the raw materials burning up and dumping.

The activity, yet, responds to society needs, to not live in contaminated lands, and to save supplies for future use. No gain comes from actual trading, since the accomplished doings are community's necessity, before, even, the individual's benefit. This leads to governmental regulations, with public services funded from taxes, and enforced targets. The issue happens, on occasion, to be questionable, especially when political interests and bureaucratic opacity cover up. Well known drawbacks are:

- no evident link between tax-collection and waste treatment efficiency, as managers

and controllers respond to political rulers, not to market;
- marked dependence on the NIMBY, *not in my back-yard*, bias, especially if sectional communities organise protests (up to sabotages);
- political biases, with national, regional and local representatives interested in short terms issues, to enjoy immediate electoral consent;
- loose involvement of the civic spirit, when the officious murkiness hinders transparency of responsibilities (and virtuous behaviours).

The administrative/municipal waste treatment/reclamation is known exercise, with lights and shadows: in Italy, the recent Naples situation has been limit case, showing the public services aberrant degradation, when the authorities collusion has no shame (with, still, electoral consent). The correct and successful balance between mandatory targets and business efficiency is not simple demand, and the end-of-life vehicle, ELV, case provides supportive hints. In the following, the RL-ELV study is reviewed, to show how the reverse logistics shall organise in case of defective local carmakers awareness.

The automotive characterises by dealing with *registered* goods. This means, in Italy: the notary affirmation of the ownership rights, the official declaration (by ACI, the in charge certifying body) for circulation liability, and the administrative governmental recording (into PRA, *pubblico registro automobilistico*). The all is a bit redundant, and exists since the time when cars were only elite products, with few involved purchasers. The changes found opposition of concerned actors (the notary class has protected rule, etc.), and carmakers (the second-hand trade suffers of high administrative costs, compared to the exchanged items value). The ELV requires *de-registration*, with three official steps before issuing the *certificate of destruction*, CoD. The enacted EU regulation requests:

- member states need to establish collecting systems for exhausted vehicles and parts, at authorised sites, where suited treatments shall grant safe and reliable fitting out, by removing noxious and harmful parts;
- withdrawal needs to be accomplished without charge on the final owners (prescription fully enabled from 01.01.2007), but included in the product-service delivery, as inherent attribute;
- users co-responsibility could be invoked, for non-conservative behaviours, when critical pieces are damaged, removed or modified, downgrading the supply original setting;
- dismantling and destruction duties ought to be certified, with assessment of reused parts, recycled materials, secondary-fuel and residuals dumping, to be notified to the EU officials.

The enforced recovery (reuse, recycle) targets are severe, required to reach the 85% of the vehicle weigh, out of 5% (similar) fuel and 10% shredder residuals dumped to landfill (modified from 01.01.2015: 10% fuel, 5% residuals). The ELV disposal is a cost, since the recovery does not repay the attainment of the enacted targets. The carmakers responsibility is deemed to embed the cost at the *point-of-sale*, because all marketed items shall include the ELV *free* take-back. The clause needs to be interpreted: the vehicle's de-registration and transport/handling to the treatment facility are extra-cost, not included by the *free* take-back obligation, but requiring further governmental regulation (out of the carmakers commitment and of the EU ruling frame). The reading in each EU member state differs, following, in the average, scattered efficiency/deterrent issues, as it occurs in the municipal garbage treatment and reclamation processes.

The «robot age» approach is highly relevant, due to the joint management of material and information flows, with the latter heavily involved in de-registration and targets certification duties. The approach provides visibility to the costs build-up, and when a member state is penalizing its residents by ineffective rules, the citizens are, at least, informed (and: *forewarned is forearmed*). Let, thus, look at the *reverse logistic* technicalities and, subsequently, at the activity setting. The backward flow, from exhausted goods, to remediation assets, encompasses:

- a network to collect the disposed items, with transportation, handling and storage equipment, with the related information processing and archival;
- the allotment of *authorised treatment facilities*, ATF, strategically located, to grant efficient, safe and reliable securing of the collected wastes;
- suited reuse flows (conditioned parts) and recycle stands (shredder, sorter, re-former, etc.), along properly specialised processing streams;
- convenient trade lines, to support the reuse/recycle flows, provisioning the manufacture and the maintenance areas.

The EU directives establish the essential diversity between goods and wastes, and only *authorised* companies are allowed to handle, process and store scraps or wrecks, making unlawful the ELV treating out of official sites. The ATF number, location and capacity, thus, are central choice in the reverse logistics efficiency, and, unless free market promotes challenging competition, the Authorities ruling is the critical prerequisite of effectiveness. The foolish distribution of ATF could compromise the all backward flow efficiency.

The *reverse logistic* activity, in the «robot age» setting, organises according to three schemes (*equivalent*, in the present EU spirit):

- *extended* enterprise, if the carmakers cover the recovery tasks, as part of the unified (forward and backward) supply chain;
- *virtual* enterprise, if the recovery is ruled by a body, or business broker, performing the backward track, as supply chain fulfilment;
- *net-concern*, if the recovery data-flow is centred on the ATF, only, using PLM-RL and interfacing for *CoD* and other accomplishments.

The first two schemes match up suitable *industrial* answers. The last is rough approach, when the «industry» effectiveness is lacking, and the «knowledge» aids of the «robot age», anyway, apply as *service engineering* instances. In Italy, due to the known administrative/municipal waste treatment and reclamation situation, factual advantage might be come from the information flow management. This is, in fact, subject to two anomalies: the excessive de-registration onerousness, and the lack of reverse logistics qualified entrepreneurship.

The ATF settling is in-progress. Most likely, they bring together three duties:

- ELV safety setting, by removal of the polluting/dangerous parts;
- wreck splitting into: pieces for reuse, materials to recycle, etc.;
- *CoD* drawing-up, with related forms to testify the recovery targets.

The treatments stands will, increasingly, exploit front-end automation (always in safety setting) and/or mixed-mode automation (in pieces-for-reuse disassembly, etc.). The ATF should be distributed on the territory, to lower the ELV collection costs, still, assuring *sufficient* economy of scale, to allow the facility investment in the suited robot technologies. Anyway, characterising duty is the third, for which the networked infrastructures are essential. To that purpose, the suited information platform needs to be developed (within the frame of paragraph 2.1.5), using suited modelling and simulation features (see, also, the hints of the paragraph 4.1.4). The platform, basically, includes:

- series of facilities/functions, one for each backward track step: collection, disassembly, reuse/recycle, secondary resources provision, etc.;
- relational data-base containing: references PLM-RL, car registration files, recovery process issues, facilities/functions inventory, etc.;
- web interchange between concerned parts: carmakers, clients, authorities, recovery process partners, certifying bodies, etc.;
- information controller, for documentation, reporting and data archival.

The information flow centrality allows to follow, steadily, the backward track from the ELV entry and *CoD* setting, up to the reuse/recycle flows and residuals dumping, providing and assessing the data to know: the car final owner (for the deregistration), the carmaker (for the recovery figures), and the national agencies (for the EU compulsory targets). The information platform shall be permanently accessed by the heterogeneous number of parties, with different interests, liability and responsiveness. The web interchange, thereafter, has to be built accordingly, generating barriers, with proper protection of the sensible data (clients' privacy, manufacturers' secrecy, etc.), warranting, nonetheless, the full transparency to the EU directives regulated targets.

In the car dismantling/recovery, the information infra-structure plays critical roles. Our separation into the *extended* enterprises, or *virtual* organisations, or *net-concern* settings, is useful outcome of the suggestions up now outlined: the issue allows, as well, to provide an explanatory example on how far the «knowledge» paradigms are from the «industry» one, notably, if the manufacture business puts off the *externalities* to *internalities* count, in planning the company engagements. This example application closes the review, showing that the net-concern, out of the entrepreneurial *extended/virtual* enterprise efficiency, still provides visibility of the technical (dismantling, etc.), and of the bureaucratic (deregistration, etc.) duties, providing clear-cut evidence of alternative reverse logistic settings/fittings, with related allocation of the profit/loss figures, induced by the administrative and organisation policies, each time promoted (and enacted).

The outlined RL-ELV study shows the typical situation, where the networked partnership appraisal (see the paragraph 2.2.2 and following) is fundamental duty, to reach proper effectiveness (or detriment). The challenge is open, and there is no reason to put off the *extended* enterprise opportunity.

Chapter 6

6. REMARKS, PROSPECTS AND CONCLUSION

This chapter, to try coherent syntheses of the heterogeneous topics grouped up now, moves, again, from the prospected generic scenarios of section 3.3, to figure out, in each of them, the positive sides that might offer valuable outcomes:

- to make globalisation work, opting the *hyper-market*, means going beyond the *trickle down* effects, and conceiving *global village* effectiveness, such to open alternative growth patterns by «industry» paradigms, as it was the case with the (many) agricultural revolutions occurred in the past;
- to make eco-conservativeness work, picking the *cautious headway*, leads to enable leanness and thriftiness, focusing on local *autarchy* and recovery or restoration practices, because the *precaution principle* forbids, as fully unsafe, the technology-driven innovation, out of the recognised traditions;
- to make economic/ecologic steadiness work, accepting the «knowledge» paradigms, is anthropic bet, believing in the science/technology abilities, to discover effective wealth sources, and in the rational *altruism* aimed at socio-political empowerment, as the wiseful benefit *best-practice*.

The three propositions are already contained by the book chapters, with topics at different levels of completeness. The section 3.2., *Mankind loyalty and world citizenship*, showed that the *global* view is request not to be avoided, face to the over-population, over-consumption and over-pollution planet earth constraints. Consequently, the *no-global* position is unrealistic or illusory or fraudulent, not assuring resources enough, to the world citizens, on the contrary, assuming very severe decrease policies, with quality of life drop. The *global village* ineluctably emerges, assembling all the peoples in narrow and interconnected spaces, so that the *post-global* habits become necessity, and the *altruism*, the only way to make understanding the difference between local (and short terms) castling benefits, as compared with widespread (and durable) innovative practices. The section 3.3, *Questioning looks to the future*, further, shows the forecasts inconsequence, when the «industry» paradigms are propagated, with no «natural capital» bookkeeping, and the growth entrusted with the *market* effectiveness, once all national barriers are removed. The «active» *post-global* approach is, then, devised, where «active» means not simply acknowledging the end of the industrialism we know, but, also, focusing on alternative paradigms, which lead to responsive innovation.

If the *global*, *no-global*, *post-global* track is properly acknowledged, auxiliary details are, here, given, to outline the intrinsic instability of the *hyper-market*, due to the lack of the democratic overseeing with too weak nation-states, in front of multi-national companies, and of the likely dangerous *conservatism*, if the local communities cannot reach the nearby wealth, but have strength enough to grab the possession, through hostile actions. The history lessons show that *autarchy* does not bring the harmony among peoples; on the contrary, quickly raises national or religious or political motivations, justifying and amplifying the mostly economic or hegemonic spurs. The *post-global* view, thereafter, ought to come out from the utopia domain, generalising the process leading at the EU, after centuries of wars, as the only way-out, towards common advantages. Of course, the history lessons are difficult to learn, and the *altruism* is, mostly, thought virtue out of men reach. We shall understand that the *global village* just covers the men shared fatherland, below which it is useless to erect barriers, because the cross-connections are thick and entangled, with no means to rescind them out. The *wiseful legality* is option to be developed. The overall mankind protection is difficult to conceive, if the today citizens egoistic habits leave impoverished and polluted lands.

Further to the additional remarks along the sketched three propositions, the chapter gives an overview of topics covered by the book as helpful synthesis.

6.1. MAKING THE GLOBALISATION WORK

The *sustainable growth* is recurrent image, somehow superficially accepted as viable solution of the mankind spendable riches generation, by enhanced *financial* efficiency, obtained by fostering the production, trade and administrative patterns through economy of scale practices. The schemes are well specified, with resort to the "*short global assent*" setting, under the USA hegemonic ruling. As previously said, the current trend progressively move towards multi-centric settings, with the emerging of several protagonists, the ones with assessed traditions, EU and Japan, and the many emerging nation-states at sub-continent size, Brazil, Russia, India, China, the BRIC assembly, and other equally concerned actors. The polycentrism is phenomenon, today, not clearly evaluated. The globalisation without leadership might, perchance, bring to the influence and supremacy transfer to nomadic multi-national corporations, in parallel, depriving the national governments' authority. The resulting entrepreneurial leaders are forced to maximise the short term return, and will, accordingly, point on the traditional industrialism, dressed with resort to financial clever tricks. The long-term innovation is, at the most, niche option.

The consciousness that the «industry» paradigms cannot warrant respite, and need revision, is still not fully acknowledged, but the *global* position, just, aiming at the *hyper-market* solution, with trickle down bits to less fortunate people, is far from the full consent. Similarly, the simple exportation of lifestyle patterns out of their original cultural frames is not less unrealistic or illusory or fraudulent than the *no-global* claims against technology-driven falls-off, with the motto "when in doubt, do nothing". The *cognitive* revolution shall not be exorcised, more than the previous *agricultural* or *industrial* ones, as *bio-mimicry* and *artificial life* happen to be in line with *artificial breeding/farming*, or *artificial energy/intelligence*, of the already experimented up-surges. The position face to the *industrial* revolution meets, maybe, larger incomprehension, being a phenomenon with time and place restrictions linking its outcomes to the western world culture. The all is too rough, to be acknowledged, without trying some deepening.

In the following, the presentation outlines the coming deployments, after the mention on how the necessity-driven (agricultural and industrial) revolutions did develop. The short remarks help recognizing the development promises fragility and the cross-border entrepreneurial hegemony, when the *global village* is drifting towards oligarchy constructs. In fact, the *hyper-market* guess is nothing more than "*economic global assent*" hypothesis, with no imperialistic ruler, and this prospect is total novelty, deserving vigilant warning, before its positive receipt. The study, here, tries to find out premonitions from old facts; relatable guesses are sought in the earlier breakthroughs, addressing the mankind survival along new tracks.

6.1.1. Trends Divides and Revolutions

The old *agricultural* revolution is lost back in ancient ages, before the history time, and only archaeological evidence is available to support our guesses. Still, consent exists that many *agricultural* revolutions occurred, with dissimilar issues. The most important origins are:

- the Fertile Crescent or east corridor, in Minor Asia, from Mediterranean, to Persia: domestication affected many plants (spelt, barley, wheat, etc., vine, olive, etc.) and several animals (goat, sheep, caw, donkey, etc.);
- the East Asia, in the south, along the Yangtze river, and in the north, about the Yellow river: the domestication covered plants (rice, millet, etc., peach tree, etc.) and animals (hen, pig, etc.);
- the sub-Sahara area: the domestication of cereals and bovines, from east to west, manly, concerned temporary settlings, to be left, when the utilization required unexploited surroundings;
- the America knows some three origins: in Peru, with the domestication of maize, potato, etc. and of guinea pig, llama, alpaca, etc; the maize farming in found, as well, in Mexico, and the sunflower in Tennessee.

Many other starting points exist, such as Indonesia, for citrus plants, or Russia planes for horses. Here, the stress is on the places and issues multiplicity, with, in addition, the widespread of the dates. By far, the Fertile Crescent is the earliest, as it starts by the 8 000 - 7 800 years B.C., with soon spreading to the Mediterranean area and to the close India and Europe territory. This has brought to hypothesise some Indo-European ancestors, as the civilisation common origin, grounded on the language affinity, without, however, clear definition and location of the shared progenitors. The Yangtze river zone is second old start, going back to the 6 500 B.C., followed, 1 000 years later, by the Yellow river and the Indochina areas. No contiguity exists between these two starts, and the cross-links shall, most likely, to be deferred, maybe, some 4 000 years later or more.

The sub-Sahara Africa start is difficult to date, but it is chiefly placed by 2 000 B.C., and entrusted to Ancient Egypt populations, migrating (south and) west, to reach better pasture lands. That diaspora is proved by the language (Peul) spoken, with allied structure, in Cameroun, Senegal and Guinea. The American zones are, likewise, dated back around 2 800 B.C. in the Andes, and about 2 500 in Mexico. The peoples moved to New World from Asia, through the Aleutians islands road, by the 10 000 year B.C., and went south (and east); this is hypothesised, because of the agglutinating language spoken, somehow, related with Mongolian, Korean and Japanese. The agricultural revolution, anyway, started original tracks.

The short hints are far from certainties, still they show local divides, when the rise of the inhabitants made survival impossible to hunter/picker populations, by merely, opportunistic economy. Due to intrinsic relevance and peculiarities, the revolutions at the two opposite Asia ends are compared to single out, again, some noteworthy facets, as it was the case, discussing the industrial revolution. On one side, we have the (today, named) Indo-European peoples. There is no evidence that a single (warrior) population made the conquest of the huge areas, from north Europe, to the India regions. The languages background, moreover, is guessed by learned glottologists, not without discussions. Much more intuitive, perhaps, is to consider the expansion of a common culture, founded on the shared subsistence means, namely, the agriculture, making possible the nourishing of greatly bigger amounts of *similarly civilised* citizens.

The common culture quickly brings in peculiar traits, such as the urbanisation, the craft segmentation and the trade organisation. The Damascus or Jericho sites are inhabited since 6 000 B.C., and, maybe, from the agricultural revolution very beginning. The farm-city link shows that the inhabitants gather into narrow spaces to assure better defence against predatory enemies; in parallel, the work division distinguishes, craftsmen, traders and peasants, making effective big communities (several hundreds citizens, or more), as compared with the nomadic assemblies (never bigger than several tenths men, to gather enough food by natural picking). The social crosslink, as observed in paragraph 1.2.1, arranges into clusters, having central hubs, to exploit representative ruling. The consciousness to need internal solidarity, together with external agreements binding the nearby clusters, is quite strong, as well as the awareness of the hubs multiplicity, with friendly, hostile or neutral mutual positions.

The each one autonomous cluster is forced by the geography, with little and specialised agricultural lands, divided by mountains and seas. The complementary produces force the trading with foreigners, at once recognised as holders of unlike rituals. The Fertile Crescent, visibly, gathered three traditions (from Noah's sons, Shem, Ham and Japheth), so that the Indo-European populations were only part of the quite diversified society. The cultural exchanges are, from the beginning, very active, with cross-profits. The Greek (the Latin, etc.) alphabet is derived from its Phoenician origin, while the Semitic (Hebrew, Arab) kept the different structures of omitting the written vowels. The habit to co-exist and co-operate, keeping nice distinction of the origins, is important feature, promoting assertive attitudes. The cultural background prefers bottom-up structures, to preserve the personal identity only mitigated, through the delegation procedures and legal contracts. The shared distinctiveness, as said, is cultural (rather than racial); in addition, the linguistic connection is rather loose, being limited to some affinities in deriving the words from roots and in sentence construction. The abstract phonetic correspondences permit obtaining unlimited number of words, combining roots and suffixes, while exploiting the timber modulation, with big variability. The link of the writing and phonemes is exploited to create and to stabilise the national languages, so that the specific characterisations are readily coded and hand down.

The situation happens to be quite different in the East Asia regions. Here, the ordered individuals are strongly connected over uniform spaces, giving rise to the spread-out monolithic communities, with scattered random bridges, to provide the general connectivity. The social horizon is flat, without the need to co-exist with less than regular near neighbours. The hostile foreigners can even act as winning invaders, but hardly modify the monolithic stability, and the (external) rulers are, quickly, assimilated, to share the assessed culture. The speech peculiarity shows the tone languages features, with monosyllabic words, coded by ideograms. The Indo-European polysyllabic words allow the understanding of the differences and the tracking of the changes. With monosyllabic words, every modification makes the backtracking uneasy, so that, very, quickly, the understanding is deferred to the context, the ideograms associated to specialise the meaning. In China, too, we have several languages: at least 13 official languages are spoken, using the same ideograms, differently read. The Mandarin (or *Han yu*, Han language, with widest spread, as well named *pu tong hua*, common language) has steady history, being coded with the same ideograms from always. At this point, it is hard to figure out the traits of changes in space and in time. Two Chinese will understand each other by reading, not speaking; no sureness exists about how the language was spoken in the past. The fact assures the society monolithic stability, not affected by spot or ethnic preferences.

Such stability is again found in the grammar, essentially, based on the concept ordering. The standard order is quite simple, enchaining [1]: the local surroundings, the verbal name and the object complement. The monosyllabic speech is, quickly, facing ambiguity risks; e.g., between *shü* (book) and *shü* (uncle), it would be hard to make the choice; saying *yī běn shü* or *yī wèi shü* no doubt exists, as *yī běn* is the numeral one for items such as tome, copybook, etc., and *yī wèi*, the one for prised persons. The context dependence allows creating new concepts; e.g., the words *yī* (*one*) and *dài* (*belt*), need the numeral specification to mean *a belt*; together, they mean *hereabouts*, if placed after spot definitions, so that *zhè yī dài* means *on this side*. Moreover, the verb is not specified with independent nature (compared with substantives); it is, more correctly, a verbal substantive, showing the «becoming». The series *tiān mèi xià lǎi* composes of four characters, meaning *day*, *dark*, *below* and *to come*; together meaning *the night comes little by little*; the action is implicit in the locution *xià lǎi*, or, usually, by the linked particles, at times, without sense out of the context.

We have already pointed out the differences leading at the *reductionism* spirit, either at the holistic *complexity*, to discern the «competition» from the «harmony» patterns. Here, we recognise several *agricultural* revolutions, which occur, when the populations cannot survive with the opportunistic economy, and are forced to devise *artificial* orders, more effective, as compared to the *natural* ones. With the time, the different inventions (and domestications) become the heritage of the all mankind: the *potatoes* or *tomatoes* are German or Italian staples, no mind of their original selection. We might assume the industrial revolution similar variety, out of the European style one. The «robot age» deployment, then, switches from the *scientific* to the *intelligent* work organisation, prising «complexity», as supporting method, still leaving unsolved the industrialism open questions.

6.1.2. The Development Promises

The «industry» paradigms, basically, request entrepreneurial competition and *global* free market to keep benefit of the manufacture efficiency. On average, the highest advantage is obtained by the multi-national companies, through the *hyper-market* deployment. The developing countries, anyway, are said to profit through the *trickle down* process, sharing improvements, otherwise un-hoped. The recent results are not encouraging: the riches gap widens apart, with poverty increasing. Different remedies are considered:

- financial capitals provision: loans allotment provides rapid benefits, with longer terms restriction, face to wrong investments, with low productivity;
- unbiased trading areas: privatisation (governments bias elimination) is nice recipe, with heavy drawbacks out of stable, well organised countries;
- human capital fostering: to aim at social and education strengthening is the new bet, to prepare qualified work-forces, as widespread resource.

The question, on how wealthy countries could contribute to the development of the others, is widely disputed, leading, on occasion, to international governance bodies, to address the public/private organisations towards optimising the decision keeping mechanisms, for the best profit of the country's citizens. The interest, in higher people spendable riches, is, surely, genuine, because this grants the market expansion, with direct profit of the clients and manufacturers. On these premises, the trade fairness is only

---

[1] The considerations on the languages are taken from: Malherbe M. (1995). "*Les langages de l'humanité*", Robert Laffont, Paris, pp. 1734.

constraint, to not enrich the few, impoverishing the less protected. At least, such viewpoint could make the globalisation work, getting rid of the distortions, fostered by the *hyper-market*, whether the set of more fortunate actors are maximally privileged, assuring widespread buyers.

That *global* approach is fought by the *no-global* one, on many bases, specially because of the position advantages of the wealthier partners. This is, chiefly, clear in the patents, licences and trademarks domain. Now, the innovation is important, and the inventions make feasible to improve the world life-quality. The discovers' significance to have their work protected and remunerated is, in addition, general interest, the technical capital being the driver of the human development. This is, nonetheless, rather tangled affair. If licences and trademarks are used to hamper technology and know-how spreading, the growth is obstructed or forbidden, with mankind damage. The fact is well know in drugs and chemicals, when the items cost is so high, to be outside the reach of many sick persons, and of (developing countries) health systems. Of course, no obvious answer exists. Somebody has linked the (today) patent rights with the «enclosures», which, in the Middle Age, permitted the land privatisation, starting again the economic deployment, due to more intensive farming exploitation, after the stagnation period, with the private ownership loose protection.

Besides, the fundamental research progresses quicker with the free and fastest circulation of the ideas. On the contrary, it remains blocked, when the corporation efforts are focused on generating the highest patent intricacy, to book up falls-off and incremental improvements, not allowing the know-how spreading. The *fair* intellectual property regulation is, surely, open question. It becomes urgent need, if the technical capital is only way, to aim at sustainable growth. The *intangibles* peculiarities make difficult to say that the «enclosures» regime might be winning. The knowledge pieces exchange is quite different, permitting infinite duplication and exploitation, so that the highest exploitation does not profit from the careful retrenching. On the contrary, the «enclosures» regime is, now, considered, face to the ocean resources expansion (seam/field prospecting, seaweed cultivation, fish breeding, etc.), today, left to the opportunistic exploitation, with stocks withdrawn without attention on sustainability. By such way, the «enclosures» would mean to fix the ownership privileges, promoting the effective and shrewd management of the individual concessions. The all is typical *global village* affair, presently rather dubious, out of the narrow costal waters.

The opposition between tangible *vs.* intangible goods exploitation rules is well acknowledged, but, today, hindered by the ownership «enclosures». The creation of endorsed spaces is long dispute, with many definitions, *legal robbery* included. Should the «natural» capital be undifferentiated mankind property, framing could never start, and the impasse of community belongs or individual enclosures is just negligible trifle. On factual issue, the history shows what has been named *tragedy of shared belongings*, when the lack of concerned persons in charge allows each one to squander and destroy the resources (by today ocean fishing). Appropriation should distinguish from fraudulent conversion, setting up legal frames, enacted by governmental authorities. The «natural» capital, as a whole, supports instrumental goods. Whether enjoyed, the withdrawal and restoration balance requires interest payment, as with any other sort of capital.

The national resources correct bookkeeping is the earliest challenge, faced by every country (with resort to the **TYPUS** metrics, or equivalent scheme), without leaving out ambiguities zones. With the *hyper-market*, thus far, the multi-national companies are severe handicap, having the possibility to exploit dominance ranks, out of the local governmental ruling. The situation is further worsened, when the lack of democratic control exists, and the corruption of a few leaders makes easy predatory profits. The *hyper-market* does not, by itself, prevent the globalisation to work; it might, on the contrary, become beneficial, whether the *global village* is fully subject to uniform and fair regulations. The entrepreneurship, in its totality, needs to consider the ecologic and social responsibility standard engagements, not to be circumvented, because the *global village* law does not leave short cuts. The worldwide government is, possibly, utopia. The *hyper-market* benefit is, however, on the average, development promise with little reliability, whether right political control on the cross-border corporations could protect the earth ensemble, issuing solidarity pacts to safeguard third (and future) people. In theory, the *hyper-market* might develop on the *"ecologic global assent"* conjecture, and few hints are given in paragraph 3.3.1; the hypothesis is, nonetheless, even less plausible than the one directly derived by the «financial» *global village*. Even so, this conjecture is again considered hereafter.

6.1.3. The Hyper-Market Self-Administration

The *economic global* scenario looks after efficiency, through getting rid of the local bureaucracies, which rise local barriers, incompatible with the *hyper-market* build-up. A multi-national corporation is free from single government rules, and enhances the financial return through productive break-up and job out-sourcing; it optimises competitiveness, transferring the activity to low wages regions, with the proprietary know-how. The market polycentrism, little backed by the nation-state capitalism, is, now, ruled on trans-national entrepreneurial patterns. Sometimes, it benefits by removing local autocracies, replaced by parliamentary democracies, open to the *free-market* axioms, with, however, imperfect authority, not to hamper the cross-border corporations' path.

The process is not without ambiguities. The localisms can be exploited to split rich or poor communities (Checks from Slovaks, Kurds from Iraqis, etc.), leading to timely better homogeneous structures, more permeable to trans-national ruling. The *market efficiency* generalisation is important feature, notably, aiming at the privatisation efficiency, dismantling authorities' ruled (and/or protected) services, replaced by communication, transport, instruction, health, etc. enterprises, capable to compete for the citizens' satisfaction. Sets of civil servants (starting by doctors, teachers, etc., and expanding to judges, policemen, etc.) will become employees, with office-work requiring clerical responsiveness, within the hierarchic reliance on the their trans-national companies (placed before the local administrations).

The *hyper-market*, freed from national ties, turns to *self-watching* practices, in an *ideal world*, with two objectives:

- the transparency of the delivered *functions*, to grant company-wide quality with clients' satisfaction and third people safeguard;
- the prospecting bigger *market shares*, to widen the entrepreneurial profit, by expanding the acquisition of potential clients.

The companies' *self-watching* capabilities are needed to shelter the business, by keeping under surveillance the internal employees, the actual/potential clients, the suppliers, the competitors, and the regulation authorities. The governmental overseeing cost's lower, most of the duties being privatised, moving, however, the power out of the nation jurisdiction. The *privacy* protection, today, fought against the local administrative intrusions, shall face transnational polycentrism, maybe, less oppressive, but out of the political polls democratic control. The nation-state deconstruction, initially, shows the *hyper-market* efficiency gains (as compared to local market biases), but the lack of direct/parliamentary democracy might turn in serious drawbacks, when the polycentric market transfers the riches, where it gets the highest profit, with lower taxes and wider autonomy. The emerging order is in polarised wealth distribution, with few high peaks, and extended poverty, so that the worldwide political instability will surface, making unlikely the *hyper-market* preservation, even, by the most powerful multi-national corporations.

Looking back at the capitalism history, in section 3.2, we have already met the *mercantilism*, characterised by weak political authorities, face to the prominence of trading set-ups. The differences are, now, more impressive than the analogies, with the present transnational corporations deeply rooted in the industrial work-styles, and the transformation efficiency paradigms applied to the non-renewable «natural» capital. The city-state capitalism has been linked with *direct* democracy ruling, surrogating too weak political authorities; the trans-national corporations, on the contrary, start with the backing of strong national settings (in, both, *short* and *long global assents*) and are, perhaps, discovering further development tracks, when the (yet to be) multi-centric *global village* might be exploited by the trans-border companies oligarchy, to get rid of the local diseconomies and impositions. The course is difficult to figure out; most likely, it could be top-down process, not the bottom-up one of the city-state capitalism, when the *mercantilism* was backed by the individual spirit of adventure. Of course, the analysis is a bit too schematic, and additional remarks are useful, for better describe the possible future scenario.

The *hyper-market*, whether the *global* track is privileged, might be temporary arrangement, at different deployment levels. Its duration is largely linked to the capability of creating wealth through the elimination of improper administrative costs. Surprisingly, the all scenario develops as if the «natural» capital is implied, tacitly presuming that the earth and the universe exist to be exploited by the man. As a matter of fact, the tacit positions presuppose the disciplined lawfulness of:

- multi-national corporations competition, to share out earth (and universe) resources: political arrangements need to be devised, through settlements ruled by privatisation agreement and/or wars between nation-states;
- surroundings entire availability to the man benefit, with the full inclusive fitness/allowableness of all withdrawals/manipulations, being the market profit, the anthropic must, before every ecologic limitation warning.

At the moment, the interplay between the cross-border corporations and local countries is open, and the *hyper-market* is only a possible picture. The second fact is, by now, accepted or rejected, with resort to fixed thoughts, maybe, political or religious beliefs, well proved by the *global/no-global* dispute. Should the set-ups turn in the outlined direction, the victory of the *market* over the nation-states, will request worldwide administration rules, all over the *global village*. Supra-national organisations (ONU, WTO, etc.) and agencies (BPM, ISO, etc.) should undertake *hyper-watch* tasks, becoming, themselves, important management business, to be carried as autonomous function, independent of multinational companies concern, under the sponsorship of the (residual) nation-states influence. Such worldwide administration will be liberated from democratic control, with no direct answer to urging voters, thus, apparently, conscious of longer terms (ecologic) demands, if the pressing cross-border corporations profit could be restrained. This issue is far from desirable, and rather difficult to be accepted into *rational* frames.

The supra-national management, based only on objective technical principles, is quite like to be utopia. Quite the opposite, for the liberty's sake, the individual freedom risks to disappearing. With the excuse to improve efficiency and widen the spendable riches, the *hyper-market* will broaden the *self-watching* practices, to choke the individual autonomy under the business productivity urgency, with, this time, no trade-unions or worker coalitions capable to exert political pressure. The only antagonism, should such global deployment fully establish, might come out of the «terrorism», and related fighting acts against the market dictatorship. The *ideal world* objectives are, surely, too abstract and the *self-administration* issues cannot lead to far-seeing *altruism*, once efficiency conflicts against these unlikely «financial» *hyper-market* needs, uniquely tied to the «economic» *global village*.

The financial wealth deregulation, with trends driven by the *self-steered* rules of the (independent) rating bodies, has, recently, shown scary inadequacy, getting to disrupt leaders' establishments. The (short term) merely «economic» appraisals are *artificial* constructs, easily to become deceitful, when the building frailty does not permit the soft landing. The fault is not in the *market* (or *hyper-market*); it is in its misuse, placing first the institution short term current benefits, as if the back frame could remain steady; we are forced to consider the driving contexts, which affect the momentary balances. And these balances, to be meaningful, need to be complete, duly covering the «natural capital» bookkeeping.

6.1.4. Auxiliary Hints and Initial Conclusion

The *hyper-market* scenario might appear as the new *free market* setting, when the *economic global consent* does not repeat known arrangements, under basically single hegemonic rulers, since no nation-state (not even at sub-continental size) is sufficiently powerful to establish worldwide control. In fact, the picture may well be accepted, with the emerging of many economic world powers, plainly involved in supporting multi-national corporation champions. At this point, two prospects could establish:

- repetition of known situations, ending the ruler's imperialistic dominance by worldwide conflicts (the *worldwide wars* picture);
- the depicted *hyper-market* situation, transferring the dominance to series of trans-national companies (the *archipelago* picture).

It is difficult to say which prospect is more probable. The former distinguishes from earlier set-ups, because the worldwide spread nation-states modifies the old Eurocentric patterns, so that the conflicts deflagration does not refer to restricted *vital space* requests, rather diffused economic necessities. The latter is pretty new opportunity, whose configuration is hard to figure out. The *hyper-market* concept is the rational exploitation of the efficient work organisation, ruled by competitive financial infrastructures. The information and communication technologies supply the supporting instruments, where the *global* attitude does not ensue from specific political hegemony (with governmental ruling), but from generic technical options (fostering the worldwide entrepreneurship). The correlation with the *mercantilism* is misleading, as already pointed out, being bottom-up organisation, which aims at surrogating the lack of political authorities, while the *hyper-market* needs to fully eliminate, or to marginalise the local national authorities, otherwise the authentic effectiveness can never be achieved.

Up now, the multinational corporations are strongly connected to hegemonic nations (today, typically the USA), and the *hyper-market* arrangement happens to be some earlier imperialisms variation (of the *long/short global assent* ways). The hegemony, now, moves outside the single nation jurisdiction, and becomes affair of inter-state companies, quickly migrating to maximise their return. Apparently, this might appear advantageous, replacing political nationalisms, by economical supremacies, ruled to the profit of financial organisations. The *global* attitude is novel experience, showing the way out of retrenching into nationally protected selfishness, towards the cosmopolitan advantage of universal effectiveness. The recipe has to be invented, and, as pointed out, severe objections exist.

Without any claim of completeness, the following aspects are recalled:

- the *economical profit* maximisation, to satisfy the short-terms return of the stakeholders, timely engaged in the corporations;
- the power transfer from the direct/parliamentary democratic control, to the *self-management* of entrepreneurial organisations;
- the trust in the *free market* ruling law, as if fully limitless steady resources feed the industrial transformation processes.

The aspects lead to already outlined objections, whose order of importance is, perhaps, to be reversed. Anyway, keeping the above order, further comments are outlined.

The *classic* economics genuinely believes in the community profit, when each individual acts in *self-interest*. This, quite obviously, leads to maximise the return on the each time allotted capital assets, concentrating the efforts on obtaining the most competitive planning, to augment the personal wealth. The general benefit is direct consequence, when every ones do their best, living in regimented contexts, which assure the individual achievements secure guardianship, private ownership included. The instability risk, here, comes, when we deal with huge corporations, having temporary stakeholders and shareholders, quickly interested to modify the activity profile and/or finance sharing, if better opportunities arise. The short-term return, then, is *necessity*, obliging the company managers to operate accordingly, with enhanced crucial outcomes in *hyper-market* arrangements, where the power follows *self-administration* issues.

The *new-classic* economics, at this point, suffers even greater hindrances, as the *macro-micro* interplay makes sense, when the timely enabled social measures exploit the inner/outer biasing to stimulate (boundary protected) manufacture and trade work opportunities. The *welfarism* of the *hyper-market* settings shall still be invented, and can only be grounded on strong (governmental-like) authorities, this time operating at the worldwide level, fixing priorities, without real possibility of the execution control and the infringement sanctioning. Indeed, the governmental influence depends on its legitimacy, through electoral polling and national loyalty. Failing these, every regulation appears arbitrary, not grounded on shared consent, and not easy to accept for the common welfare. The ethical/social multi-national corporation is abstract entity, having welfare practice and policy out of structured democratic jurisdiction, unless the (possible) sponsorship of famous philosophers, eminent academic/religious authorities, or distinguished civil servants. The aim of such organisations is important to encourage dialogue and debate across personal, collective, intellectual and international boundaries, on the mankind issues, tying the professional interventions with the citizens' life. The actual political falls-off are to be discovered, once the *global village* loyalty raises as common glue of the tomorrow populations. Whether found, they logically lead to set-ups, akin to the *altruism*, ruled by some *hyper-democracy*, sheltering the mankind lineage.

Then, the discussion turns on the third aspect, the ecologic concern, which is in full contrast with the *free market* rule. The *supply and demand* law is beautiful short cut, if unlimited resources are available to exchange, just subject to rate or transport limitations. This is not the case, when non-renewable stocks are handed, with the further pollution burden. The *hyper-market* setting does not imply viable solutions, and requests switching to the *no-global* or *post-global* viewpoints. This one, at least, is fine initial conclusion, coming from the just devised scenario. The multi-centric world spreading, with several countries at dominant economical and political range, is *hyper-market* alternative, but with more shades than lights, yet again discussed in the following (as *archipelago* picture).

6.2. THE ECOLOGIC CONSISTENCY PLEDGE

The theoretic limits of the development are well established in physics by the entropy principle, stating the unidirectional trend of the transformations, towards the undifferentiated chaos. In the «industry» paradigms, the reversibility wants are ignored, as if the waste and pollution become part of the productive process, with the closure of hypothetical regeneration and reclamation cycles. The industrialism economics is pervaded by such nonsense, as if the mechanics laws are static truth, not affected by progressive decay. Now, being the ceaseless growth claptrap, the conservative evolution is the only reasonably alternative, looking after the human behaviours, assuring the conscious limitation of the intensive downgrading. Such approach is named the *cautious headway* (to oppose the *durable growth*), because the absence of development is mastered by series of careful actions, enhancing the preservation of the existing resources. The attitude can, in some cases, lead to the bio-economics, in the restricted sense to look at economic systems consistent with the bio-sphere, we know, which is life-friendly provided that its stability is kept.

The *cautious headway* does not, strictly, mean decrease. It clearly opposes to the *affluent* practices, and coherently sponsors the *thrifty* habits, while the series of cautions spread out, focusing on the bio-diversity, instead of the massive resort to the (industrial) farming efficiency. The precaution principle «when in doubt, do nothing» is fundamental motto, with proper attention on the restoration of the old traditions, having proved safety margins. This summarises into the *8R-changes*:

- to redefine, turning the value scale, towards *cautious headway* scopes;
- to revalue, fixing priority benchmarks, headed to *thrifty* behaviours;
- to reassign, modifying the riches allocation, charging the *consumers*;
- to relocate, fostering *autarchy*, prising the local products/services;
- to reinstate, poising the production facilities at their maximal leanness;
- to reduce, lowering the bio-sphere footprint, by mandatory targets;
- to reuse, expanding the items lifecycle, with second-hand efficiency;
- to recover, generalising the reverse logistics to the traded items totality.

The first four *R* aim at deep modifying the people habits, by *parsimony mind*, specifying the eco-consistency imperatives: conservative safety value, thriftiness priority, consumption taxes and work/trade centring. The subsequent four suggest paradigm shifts in the earlier manufacture flows, with: internalities specialisation, externalities betterment, conscious market policies and backward tracks support, expanding the business, by *servicing mind*. In abstract, the *8R-changes* are suited practices, and, in the following, additional aspects are recalled, prior to entering in more details of the *no-global* spirit. The *cautious headway* is nice answer, to face *over-consumption* and *over-pollution*, not to deal with *over-population*, and, more over, with quality of life enhancement.

The *no-global* approach, it should be said, shows clear-cut positions, when the opposition to the «economic» global mind-sets are dealt. At the proposal step, the lines are considerably evanescent, and the discussion of the next paragraphs might slot in series of concepts, broadly reflecting the plain *post-global* views, when the end of the affluent society is acknowledged, but no active alternative is conceived. The topics are, nonetheless, useful, since the *global*, *no-global*, *post-global* course profits of joint analyses, to work out helpful hints.

6.2.1. Planning with Parsimony Mind

The four *R*, with *parsimony mind*, aim at moving off the affluence face of the «industry» processes, to come back to previous more conscious thriftiness habits. The people mentality change successfulness is hard labour, and its usefulness is, for sure, questionable. It stands for reinstating the hunting/picking *opportunistic* economy, when the neighbouring populations enjoyed *agricultural* options. The evolution is unidirectional, and the cultural background is inherited, summing up the older generations knowledge. However, the *cautious headway* scopes have to be considered for their ecological concern. All four scopes are especially relevant. The first two, for their educational purpose, help to orient towards eco-conscious behaviours. The third is noteworthy change in the taxing systems, today, oriented to redistribute the «financial» capital, to less fortunate citizens; tomorrow, aimed at charging the consumers, in proportion to the withdrawn «natural» capital. The variation is not simple, because it requires an objective metrics, built on absolute scales, so that the earth resources are referred to unbiased standards. The **TYPUS** metrics is example reference, requiring, however, worldwide consent, to become legal metrology standard.

The fourth change looks at setting (or re-setting) autonomous local economic frames, with the production of self-sufficient resources, so that the transportation of supplies and transfer of people are fully removed, totally stopping the related wastes (and costs). The efforts in that direction are documented in some studies, to sketch, e.g., the Netherland project, devising full *autarchic* cycles, based on the intensive greenhouse cultivation, sufficient for that highly inhabited country. The study is, as well, suggesting innovative goals, to look at bio-diversity enrichment, prising *natural* processes, to avoid phyto-drugs and grant wide produce variety.

The *autarchy* outcome is fundamental step in the leanness build-up. The local breadth assures fresh foodstuffs, breaking the mass-distribution hegemonic rule, with widespread front-end occupation into the self-supporting communities. The all looks at reinventing the city-state, with, however, fair solidarity conditions, to crosslink the autonomous communities, to grant equal-opportunity headways to all people. The local dimension, within otherwise worldwide monolithic societies, permits the peripheral direct democracy, with the advantage of unbiased govern, and of balanced solidarity, if the local entities interact without the intermediation of national selfishness. Indeed, the nation-state capitalism already proved to lead towards violent policies, with the leading nations exploiting the less aggressive, to obtain more resources and to start out-of-control spoilage. The *autarchy* of small districts is meant to warrant spread autonomy, so that the belligerent antagonism is avoided, and the *cautious headway* becomes safe and reasonable prospect.

The idyllic picture is completed by two side effects:

- the promotion of *friendly relational* supplies;
- the *precaution principle* strict application.

The first means that the cultural *R*-changes lead to the massive production of intangible provisions, as such, indefinitely duplicated at no charge, with twofold outcomes: to grant the local identity awareness, fostering the self-independence; to back-up the worldwide monolithic responsiveness, by the transparent portrait of the *cautious headway* advantages. The cultural exchanges promotion is direct opportunity of the information technologies, making possible the worldwide web use, and the networked cross-assessment-and-control of the local communities to make all people aware of the balanced fairness membership.

Today, the information sharing is acknowledged as powerful way to move the value chain along conservativeness, while the resort to bio-copying mechanisms are rich of charm;

but also feared as potentially dangerous. The perception of the life principles highly puzzles individuals. Bio-mimicry, e.g., is kept in quarantine according to the *precaution principle*, which assumes, as potentially dangerous, every (artificial) bio-genetically modified product, until the experimental proof of its full safeness is achieved, as for future and indirect falls-off. The (potentially) positive outcomes are delayed far away in time (to unpredictable horizons), since the life-sciences do not possess systematic recognition, and require wide testing before reaching empirical evidence. The question is, by no means, subtle, and if the *precaution principle* applies drastic limitations into, basically, conservative frames, very little chance exists to accept the *artificial* bio-genetic tracks, for high productivity re-materialisation processes.

The planning with parsimony in mind, as general precept, is valuable practice, with, possible warning, only, when the localism (and *autarchy* closure) is prised, at too idyllic pictures, not directly acknowledged in the current practices. Besides, the enhanced localism and conservativeness are dangerous attitude, at any rate, on the known past history experiences. The enduring thriftiness is virtue, coming up close to the fatalism, not necessarily tolerant face to some one else's success. The all brings to conservative vigilance, by itself surely not harmful, unless if the lack of spirit of adventure forbids looking at effective ways-out.

6.2.2. Planning with Servicing Mind

The four *R*, with *servicing in mind*, aim at creating backward track options, in addition to the traditional forward ones. Actually, the first two are general scopes, leading to lower waste and pollution, suppressing useless duties and redundancies, and to enact compulsory targets on the bio-sphere impact. The other two, openly, refer to the reverse logistics business. The four together foreshadow the new way in the material supply chains, through which the more elaborate provision of any items, embed the parallel information flow on the used resources bookkeeping. It is vital to spare the «natural capital»; equally relevant the value chain transfer into intangibles, so that, wealth being equal, the value added is mostly incorporated in the enabling *functions*. The *function* trading gives the rise of several opportunities:

- the selling of *product-service*, when the provision include lifecycle support for *conformance-to-use* certification and for *end-of-life* take-back;
- the supplying of *product-generated-function*, so that buyer's satisfaction is reached by intangibles allowance, without items' ownership transfer;
- the timely allotment of *facility/function* aids into entrepreneurial set-ups, in order to enable provisional (project-driven) productive-organisations;
- the generalised resort to *relational allocations*, in diversified fields, from education, to entertainment, etc. to broaden the intangible supply chains.

The trading of *services*, in addition or instead of products, is not new, unless that the provision is deemed to become knowledge enterprise market staple. In the past, the manufacture products did not replace the land produce, even if the work ways deeply changed the farmers' activity. The knowledge entrepreneurship shall not supersede the traditional shop-floor; it will totally modify the ways to compete in the market, with emerging eco-responsibilities, leading to the *service* delivery.

It is obvious, that the life-quality deployment profits from *soft* technologies; it is inconceivable to entrust to these, the potential for eco-sufficient solutions, when exclusively exploited, as already noticed, along the «to de-materialise» ways. The prospected market staples entertain all four co-operating capital assets: technical, financial, human and natural, with, moreover, the warning to distinguish between *native* contributions (the last two), from *artificial* additions (the other two); this, at least, is the *cautious headway* strategy counsel, because of the artificial quantities (technical and financial) inborn limitations, which might result deceptive.

Anyway, the enterprise competitiveness turns from the capability of offering new products (fit-for-purpose according to individual needs), to the capability of providing (intangible) services (offering lifecycle support or replacing functions, with full satisfaction and better non-renewable resource balance). The eco-value implies to measure the bio-sphere footprint, as for the raw materials depletion and dispersed pollution, and to promote the eco-conservativeness by:

- concerned measures: regulatory agreements, which expand the alternatives to non-conservative practices: to require frequent conformance checks, to impose end-of-life take-back, etc.;
- peremptory acts: mandatory collection of taxes in proportion to the caused impact, with solid responsiveness of the consumers (producers and users) and with resort to acknowledged measurement standards.

Proper technical solutions exist and answer to these goals, deeply rooted in the mankind deployment; the cultural requisites will soon follow, leading to the listed *8-R* example changes, integrated, nonetheless, by their *action* matching parts, not encouraged or, even, rejected by the conservative fear of adventures.

The natural capital centrality remains obvious reference since old times, as the men received their subsistence from what they find or produce in the land. The proposition, by which the (local and temporary) wellbeing is artificial constructs issue, is modern time result, and it leads to recognise the technical capital role, as the origin of the agricultural and of the industrial revolutions. The options made possible to devise refined models, with, more or less, coherent causal schemes, showing the handed down knowledge primary (even if implicit) contribution, the cumulated financial means support, and the human capital enabling intervention, permitting, as previously pointed out, to historicize the becoming. The claim to distinguish *natural* (safe) and *artificial* (risky) contributions, according to *a priory* judgements and scales, is far from rational. The *precaution principle* shall work on *a posteriori* empirical evidence, and *natural* poisons provide little trust, as safe produce, if *a priori* truths only are enunciated.

The other way, the *cautious headway* path suggests important measures, when the recovery (reuse, recycle) targets are detailed. Surely, the waste of goods (and people) of the *affluent* society ought to be removed or drastically reduced. Today, the reverse logistics is at the *opportunistic* chance level, and more conscious acts should be devised, without, inevitably, demonising the industrial options. When looking at the civilisation *artificial* side, one is deemed to express trivial truths. The *artificial* intelligence is safe and sound aid, more than the *artificial* energy of the old manufacture capitalism. It, clearly, switches on (mostly) intangibles, after centuries of wealth, built on drawing on material stocks, with no restoring. As an all, however, it contributes to the «to de-materialise» track, with only negligible falls-off on the material resources strengthening and recovery.

The planning with servicing in mind is, again, as general precept, helpful way to spare natural resources. It is not, for sure, sufficient to create eco-conservative frameworks, but the commitment, of transforming *opportunistic* reverse logistics, into compulsory obligation, is fundamental claim, not included by the *supply and demand* law, whether the only economic side is investigated, without paying the «natural capital» real value. In short, the trust on relational allocations helps, but it does not produce the basic resources for the mankind necessities.

6.2.3. The Locally Conservative Practices

The ecologic widespread scenario looks after *cautious headway* strategies, by bottom-up policies, based on the regional self-sufficiency. The frame is not at all new, if one considers the European history (after the dissolution of the centralised Roman Empire), from the city-state, to the nation-state paradigms. The *no-global* interest in looking at *old* certainties, disregards, not only the industrialism refusal inconsistency, as well the transport/communication exchange importance for the building of shared comprehension and acquaintance. The empires' rise and death are fascinating events, based on how the leadership ruling establishes and exerts their hegemonic functions. The local dimension *autarchy* is stable, when wealth and strength are equally distributed, making feeble the envy, and risky the profit. The hypothesis is rather improbable.

On the contrary, the anger of less fortunate communities, against (the *market* order and) the richer peoples, will quickly gather the coalitions of the desperate individuals (leading to the terrorism actions) and countries (blocking the *cautious headway* equilibrium). On these conditions, the chance to expand monolithic nets between equals vanishes, and the national (religious, ethnic, etc.) divisions bring to gather sufficient power for more or less local struggles, up to generalised wars. The conflicts' extent depends on the availability of armies and weapons. The rage against (real and supposed) privileges is enough to motivate the fighters; the new technologies are fit for more effective and lethal arms.

The preliminary targets, surely, are the rich countries, even if the direct attack has to be postponed, until their strength is lowered, through the combined effects of the internal lack of cohesion, and external deficiency of vital resources. When the retreat on the *cautious headway* local dimension is unilaterally performed on regional (or even, continental) scale, the (spot or total) attack risk rises, notably, if the watch lowers. The aggressive regimes containment or, even, the precautionary war gives little respite, when the fight concerns the acquisition of vital resources. The policy towards parsimony and thriftiness has little chance to be winning, if, mostly, endorsed on the passive fairness of fatalism conditions. Here, again, the supra-national resource management and conflict resolution organisations should be addressed, though without, the multi-national corporations force, in this case, replaced by the local dimension businesses. The *autarchy* helps fixing barriers, with the principle of precaution turned to worst-case scenarios, each time giving dark portraits of all changes, assuming "better safe than sorrow".

The psychology "when in doubt, do nothing" might result winning in narrow political context, but with long terms damages, due to the irresponsible riding of the sloth conservatisms irrational side. The kind of decision keeping, when risky conditions appear, is strictly related to the institutional frames and cultural ranks of each given community or nation. The political class aiming merely at the easy consensus, rather than at the hard effectiveness in view of the people wellbeing, has better way along the "do nothing, when in doubt" track, just referring to the precaution care, in front of risky negative outcomes. With the «robot age» mind, the anti-scientific approach should disappear: it is surely unsafe, with the unveiled threat to lower the future populations' wealth. Squandering and wastage are rather different affair, however, harmful, not less than the innovation refusal.

The *cautious headway* concept correctly opposes to the *durable growth* one, if the focus is on the entropy law, and consequent physical world decay. The simple *8R* changes do not provide viable solutions, out of the worthy leanness habits and the limited recovery and reclamation issues. These advantages are stoutly paid by the *autarchy* localism, very difficult to generalise, if the eco-ruled self-sufficiency is only programme, under the imperative, narrow sense, *precaution principle*. The *no-global* views have the merit to single out the *global* deficiency, when bounded to merely economic assessments. The so called «glocal» advance, joining local to global dimensions, is rather unstable, and does not propose spendable riches at the level of the future world increased needs.

The presentation of the *no-global* position is affected by ambiguities, because, at the moment, several political movements recognise themselves in the area, but with diversified claims and objectives. The *no-global* only unifying views sort out *in opposition*, even if they cannot fully disregard the otherwise shared knowledge. The «glocal» concept, with its fuzzy outlines, is example oxymoron, on occasion, used to distinguish different aims in the *cautious headway* surroundings. The tiny outlooks here given are just interested to discriminate alternative scenarios; not at all to give exhaustive descriptions. The conservatism alternatives do not appear to offer, in the present situation, constructive suggestions.

6.2.4. Added Remarks and Close Updating

The *no-global* scenario might emerge as the negation of the *global* picture; in fact, it mainly follows such attitude, with, nonetheless, some exclusion. The clear opposition involves two issues:

- the focus on the local realities, giving the way to *self-sufficient* districts, in which the mobility, transport, etc. squandering disappears;
- the parsimony planning, leading to *thrifty* societies, opposed to the affluent ones, avoiding most disposables and use-and-dump habits.

The exclusions are a bit entangled, achieving formally similar issues, however through rather dissimilar approaches:

- the *servicing* planning assures centrality to *relational* entrepreneurship and *intangible* value added, as instrumental (not standalone) option;
- the «glocal» concept allows exploiting the *net concern* support, mainly, as the *local* dimension districts' *worldwide monolithic* complement.

The business, education, entertainment, etc. *relational allocation* is expansion of the «robot age» mind, obviously, surfacing in all future societies. Similarly, the networked infrastructures permit to modify the interpersonal correlation, included the political sphere, enabling sorts of direct democracy, to bring the peripheries in touch with the ruling authorities. Hence, the *global/no-global* common issues are, basically, technology-steered, and another way shaped, depending on the implied socio-political preferences. It needs to be noticed that the *global/no-global* fight is typical western countries phenomenon, and the dialectic opposition is not without dogmatic (or childish) pettiness, when the dialecticism might be wiser.

Back to the evident oppositions, the *no-global* position is gravely conditioned by the negative policy, and heavily biased by *precaution principle*, as if old habits are automatically safer. The battle against innovation, on the contrary, is not safe at all, as we know that, face to the present over-population, consistent amounts of new resources are necessary, not to worldwide expand sorrow life conditions. The motto <in doubt, do nothing> is the simple *cautious headway* slogan, which takes no heed to the people welfare. Of course, the other industrialism outcomes, over-consumption and over-pollution, are real dangers, not simply solved, however, by the mentioned motto. Thus, the *no-global* position correctly pays heed on *ecology*, and prospects responsive palliatives (e.g., parsimony plans), but prefers avoiding the whole question. Even worse, the biology laws are interpreted, distinguishing the (apparently safe) *natural* evolutionism, and the (surely harmful) *artificial* one. The agricultural revolution, also, is regarded as doubtful, being the *domesticated* vegetables/animals origin, and the promoter of (many) species extinction.

The negative policy, of course, cannot be expanded up, by coming back to the primigenial *opportunistic* economy, and the *cautious headway* accepts exceptions as well established safe products/habits, more or less as the *global* position is now open to subordinate the *economy* to the *ecology*, especially if the over-pollution is becoming impending danger. Hence, the *global/no-global* overlapping is regular, out the extreme positions. Nonetheless, even when the common sense leads to not too afar factual behaviours, the basic socio-political backgrounds are far-off, with the *hyper-market* the exact opposite of the *autarchy*. The fact might be understood as the *post-global* position reference start, namely, the realistic acknowledgment that both scenarios lead to untenably biased situations. The *hyper-market* trend is based on the competition efficiency, and requires pulling down the local barriers, typically, leading to highly lopsided wealth distributions. The local *autarchy*, the contrary, aims at beautiful smallness, accepting parsimony plans, with accurately smoothed wealth distributions. The wealth concentration, in the first case, is not localised, strengthened through supra-national legitimating. In the other one, the individual local district is the author of its own wealth, sometimes, quite different from the nearby districts one. In both cases, less fortunate citizens exist, without equalisation (welfarism akin) mechanisms. The solidarity is positively valued as individual/community option, again, looking back at the old charitable practices, thus, entrusted to the volunteers, and bounded to the territory scale.

At this point, additional conclusions might be proposed, putting in evidence the drawbacks of the prospected *no-global* scenario:

- the (proclaimed) scarce/null interest in *growth* does not permit riches build up, to carry on/spread out the (today reached) quality of life;
- the *cautious headway*, with (local) closure is conflict unsafe presage, when the rivalry opposes countries, at the national interest range.

The ecological concern becomes, today, priority at the industrialism stage, and linked over-consumption/over-pollution ranges. The *hyper-market* set-up does not offer solutions, as said, because the ecology constraints cannot be ruled, according to usual competition schemes, among enterprises on short-term profit horizons. If this is clearly stated by the *no-global* opposition, the offered solutions are lavishly biased by assertive preconditions, making the *precaution principle* rigid obstacle to technological innovation, and, there through, to sustainable growth (at least, up to the man history range). Moreover, several side-effects, perhaps, still not fully acknowledged (from the «glocal» concept, to the *autarchy* risks), suggest looking at a possible further scenario, here, referred as active *post-global* position. Again, the outlook suggests a series of hints, requiring proper deepening, assuming that it is necessary to modify the current trends, and that positive falls-out have desirable chances to be devised.

6.3. THE ROBOT AGE CHALLENGE

The men have the unconscious belief, to play preferred role in the universe, as if the intelligent life does not simply reduce to the more or less curious outcomes of anomalies, leading to carbon-driven life compatibility conditions on the earth, thereafter starting the astonishing gratuitous evolutionism towards us, but, on the contrary, as if proper pre-conditions existed (since always), to make the build-up of the humanity, and to grant meaningfulness to the world. Indeed, when the life started on the earth, perhaps, some 4 billions of years ago, no notice of the fact could have been detected (by an hypothetical observer, somewhere located in the universe). Little by little, the vegetation started to incorporate the CO_2 from the surroundings, and to release the oxygen. This latter becomes dominant element in weight on the earth's crust, and it is the 21% of the atmosphere, in the molecular state, thus, assuring strong oxidising action. The limit, however, is critical figure, not to be modified, not to reach spontaneous combustion conditions, making the carbon-based life impossible. The fact that *natural life* and *intelligence* exist, does not give certainties about their safety, out of the known space and time spans. The opposite, as well, is arbitrary. It is not sufficient the preserve the *natural* evolution to be sure of favourable future conditions (the earth experienced past life-hostile situations), and there is no reason to believe that *artificial* transformations are, by themselves, intrinsically noxious.

On such premises, the physical world decline, because of the entropy law, is not disputable truth, on the cosmologic time span. On the evolutionism periods, it is likely recognised, to experiment typical *becoming* statements, exemplified by the *three facts*: death, sex and speciation. The death shows that the future differs from the past, suggesting renovation patterns. The sex permits to mix and spread out the genetic characteristics. The speciation creates diversity, by surroundings adapted species. In lieu of decline, the *life* phenomena suggest progression. The impression is fortified at the *intelligence* range. Again, we might recognise *three facts*: revolution divides, cultural inventions and language varieties. The up-turns are tragedies, forced by the survival impossibility. The cultures create cross-types, strengthening the original inheritances. The languages permit community singling and identification, granting the social richness.

The man is development and growth, not lessening and decline symbol. The *post-global* views, then, aim at integrating the *economic* improvement, along the *global* approach, with the *ecologic* conservation of the *no-global* warning, to look at feasible policies, more balanced, lowering the menace of the certainly instable socio-political organisations. The section considers in sequence:

- the new robots' opportunities face to the existing overall constraints;
- the options, to helpfully mitigate the *hyper-market* inconsistencies;
- the options, to profitably avoid the *cautious headway* dead end halts;
- some example outcomes, explaining the «robot age» spirit potentials;
- the basic *post-global* value chains of the devised *four* capital assets;
- brief remarks on the socio-political plausibility of the *altruism* bets.

The section concludes the book, without offering substantially new ideas. On the contrary, the continuity of the «robot age» spirit is essential guess. During the "*short global assent*" parenthesis, the value creation primarily moves towards the *technical* capital, with the further *natural* capital enhanced productivity, joint to the transparency provision. The end is break, not defeat, leading to the knowledge society, and related ecologic potentials. The *global village* is fostered by the net concerns, and, in same time, is split into districts, as somebody prefers to have a local economy, being reluctant to share a temporary wellbeing. Nonetheless, if the current options provide limited and unfair advantages, or, even worse, convey to dead end, the way should be modified, accepting breaks, not to incur into defeats.

The resort to the «robot age» concept, and related changeful «knowledge», is deeply linked with the thought that the western world style industrialism does not have future. The simple rejection of known paradigms, however, is totally devoid of sense; these ought to be revised and amended if defective. The responsiveness spirit and adventure character are, notably, permanent help. In the revisions, the attention on alternative cultural backgrounds is, especially, relevant, because this permits to understand that multiple-cultural tracks exist, and the wise man should profit of the totality of them. In the sketched analyses, the *global* spirit is clearly referred to the western world style; moreover, also the *no-global* attitude is, also, (and even more) issue of the same culture. The conclusion that both positions are only partially true (or wrong) should not be rejected.

6.3.1. The in Progress Changeful Knowledge

The *global* views, leading to the *hyper-market* efficiency ruled by prevailing multi-national corporations, are confident of the anthropocentric *economics*, with the belief that the *industrial* revolution (as the earlier *agricultural*) could take new forms, discovering different staples, each time adapted to the emerging situations. The *no-global* views, leading to the *cautious headway* continuation by *autarchic self-sufficient* communities, are conscious of the *ecologic* limits of the spendable riches obtained by transforming the earth resources, into waste and pollution, and require drastic changes, towards consistent behavioural thriftiness. The opposition observes that the *sustainable* (or *durable*) *growth* concept is cheating, or, at least, an oxymoron, the growth being unidirectional phenomenon, typically reaching its final state, before decay and end. We shall get persuaded, therefore, that we need to start relinquishing the absurd *affluent* society style, and to adopt eco-balanced habits. The *economic/ecologic* dilemma is central concern of the new millennium, having both, the *global* and the *no-global*, theses many convinced defenders. The big drawback, at least in the analysis tried through the book arguments, is that the two views are awfully partial, and lead, in more or less short time, to conflicting situations, both the *hyper-market* and the segmented *autarchy*, bringing to biased societies, because of the highly unfair wealth distribution, either, of the non-even opportunity local communities.

Thus, the analysis of hypothetical *post-global* views is thought deserving high relevance, and is performed, moving with coherence along the industrialism track, to understand doable eco-conservative alternatives and desirable ways-out, which allow mitigating the «industry» patterns negative falls-off, and, perhaps, devising new patterns for revolutions to come. By now, the entropy principle is surely well assessed, and, on long period issues, the physical laws teach us that the mankind life (and intelligence) shall disappear. The time orders are the problem, here. The man history spans are, perhaps, three thousands years, to be roughly doubled up to the agricultural revolution pre-history, or widened, maybe, three times or slightly more, to find the *intelligent* living beings existence. The figures are negligible, if compared with the cosmology spans, at least, whether the current explanations are accepted. The anthropic reality, accordingly, is totally insignificant phenomenon, confined within a trifling planet, for an irrelevant while. The *precaution principle*, then, needs to be assessed on the earth life conditions effects, with little worry on unlikely falls-off. The *sustainable growth*, thereafter, assumes accordingly limited spans, where *sustainability* means to obtain the balanced *economic/ecologic* mix, permitting suited life-quality standards to current earth populations. The horizons, surely, are important figure, and alternative issues ought to be weighed against the riches amounts and ranges, each time, devised.

The *post-global* views, thus, take due account of the *cautious headway/hyper-market* opposition, trying to detect scenarios with lower intrinsic instability, on, both the ecologic and the economic sides, to permit comparatively steady quality of life conditions, on the few lifespans range of the generations to come. The all can be summarised with a set of statements, moving around the «robot age» ideas:

- the *industrial* revolution is recent history phenomenon, with well localised geography and period, where the «robot age» ideas show a divide, among the *scientific* vs. the *intelligent* work organisation patterns;
- the «knowledge» paradigms, replacing the «industry» ones, have falls-off exceeding the *manufacture* market: directly, addressing *products-services*; on the whole, aiming at «robot age» *intangible* staples;
- the *artificial energy* provides autonomy from the *natural* emergencies; the *artificial intelligence* gives conscious autonomy, through the «robot age» transparency of the enabled transformations;
- the «robot age» stands for the *artificial energy* and *artificial intelligence* operation synthesis, and brings about new bets, towards the *artificial life* (and *artificial man*), supporting *cognitive* abilities;
- the «robot age» allows overseeing the (*cognitive* revolution) *bio-mimicry* processes, broadening the man steer/control potentials, after the *artificial* domestication/manufacture of the *natural* (living or not) resources.

The above listed are example propositions, showing some implications carried over, from the «robot age» changeful knowledge. The future scenarios, right now, are hardly figured out, and the listed propositions might be differently understood, whether: we believe to be at an upturn, opening the *cognitive* revolution times; or we prefer to prise the industrialism continuity, through the «robot age» staples up-dating; or, as well, we conceive *8-R* changes alike patterns, discovering *artificial* farming/breeding processes, precaution principle consistent, supplying foodstuffs the worldwide around. In the three cases, the focus principally turns on biological transformations, to aim at (mostly) renewable resources, so that the effectiveness does not immediately speeds up the earth habitat entropic downgrading.

The «robot age» long terms issues cannot by forecast, as the new technologies are bets. This clangs against the caution mode, with the aggravation of operating on *artificial life*. The pragmatic approach inclines to imply correspondences with the ancient agricultural revolution, when the man tamed animals and vegetables, according to their usefulness. No evidence exists that the *artificial* evolutionism is worse than the *natural* one. Anyway, opinions about in-progress technologies are open to contrasting interpretations, beyond the present remarks, where coherent decay scales might be devised, to slow down or to speed up the entropy trend.

The focus on renewable resources means that, on suitable periods, the natural capital is guaranteed, and the growth verifies durability checks. The «robot age» growth by *artificial energy/intelligence/life* lines comes with transparent process overseeing. To achieve effective outcomes, the accounts balance has to show that two goals occur:

- the value chain expands, by no (or negligible) tangibles consumption, and (man-safe) alternative stocks refill is, in any case, granted;
- the pollution is removed, by equivalent remediation measures, restoring life-friendly conditions, with in-progress eco-impact monitoring.

The «robot age» approach is neutral about the two goals achievement. It aims at providing the instrumental methods, being based on the incorporated *artificial energy/intelligence/life* means, but it simply stands for the enabled transforms and controls, assuring the required steering and monitoring functions. Moreover, the *cognitive* revolution, should ever happen, is outcome of new technologies, still to be acknowledged; on the contrary, we already live in the «robot age», even if, at the moment, we do not fully exploit the provided potentials. The co-existence of manufacture products and land produces shows the continuity of the men needs, face to the deep changes in the spendable riches generation. Similarly, the «robot age» will not supersede the traditional lifestyle habits, face to new transformation techniques (whether discovered).

The «robot age», in the outlined reading, entails managing the «knowledge» paradigms, in lieu of the «industry» ones. At the moment, the change adds the supply chain visibility, specifying the emerging eco-responsibilities. The *product-service* delivery, accomplished by *virtual/extended* enterprises is example issue. The life-quality deployment profits from technologies; still, it is inconceivable to entrust to these, the potential for self-sufficient outcomes, without exploring the *cognitive* processes duplication opportunities. The resort to the technical capital means accepting the prominent changes that it induces in the surroundings. Now, the balancing of the four co-operating capital assets: technical, financial, human and natural, finds in the «robot age» full transparency, granting *cautious headway* to the *artificial energy/intelligence/life* paths. The sustainable growth cannot exist, unless safe and effective answers are found. The *no-global* approach appears too conservative, and dangerously *autarchy* localism biased. But, the known paths do not bring to ahead issues, and are risky for the mankind stability. The resort to the fossil fuel combustion creates carbon dioxide, and the in-progress environmental deterioration. The existing fission nuclear plants are treacherous, because of the radioactive wastes. Why not looking at new options (free from the said risks) ?

6.3.2. The Responsibilities Framework

The *agricultural* revolution changeful knowledge, surely, took long to settle in many sites, with autonomous inventions. Then, the opportunistic economy was no more competitive, with the agrarian societies' huge advantage. In the 20th century second half, the enterprises, even in conventional manufacture, moved, from the *scientific* organisation of the work, for the mass-production by the scale-economy, to the *intelligent* one, for the customer-driven quality by the scope-economy. The evolution replaces the big flow-shop plants, by scattered job-shop facilities, where the operators are required interactive collaboration, as the robots accomplish the front-end jobs. This is said to be the initial step of the «information» capitalism, due to explicit resort to technology, with the knowledge society start-up or post-manufacturing growth, prising *intangible* value added delivering.

The *artificial* factors independence (the financial not less than the technical), by respect to the originating human factor, might be factual trick (to describe the trading transactions, or to explain the transformation economies), or recognised as real expansion, with validity disjoined from individual persons direct intervention. In the latter case, only, the concepts of financial, either, technical productivity is coherently defined, and remuneration of one or the other contribution deserves independent evaluation (out of the functions performed by the engaged operators). The doubts, around the correctness to lend money against interests, are long story, with religious flavours, and, at times, surface bringing on, e.g., the usurious rates rules. Today, the entrepreneurship risk reward has standard market, with explicit and implicit transactions. The technical facts ostracism is quite different truth; it occurs when the issues are blasphemous, compared with accepted traditions, such as, in the past, for the celestial mechanics or the evolutionism of living entities, and, presently, for the genetic manipulation or the process bio-engineering. The critical positions can be understood, until the innovation worthiness has reached proper acceptance (or falsification). However, the technical capital opportunities, offered to look at wealth expansion, cannot be rejected, as blasphemous trends. The metrics assessing the «financial» either the «technical» capital are quite more entangled affair, and the plain reference to the «demand and supply law» is fairly simplistic, neglecting the fact that the market law presumes external conditions, at least, at two levels: highly structured legal backdrops, supporting the value of the *artificial* provisions; balanced remuneration of all *four* capital assets, not allowing the unfair abuse of any of them.

The *hyper-market* deployment is, certainly, critical issue, shifting the power form the nation-states, to the cross-border corporations. Few facts deserve notice. The lesson, coming from the companies evolution, deals with the organisational *externalities* setting; this entails changes vis-à-vis of the employees, to show how the human capital role transforms; and in terms of the web opportunities, leading to net-concerns, or knowledge entrepreneurship structural innovation. Remaining in the social interpretation frames, it is beyond doubt that the large corporation is instrument to exercise power, by sets of practices, such as the following:

- over the employees and related salaries;
- over the prices recognised to the suppliers and imposed to the dealers;
- over the market policies of consumables and durables;
- over the standards (conformance assessment, fiscal treatment, etc.);
- over the choice of the trans-national production facilities allocation;
- over the trade organisation settings, and promotion or charges sharing;
- over the corporate/administrative policies, with regards on local easiness back-up, through lobbing or other means.

This requires re-interpreting the market, when it is hypothesised its strength in governing hiring, wages, prices, standards, etc., in front of the biased steering of the assessed offers, and the competition with better products aims at replacing old trade-marks. To such purpose, simple considerations lead to recognise that:

- the multi-national set-ups operate with decentralised plants, befit for local supremacy; the interstate oligopolies establish with influence exceeding all local authorities; the competitiveness is built by the central management, maximising the return on investment by fiscal planning, productive break-up, cross-border ruling, etc., through nominal market self-regulation;
- the large corporations enjoy factual separation of property (shareholders, etc.) and administration (managers, etc.); the supremacy focuses on the interest of the structure (out of current citizens), with priority to the profit of the internal staff and leaders (by respect to the nominal capitalists) or, generally, of the stakeholders with high lobbying possibilities.

In both cases, the *free market* appears fuzzy concept: from below, as the inter-state horizons hinder factual controls; from above, because the schematic dualism «proprietor-workers» becomes meaningless, even without the revision operated by the welfarism, as the power appropriation addresses the steering stakeholders field, out of the class struggle, with, in lieu:

- the establishment of trans-national managers, with the cross-border micro-economics options;
- the interfacing with civil servants, entitled of the nation macro-economics

intervention duties.

The macro-economics measures loose effectiveness, or, even, lead to opposite effects, with financial resources flight, based on the *hyper-market* style. Looking positive, in the *altruism* responsibility structure, the economic complexity allows value incorporation, from network co-operation, by flow continuity and stability. This permits the eco-impact mandatory overseeing, as set in data, included by the value chains, over the four capital assets (**KILT** models), with the comprehensive assessment of the tangibles, actually, used by the provision (**TYPUS** metrics).

The *local/global* interplay re-establishes the socio-political stability, when the *global village* has common regulations, shared solidarity, uniform overseeing and combined liability. The localism cannot create castled counties; on the contrary, is forced to conform to external laws. The multi-national corporations are subject to identical burdens and facilitations, and cannot dodge or overwhelm the identical governmental charges. The net-concern technicalities are main reference to deal the ecologic constraints. In the *virtual* set-up, the organisation goes behind series of rigid taxonomies, such as: cluster, nodes, etc. ownership; proprietary hierarchy; links with *local/global* authorities; dependence on owners/stakeholders; relations with internal managers; internalities/externalities interchange; responsiveness lay-outs; etc., so that the bottom up constraints are remedial against the *hyper-market* drifting towards oligopoly positions, because of the *hyper-democracy* superposed control, to be established with worldwide consent.

The *global/post-global* transition, according to the shortly sketched outlook, is straightforward, when one look at the mankind as a continuum, not only to cover the *global village* populations, but further including in them the whole descents of today citizens. The responsibility framework is, perhaps, conceivable, if the focus is on heritage transmission, as wealth shall be protected by solidarity laws.

6.3.3. Robots in Eco-Footprint Guardianship

The «robot age» prospects are useful technicalities, to enable the *post-global* eco-balanced society, and to support the *altruism* socio-political deployments. On the technical side, the instrumental aids, mostly, appear at two levels:

- the supply chains monitoring, and eco-impact assessment, to provide up-dated picture of the wasted resources and pollution;
- the materials balance overseeing, and earth safety watching, to bring in clear natural capital and «to re-materialise» records.

The two opportunities, both already discussed, are here shortly recalled. The supply chain transparency is based on two knowledge frames:

- the manufacturing activity grounded on the **KILT** models, making explicit the dependence on four capital assets, each requiring separate balance;
- the natural capital contraction/expansion measured through homologated standards (**TYPUS** metrics, or equivalent ones).

The ecological bookkeeping becomes constant «robot age» accomplishment, inserted in the information infrastructures complexity. The **TYPUS** metrics has to be stated within the legal metrology schemes, with worldwide acknowledgement. Its application needs to reach self-sufficient consistency. The today EU law, e.g., with mandatory materials recovery targets, but undefined energy constraints, is misleading, prising *legal* backwards paths against, perhaps, less eco-conservative *illegal*, but lower energivores ones.

The sophistication of the information infrastructures is deemed to grow, when performing the monitoring of all consumables, at least, as aggregate figures, but without biased, unfair trade effects. The manufacturers' responsibility and the free call-back scheme are, quickly, creating entangled forward and backward logistics, not easily managed, out of advanced *extended* enterprises. Moreover, the trade-off «monitoring complexity *vs.* sustainability achievements», only, aims at sectional (durables, not disposables) or partial (material recovery, without energy checks) regulations, and the issues are positive, on condition to refer to coherent frames for programmed scopes (delayed, to follow smooth, pace-wise progression). The «robot age» is at the embryonic implementation stage, and only the information structures are already properly specified.

The **KILT** model, further, specifies that the manufacture company output Q includes the environmental impact (pollution and consumption), to dress coherent balances, with the natural capital productivity transparency. The imbalances shall be charged to the consumers, according to all comprehensive assessments. The T-factor is ecology bitter surprise; the K-factor is robotic good chance. The current balances estimation is, still, out of the traditional habits, principally, as the natural resources are thought implicit mankind property. Nobody bothers measuring the actual impacts (withdrawals and pollutants), assuming unlimited earth stocks and fully automatic natural recovery. These two facts are not true, and the ecology is critical concern. Besides, monitoring is standard «robot age» opportunity, and the supply chain visibility, straightforward enabling instrument.

On the technical side, the «robot age» first instrumental level offers aids along the «to dematerialise» path, through intangible value added. It modifies, as well, the people responsiveness and commitment face to the growth sustainability, by the transparency provided to the manufacture transformation (the <*forewarned, forearmed*> saying). The steps ahead, from the *hyper-market* disinterestedness or multi-national companies' self-lessness, or from the heroic thriftiness or *cautious headway* conservativeness, are, at this point, moral achievements, providing clear and objective assessment of the eco-footprint left by the individual citizens, due to their current behaviour. The moral fairness or iniquity is translated in the taxing system, charging the individuals in proportion to the eco-footprint, in view to aim at *altruistic* compensations, to the future generations' factual benefits. The rising charges are spurring to eco-consciousness.

The «robot age» second instrumental role offers support more tricky to define, being dependent on technologies to come, centred on the «to re-materialise» path. The considerations up-now presented incline to think that divisions should occur, showing discontinuities from the *industrial*, to the *cognitive* breakthrough, on the two standpoints:

- the focusing on *renewable resources*, therefore, primarily pointing on bio-processes, incorporating, as much as possible, energy from *outer space*;
- the discovering of *artificial life* opportunities, suggested by *bio-mimicry* or *genetic engineering*, to operate under *artificial intelligence* control.

The *cognitive* revolution, if actually close to come, characterises the mankind answer to the existing demands, as it was the case to overcome the opportunistic economy, and it is now the case with the conventional manufacture economy. Of course, its factual occurrence presumes believing in suitable anthropic principles, maybe, saying that: <God helps those who help themselves>. The man, hence, is responsible of his future. The resources misusage (the squandering, with benefits spoil; the injuring, with damages falls-off) depends on lack of diligence and hard work, and has little to do with the *precaution principle* and conflicts with the pre-established order. The «innovation» is basic option, allowing the discoveries (or inventions) of opportunities somehow pre-set, assuring the men supply.

The approach enhances the man concern about the research of the better life-quality. It makes obvious the sustainable growth research, as intrinsic aspect of an *artificial* evolution. The *natural laws* are, perhaps, shorthand records, expressing the acknowledged empirical regularities by models; anyway, they can be used, as meaningful explanation of the real world. The model successfulness means the coincidence between the extrapolated guesses and the perceived assessments. The «innovation» permits to exploit the technical models, for the sound and enhanced resort to the reachable reserves. The resource expansion is possible, by efficiently exploiting the land (agricultural revolution), the labour (industrial revolution) or, from today on, the intelligence of the living beings (cognitive breakthrough). The life-quality increase is innovation outcome. Certainly, we have no evidence that the sketched approach might result successful, but it is difficult to accept that the humanity is without future, and no technology-driven path is value the challenge to establish nice set-ups, having valuable issues. The bet, in actual terms, is worth some trials, and in such active spirit, the remarks are assembled.

6.3.4. The Robot Age Knowledge Promises

Going somehow forward into technology deployments, the «robot age» spirit suggests example possibilities in the offering of valid development chances. The outlook might start, referring to *ideal* experimentation studies, using robotics as supporting means. We can begin, looking at the twofold role the robot has to play:

- the self-sufficient automaton, capable to accomplish the allotted tasks, up to desired accuracy and efficiency; the self-reproductive automaton is case example, assuring scheduled duplication and deployment;
- the surroundings watcher, capable to monitor the assigned region, joining empirical information, with domain knowledge; the diagnosis and govern duties permit to safely manage and govern the local space.

The *self-reproductive automaton*, according to von Neumann, builds on four issues:

- the coded instruction, containing all the data necessary to make the task (namely, the *self-reproduction*) feasible;
- the *self-sufficient* productive facility, able to collect materials and energy, and to fulfil the (*reproductive*) manufacture job;
- the interpreting device, capable to *duplicate* the instructions, to recognise each step, and to *enable* the job progression;
- the control device, connected with the other ones, to oversee the all job, so that the robot *self-reproduction* task is fulfilled.

In the biological processes, the instructions are DNA and RNA genetic codes; the productive facility goes with the ribosome; the interpreter, with the enzymes; and the controller appears in the current build-up. We reach two results: the logic is common to *artificial* and to *natural* processes; the *self-reproductive automaton* is within the engineering domain.

The second issue might suggest the following *ideal* experiment. We look at a small robot

[2], basically, made in silicon and oxygen, with other additions, directly found in desert lands. It is programmed to draw, from the air, the tiny amounts of water, ever existing. The sun is only energy source, converted in electrical power, for its internal needs, otherwise stored. Once the first specimen built, its use in the earth deserts could just be opposed by the traditional energy corporations. Being self-duplicating, it will proliferate, basically, in inhabited lands, providing energy and water, exactly the stuffs most locally needed. The project cost is limited to the first item initial invention, design and build-up. Is it too much ? This sand-eater robot has eco-consistent lifecycle; after the grow-up, it begins the energy/water steady production; at the end-of-life, it comes back to sand.

The focus on the biological processes, precisely, aims at circumvent the basic engineers' problems, addressing domains, where the *self-reproductive* abilities are standard feature. The equivalent of the just sketched *ideal* experiment would look at *artificially* modified trees. The trees are beautiful natural machines, capable to take air, water and solar energy, and to transform the all in fuel and noble stuffs. Only hindrance is that we are unable to take profit from the natural achievements, unless destroying the machine. Thus, we have to modify the programme (the *old* genetic code), to create, e.g., trees which do not synthesise cellulose, instead some liquid compound (alcohol or hydrocarbon), to be directly collected through simple contrivances (as for the natural robber), and forwarded by proper draining pipes. This way, the mountainous regions, badly exploited for nice living, will transform into effective «to re-materialise» resorts, with, only, the rangers guardianship.

Today, the bio-fuel is obtained by destroying noble foodstuffs (maize, soy, or the likes), with next generations advantage of lowering the non-renewable sources consumption, but consequences on the world hunger. Better results are proposed by bio-masses, when wastes and scraps are processed. Clearly, the creation of the *artificial* bio-machines is the right procedure. If we look back at the *agricultural* revolution, the, then reached, achievements are astonishing. The cereal selection, assuring grain-holding ears, to permit safe harvesting, requires the highest ability, nearly unbelievable, when the final characteristics are still unknown. The maize *artificial* farming is obligation, due to its closed cobs and heavy grains, without *natural* spreading chance. The segregation of the candidate weed is shocking, if the today Mexican herbaceous plant is the real progenitor. The *ecologic* horror of the old farming/taming outcomes is in proportion of the man-driven selection and promotion of given items, by respect to other ones, starting the in-progress earth anthropic transformation. Shall we demonise old, recent and future revolutions ?

When looking at the natural evolutionism, one finds that the process is pace-wise, by tiny changes, remaining to characterise the issued species members. The diversifying characters are strictly monophyletic, leading to *clades*, keeping the originating genealogic branch, up the final specimen, adapted to live in the given surroundings. The biodiversity richness shows that many species are fit to survive on the earth, providing mix of alternatives for the natural beings continuance. The *agricultural* revolution created the *useful* either *harmful* plants hierarchy, aiming at reducing the latter, and highly concentrating on the former species. The genetic engineering is moving in the same direction, looking at goal-oriented species, by *artificial* enhancers. The *clades* oppose to *clones*, which fully replicate the genes of the parents. The heterosexual lineage is safe process, to permit gene mixing, to avoid dangerous regression. The *clades* are open to further adaptation; the *clones* are, basically, sterile.

Coming back to the robot twofold role, the *clade/clone* interplay is right away evident. The *artificial* evolution permits to orient in the looked-for direction and to speed up the desired processes, in view to obtain *clades*, having all the desired properties. Then, the economy of scale turns to the duplication of *clones*, to reach proper productivity. Both the

---

[2] The hereafter suggested ideal experiments take inspiration from: Dyson F. (1979). *"Disturbing the universe"*, Harper & Row, New York.

operations need to be overseen and governed by the watching robots, through continuous monitoring and reporting, to make easy the prompt protect interventions. As compared with the old *agricultural* revolution, the *cognitive* one takes under total control the *clades'* generation and the *clones'* proliferation. At the moment, the two processes are technologies to come, and the shortly mentioned *ideal* experiments, just, outline the philosophy, and not actual engineering issues. Of course, the *post-global* viability heavily depends on how the *ideal* experiments are transformed into real projects.

These short outlines suggest some conclusions. The anthropic polarisation, in the world facts account, has proved motivations, if we analyse, on history basis, the socio-political relationships among individuals or nations. When the analyses are further pushed, the cosmological conjectures lead to discover that the earth is totally negligible entity, and the evolutionism leads to unveil life generation and intelligence build-up processes as local anomalies, without justifying reasons. The men innate belief, to play the main role, can be used, turning the approach around, say, looking at the scientific conjectures as factual proofs of the timely prospected physics, chemistry, biology, etc. laws, until their falsification is carried over. The truth does not exist by itself, outside the researchers; it, simply, means that suited models are discovered, useful to obtain abstract duplications.

This leads to say that <there is no physics, unless the physicists, etc., exist>. Now, the man is proved to exist, with proper intelligence, and actual abilities to interfere with the surroundings, fulfilling *artificial* transforms and constructs. The anomalies, leading to the earth biosphere and to man-like conceptualisation skill, have very low *a priori* probability, but are *a posteriori* truth, which guarantees the trustfulness to proficient discoveries or inventions. The man civilisation progress is thought to be technical capital issue, and the knowledge build-up is heritage, to be transmitted for future advancement. This is, more or less, the guess behind the weak anthropic principle, authorising the resort to *artificial energy*, *intelligence* and *life*, when these help in providing worthy benefits.

The economy *vs.* ecology dilemma or the artificial to natural assets dialectics is today challenge, to face over-population, over-consumption and over-pollution problems, affecting the life-quality forecasts. The analyses lead to acknowledge the gigantic progresses provided by the «robot age», to start from the instrumental infrastructures (net-concerns, ambient intelligence, etc.) pervasiveness, up to the proficient opportunities of the knowledge entrepreneurship (relational companies, functions provision, etc.). Nevertheless, it is evident that merely intangible value chains cannot suffice to satisfy the people needs, from the every day life material goods and commodities, up to the utilities and energy supplies fit for our habits.

6.3.5. The Proposed Post-Global Outlook

The new opportunities of the lately devised knowledge paradigms (paragraph 6.3.1), the trans-national responsiveness about the future world (paragraph 6.3.2), and the robot technologies potentialities in guardianship (paragraph 6.3.3) and in keen innovation (paragraph 6.3.4), all together, justify aiming at (duly accredited) sustainable growth, as the rational outcome, within the civilised *global village*. In fact, the problem here is to give good reasons for the *post-global* hypotheses, by gathering objective suggestions. The searched way-outs are not in opposition, as the conflict means to selecting partiality, perhaps, in view of coherence, however, risking ineffective and biased paths. Indeed, the engineering perspective favours factual constructs, smoothing all extremisms, and taking out proper useful hints whenever recognised.

The *post-global* outlook, summarised with the previous four paragraphs, is a mix of socio-economic and technology-driven concepts, developed in the spirit to make the growth sustainability (more than desirable) the compulsory path, which, only, allow fair mankind progression. Of course, these concepts are by themselves not sufficient to manage the *post-global* attainments, and the next paragraph 6.3.6 tackles more politico-legal concepts (e.g., the yet to be *hyper-democracy*, or what else might be imagined, guiding the *global village* by suited «government through discussion»), just, providing quite a rough sketch of essential topics, nonetheless, completely out the book reach. Hereafter, it seems valuable to come back on the leading ideas of the all arguments motivating the study, namely, the focus on *four capital assets* that cannot be disregarded to create spendable riches, for the human civilisation development.

The considerations on the subject are, perhaps, implicit in many treatments. In the present analysis, the assumption is explicit, and becomes motivation, to study *economy* and *ecology*, under unified contexts. Because of the assumption critical role, further details of the analysis are, again, discussed, moving backward, from the *natural*, through the *human* and *financial*, to the *technical* capital. All images are questionable, and are collected to promote discussions. Also, the *economy* and *ecology* merging is questionable, excluded the common etymology (*oikos nemein*, homeland manage, or *oikos logos*, home description); in reality, the *global village* concept suggests fascinating connections of the world populations challenge. The strictly *economic hyper-market* is, quite surely, unstable achievement; it happens to foreshadow the *ecologic* common market, critical prerequisite, to make obvious looking at the *rational legality* of the *altruistic hyper-democracy*.

The «natural» contribution is necessary, and omitting its *explicit* account is the economics anomaly, having serious consequences, included the confidence on the *supply and demand law*, exploited without inspection, out of its valid conditioning context. The term «natural», itself, is not without ambiguities. One should notice, for instance, the opposition between «produce», *yield or amount generated by the agricultural processes*, and «artefact», *object made by the man with skill, chiefly, with a view of subsequent use*. In the present analysis, both are *natural* capital, to be, possibly, distinguished because their value added comes from the *agricultural*, either, the *industrial* revolution. The conflict is reasonable, when considering their content in renewable, either, non renewable resources; in both cases, the products are *artificial*, meaning by that the man involvement, needed to modify the *natural* order, if this should be defined, strictly, out the human interventions. In my view, the ambiguity is, explicitly or implicitly, recurrent in many ecologic debates, as if our anthropocentric world could (advantageously) be set free from the man. In my belief, the humanity is part of the *natural* order, and all interventions do not alter the piece of evidence, being, only, to be classified as advantageous or harmful, by singling and assessing the related (short and long time) eco-footprint. The *natural* capital value is the today big puzzle. The twofold duty: resource bookkeeping and assets revenue, are accomplishments far from settled, on which to stake in view of the «active» *post-global* approach credibility.

The *human* capital distinguishes from the *natural* capital, because of the man *intelligence*, phenomenon with relevant consequences, principally, the reasoning ability, which permits the organisation the human knowledge, and the planning of finalised and co-ordinated actions. This leads to recognise the man active role, in the earth transformations, with positive or negative falls-off, still, putting in clear that the habitat is, to great extent, the issues of *artificial* changes. The statement is equivalent to hypothesise some convenient *anthropic* principle, strong or weak, in function of how deeply the surroundings will be transformed, by the man action. The forefront manipulation abilities are clear proof of the transformation issues, but the man workforce is only temporary instrumental side, suitably, replaced by equivalent robot. The background work-organisation capabilities are permanent enabling contribution of the *anthropic intelligence*, making feasible the *artificial* build up of civilised communities. When talking about the *human* capital, focus is on the attendants' qualification, and the front-end prerequisites are, mostly, weird legacy of outdated positions. Anyway, the *human* capital value vagueness is, now, considerably lowered, due to the trade unions settlings and the welfarism policies. The *free market* law is reminiscence, slightly affecting the actual wages. The huge question mark is on how long *sufficient* resources are available for the all citizens of the developing and developed countries.

The *financial* capital is subtle issue of the life in the civilised communities, in which the *money* (directly, and through goods, services, etc. labelling by means of conventional values), plays primary roles, to assure legal transactions and political formats, bearing recognisable worth. Today, the *financial* capital evaluations are totally conventional, having, even, lost all the residual links with the gold bullion. The fact is not without haziness, when dealing with the on the go trade affairs, but it allows wide autonomy, in creating economic instruments, diversifying from the restricted monetary functions. Here, special notice deserves the *hyper-market* set-up, and the interplay giving power to cross-border corporations, with detriment of national governments. The scenario is only possible outcome, perhaps, consistent with multi-centric political realities, without the hegemonic power of to establish the imperialism, under unifying leadership. As first guess, the *hyper-market* is bet to subscribe, at least, because of the efficiency enhancing capabilities. The worth generation mechanism requires, in the modern times, rising sophistication, having the plain stocks and bonds little relevance, as compared with the linked openings to the asset/mortgage backed securities, managed futures, venture capitals, private equities, hedge funds, etc. that set up the crucial part of the *financial* market. The *money manager* is specialist, playing with the interfaced figures of prime brokers, fund raising operators, risk rating agencies, etc., and, the other ways, the general or limited partners, to create value, through the trading of equities and futures, on the worldwide frame. The overall picture rises at dramatic level, after the troubles recently experimented.

The *technical* capital is, perhaps, more evident benchmark, as compared with the *financial* one, permitting to classify a society by the technology level at hand. The yardstick is known since the archaic *agricultural* revolution, when series of peoples started distinguishing their quality of life, beginning the process towards the increasingly *artificial* generation of resources. The numerous conquests are, now, heritage of the mankind all, and it is silly exercise, to claim exact priorities, out of the traditions locally established, from, more or less long, times. The fact is suggesting a similar approach with the *industrial* revolution. This is quite recent phenomenon, and we have the vivid picture of how its different developing rates yield big changes, in the world life-quality distribution. At the moment, the book describes the already occurred switch, from the *scientific*, to the *intelligent* work organisation, later leading to recognise the emerging «knowledge» paradigms. In the earlier «industry» paradigms, the prevalence is on the western world methods, where «industry» is the same as *diligence*. The scope economy in manufacturing, and the intangibles «new economy» value chains, now, modify the entrepreneurial profit, according to the *technical* capital remunerations schemes, shortly summed up in the section 2.2. The intangibles contributions evaluation, as already pointed out, is rather entangled, with (real or presumed) risky ties involving the *financial* capital. An iffy practice resorts to the market *active* steering, to affect the business resources, by the emission of standard or reverse convertible, equity linked, etc. bonds, to foster assets, based on promised (technical) growth. Basically, links are created between the enterprise strategy and the corporate capital, to increase the investment trend. By these ways, the emerging intangible fortunes appear in the accounting records, and become current practice to evaluate the «new economy» business, with positive (leading to previously not-existing offers), and negative (build-up of speculative bubbles) falls-off.

The independent role of the four capital assets, especially, when bookkeeping and balances are recapitulated, is fundamental attainment, before making the *post-global* value chains work. The engineering part in the all business is, surely, huge, but does not become effective, unless the economical instruments and the political consent are clearly established. Thus, to make the picture plausible, the extensive digressions out of the basic technological topics are so important. In between, the positive metrics, making possible the evaluation of the four capital assets, need to be rightly understood, out of the intriguing «demand and supply law». The resort self-consistent balances of exchanged quantities, through the loan at interest or the assets reward, brings to recognise proper *productivity* figures, never to be disjoint from the capital investments. The way permits to establish the **TYPUS** metrics, or other equivalent balancing record, to deal with the «natural capital». Basically, the procedure makes use, as said, of the legal metrology rules, enacting conventional standards, acknowledged over the whole *global village* (today, through inter-state agreements). This point cannot be forgotten.

6.3.6. The Knowledge Society Guides

To end the devised framework, the plain technicalities are, only, limited view, out of the socio-political features, making the sketched scenario plausible. Here, the discussion becomes exceedingly fuzzy, since grounding the *altruism* on sound evidence is hard job, even more hard than thinking up the «knowledge» divide, by respect to the «industry» habits. Nonetheless, some embedded motivations can be figured out. The breakthrough prospect requires acknowledging its urgency, and discovering motivations and protagonists. In that sense, the interest of preserving habitable earth conditions gathers shared consensus, rush adjusted, only.

From the «human» capital standpoint, the lessons taken from history are huge help. The man action in the climate upsetting is accredited, with some confusion, however, as reciprocity should exists: the reviving remedies affect the *artificial* trends, not less than the induced decay. Then, the ecologic calamities foster up-surging actions, at least, if we believe in shared rationality. At the end of every war, the *wise* man takes lessons from the past, and looks for durable antidotes, to plan out more stable political set-ups. Similarly, once recognised the harmful man effects on the environment, the *wise* man singles out the pertinent cures. The new world equilibrium shall, necessarily, be *global*, as the *local* issues little modify the earth future. It shall, as well, be durable, say, with effects lasting for the citizens to come, because temporary benefits are false, and their deceitfulness is bigger, if the momentary advantages are obtained to the expense of larger future damages.

The multi-national corporations' *hyper-market* attitude, in the sketched views, provides untruthful advantages, on the short spans, being urged by stakeholders, and no or slightly controlled by empowered democratic governments. Apparently, as previously remarked, little can be done, if the efficiency achievement is simply deferred to the *market*, up to worldwide size. The past history teaches us that the nation-state capitalism of the "*long global assent*", already, failed, and even at the sub-continent size, permits the ephemeral "*short global assent*" outcomes. The ecologic concern is insurmountable barrier, and the *wise* man ought to realize that adequate interventions are needed. These have, previously, been identifies in the strengthening of supra-national organisations. The result can, chiefly, be obtained by means of the democratic will delegation for the *global village* administration, to obtain balanced *solidarity*, in view of the rational acknowledgement of the best profit sharing.

The issue leads to trans-government authorities, enabling what shall be named *hyper-democracy*, or similar equivalent concept, which characterises on two facts:

- it permits the unified government of the *global village*, at least, so far as the ecologic

regulation is concerned;
- it operates for the safeguard of the generations to come, thus, responding to *virtual* (non-voters) peoples.

The first achievement is not utopia, when we look how the EU is established, gathering countries since centuries in perpetual war, to obtain some advantage, at other people's expense. Evidently, the *altruism* is deemed beneficial, centralising the *economic* policy at the sub-continent level. As well, the *altruism* is mandatory, centralising the *ecologic* policy at the worldwide level. The central authority and local governments interplay is ruled according the *subsidiarity* rules, which do not avoid ambiguity margins. The *hyper-democracy* vs. *autocracy* legitimacy is basic puzzle, because only the member-states parliaments directly respond to voters. In reality, the ruling is deferred to the EU Commission, basically, administrative set-up, yielded by the local countries weighed representation. The democratic control applies at the low level; the upper one is obliged to satisfy the preset bureaucratic rules. To switch to the world dimension will, necessarily, broaden the constraints, making the administrative accomplishments rather entangled: the prospect is bet, worthy to be tried. The *hyper-democracy* is institution open to alternative settings, not to be here discussed. Actually, different democracies have been experimented, even, each one exclusive, e.g., opposing the parliamentary to the people's ones, in which the rulers granted the country's benefit *authentic* interpretation, unfailingly ratified by the (Bulgarian majority of) citizens. The word democracy is, somehow, misused, and the «government through discussion» is well-matched with control forms fixing future targets.

The second achievement is more subtle, because, here, the non-voters' will is involved, being fundamental to look at durable falls-off, in the *altruism* spirit. The argument might turn on side enhancers, to make the guess plausible. The idea is to check the usual lifestyle elements, with their in progress changes:

- the *hyper-democracy* entrepreneurship: the *relational* corporations deserve special attention to the inter-personal interests; the abstraction will quickly turns on the future generations, understanding that the human happiness is in the stability, leaving safe life conditions;
- the *hyper-democracy* infrastructures: the hierarchic clustering of the local, national, continental and worldwide structures is helpful step to assemble ordered organisations, with assigned administrative functions; the political control establishes on bottom-up architectures;
- the *hyper-democracy* market: the local *autarchy* ought to be mitigated by the outer complementary specialisation; the common welfare will exploit the *collective intelligence* (not to be patented), trading the *altruism* value added, which assures wide nomadic spendable riches.

The *altruism* spirit should be the rational answer, once the *hyper-market* and the *cautious headway* scenarios do not, both, provide safe ways-out. The *hyper-democracy* image is quite fuzzy, being loose concept especially in understanding how the political control of non-voters citizens (to come) might operate, for the humanity benefit. Nonetheless the *simple* alternatives are hopeless, and *solidarity* only is well-matched with future targets, involving the today citizens in the build up of more stable frameworks.

At the level of abstract principles, the factual definition «government through discussion» should be extended to the *hyper-democracy* settings. This entails the citizens' freedom, with the capacity and competence to take position into debates, grounded on the subjects' pertinent knowledge and judgement's proper autonomy. The reaching of the shared agreement is, already in *democracies*, elusive question, as it is documented by the NIMBY effect, and the "*conflict of interests*" limitations, making difficult the endorsement of resolutions clashing against the voters' profit. With the ecologic restrictions on the indiscriminate *natural* capital, the attainment of pooled approval requests high civic mindedness and documented awareness of the obligation towards the future generations. How the *hyper-democracy* will deal with immediately acting restrictions, is big question, maybe, aiming at subtle split between «discussion» and «approval», bringing to the *direct approval* the general principles, and enacting *mandatory regulations*, expressly entrusted to inter-state commissions, or experts panels. This way the «discussion» is open to minorities, but the local or sectional interests will not prevent the fair choices, from reaching the general benefit.

The law-making is, maybe, knowledge society duty, not to be further chewed over. The political plausibility of the *altruism* bets adds to the inbuilt soundness of the *post-global* value chains based on the *natural, human, financial* and *technical* capital assets, to the reactive clades/clones «to re-materialise» management, to the robot eco-impact assessment for cautious advice, to the responsibility framework in the market deployment, and to the economy/ecology adjusting knowledge. The all combinations, together, provide rational motivations, to *invent* administrative settings and government systems assuring the mankind future. The all paragraph, anyway, leaves open several basic demands, and, even, seems to contrast current thoughts, requiring the next paragraph digression.

6.3.7. The Post-Global Perilous Uncertainty

In the prospected third scenario, the *post-global* view assumes the active deed of the *world village* citizens, accomplishing responsible measures, in view of the mankind safety. The active awareness does not, simply, mean to recognise that, both, *global* and *no-global* positions fail, but need to give rise to the subsequent socio-political organisations. It means, moreover, to find out and to make possible the suited interventions, keeping, basically, the twin commitment:

- the ecologic concern, based on the assets bookkeeping (paragraph 6.3.5), by **KILT** models and **TYPUS** metrics (paragraph 6.3.3);
- the civic mindedness of the *altruistic* global village inhabitants (paragraph 6.3.6), under fair trans-national ruling (paragraph 6.3.2).

The active *post-global* position is suggested, in the book, to look at ways-out, once the failures are analysed on their critical evidences. Besides, as before noted in the paragraph 1.1.6, the world politico-economical situation offers a long list of intrinsic reasons, showing the end of the «industry» age, even before the ecologic motivations. The listed reasons are likely showing the *global* frailty, making clear that the related epoch is over. On the historical viewpoint, the *post global* age is dated from the XX century end, when the "*short global assent*" parenthesis stops, with the «new economy» crack, and the growth trend reversal, vividly marked by the September 11, 2001 slaughter. On these facts, the economics wobbles, and the forecasts cease to trust in simple mathematical algorithms, too much grounded on steady driving phenomena, as if the universe imperturbability could be axiomatic truth. This means to bring «complexity», where for a long while, the *reductionism* allowed to devise the industrialism steady economical growth.

The mentioned socio-economic intrinsic reasons are plausible starting date of the plain *post global* age beginning. They are not sufficient, however, to call for a drastic divide, the one compelled by the world over-population, over-consumption and over-pollution.

Afterwards, the plain reality happens to be quite different, so that the related *post global* views hang back, without active awareness. In similar way, simple trade and organisational upgrading, along what said in the paragraph 3.2.5, only, provide limited and temporary relieves, likely, to be later recognised as deceptive measures, induced by the current perilous uncertainty.

In fact, by the piling up of dates, the plain *post global* age begins with perilous economical and political uncertainty, even out the impending ecological distress. Lacking the glue to fight the terrorism regimes, the USA-EU alliance is forced to leave out the dreams of the peaceful economic integration; the twin uncertainty, instead, shows the political weakness of prospects, where the economical growth totters. The moral glue against the «terrorism» has loose and passing effects, not providing certainties. On these premises, the *hyper-market* is awkward surrogate. It is the theoretic extremism of the "market", at the worldwide range. Up now, the two "*economic global assents*" developed on imperialistic bases, with the nation-state capitalism, to which all corporations belong. The multi-centric scenery, with no dominating rulers, is, perhaps, captivating hypothesis, on condition, however, to grant the continuity of the expanding industrialism effectiveness. If the long list of intrinsic reasons shows the end of stable growth, the scenery has little respite.

The twin economical and political ambiguity is something going farther on the typical entrepreneurial *risk*. The challenge, at present, is, not only, to construct the business in the known governmental set-up, rather, in the same time, to re-shape the legal contexts, according to the «market» founding principles; only if both the goals are attained, the *hyper-market* effectiveness befalls true. However, the issue defies all accepted entrepreneurship paradigms, entailing the sharp *risk* appraisal, while the yet-to-be multi-national company manager is forced to face trans-border ambiguities, not be dealt with the known evaluation standards. In such conditions, the companies' policies choice becomes fuzzy exercise, and it is rather difficult to devise whether in the *world archipelago*, the political or the economic power will prevail. In the former case, no "*global assent*" establishes at all, otherwise quickly it brings worldwide conflicts, between the incompatible political governments.

The passive, or plain, *post global* age of the *world archipelago* is the picture, at times, evoked, to describe multi-centric sceneries, where self-ruled islands co-exist, covering the whole world. It is factually little different from the *no-global autarchy*, even if with the initial principles in opposition. Anyway, assuming that the *hyper-market* multi-national companies' dominance will not set up, the plain *post-global* age upshot brings to several sceneries, e.g.:

- the consolidation of hegemonic rulers, replicating the "*long/short global assent*" set-ups, each one competing to keep/expand its dominance;
- the world division into equivalent power zones, freezing opponent groups, in multi-centric stand-by, which accept co-ordinated agreements;
- the worst possibly scenery, with the deflagration of internal, regional and international conflicts.

The first scenery is axed on reinstating the USA hegemony, alone or with the side EU help, to fight the *evil*, and to establish the world order, according to well acknowledged roles. On the merely military side, this has, perhaps, credit, not on the economic one, due to the set of emerging industrial countries, with leadership in many critical domains, starting from the communication and information ones. The coalition, based on civilisation fights, could face emergency harassments; it would quickly vanish, when the actual returns cannot repay the efforts. Moreover, here, the «identity» weigh largely outruns the «market» one, and this *post-global* scenery has little chance to last for long. It has to be noticed that, in the western world, the "civilisation conflict" scenery is addressed by several people as forced occurrence, not to drop away, unless sufficient prevention makes clear the on-the-field strength. Then, the military hegemony and related war missions are necessity that put in second line all the ecologic menaces.

The second scenery leads to a (somehow) *hyper-market* alternative. It aims at the integration of localised areas, into clusters, with ruling hubs. The outcome is conceivable, widening the already existing trade agreements, up to fair political unions, thus, solidifying the EU-like experience, into structured national powers. The *world archipelago* image is, possibly, consistent with the *autarchy*, whether local aggregations build-up, on common economical and political affinities. The multi-sectional arrangement presumes: the lack of hegemonic powers, so that the imperialistic hierarchies cannot create; and: the establishment of balanced trading conditions, assuring fair wealth distributions, through voluntary agreements. The two conditions need to continue, maybe, after modifying the archipelago partners, but persisting in the mutual influence assets and the fair commercial steadiness. It is uneasy to figure out the *world archipelago* soundness. Yet, some studies show the today market *partial* integration, carrying commercial flows, concentrated in prevalent domains and areas, and such alternative to the *hyper-market* is, possibly, more desirable outcome. The all is particularly desirable, whether eco-conscious behaviours are coherently added, following proper solidarity (*altruism*) schemes.

The third scenery appears as nightmare, and is, of course, to be acknowledged, to push all populations towards more safe conducts. The plain *post global* age, on the outlined political sceneries frames, does not, however, describe the urgency of the ecological demands. The perilous politico-economical uncertainty, even out of the worst possibly scenery, cannot excuse the lack of perception, face to pressing mankind survival menaces. The mention of the above plain *post-global* sceneries is recalled, to point out that the analyses developed all along the book are far from usual, and are, even, sometimes, underrated or neglected, face to relevance given to the politico-economical questions.

At this point, the comparison between the passive and the active *post-global* views suggests two levels of reflections. The plain (passive) acknowledgement of the *global* approach failure is already implicit in the twin economical and political current uncertainty. There is no need to look at ecology constraints, to understand that the civilisation conflict will prospect dark outcomes to the future generations. The active *post-global* approach, then, is obvious (higher level) reflection, to try the joint solution of the political-economical and technical-ecological threats. The *world archipelago* picture, on that premises, is only partial answer, to be linked to the strong *hyper-democracy* overseeing authority. Merry districts cannot build up, and the all world village populations need to be subject to exactly identical eco-regulations, under fair «government through discussion» settings.

The severe politico-economical uncertainty brings about the unsafe attitude to reject eco-conscious behaviours, as if the natural capital bookkeeping, with resort to coherent metrics, could be ignored. The short considerations of the paragraph deserve especial notice, because such uncertainty is proof of the mankind serious danger that exceeds the traditional «industry» paradigms and obliges to abrupt up-turns. Thus, we cannot confuse side issues (the politico-economical uncertainty), with initiating causes (the natural capital depletion). The *global*, *no-global*, *post-global* parable has to be tracked in its integrity, without stopping at illusory half-way perilous stages.

The active *post-global* age merits to be tackled with absolutely different spirit. The plain *post global* acceptation is passively registering the unsuitableness of the *global* and the *no-global* positions on merely traditional figures, as if the current industrialism could be propagated. The blindness in front to the ecological threat is amazing, even more shocking than the free trade faith in the *hyper-market*, or the parsimony loyalty in the *cautious headway*. Both, these suggest solutions, and *active* positions, when menaces occur; not the plain *post global* acceptation of the on-going trends. That spirit is recovered to reach some conclusion, leaving out the digressions of the present paragraph.

6.3.8. Illative Prospects and Final Conclusion

What can we infer from the *hyper-market* or the *cautious headway* scenarios ? Is it reasonable to look at reliable alternatives, granting sustainable growth (within the above mentioned restricted sense) ? The scenarios reasonableness shall simply refer to the mankind survival horizons and to the life quality levels to come. If one or the other figure is too low, the way out reliability is not sound, and the vision is far from safe. The *post-global* scenario might appear as sturdy issue in the «robot age» spirit, say, as technology-driven solution. The socio-political side happens to be much more questionable, and the *hyper-democracy* setting of the *global village* is, for sure, out of the today world nations and citizens mentalities. My conclusion tries to look positive, and hints (inferred from the past events) show that the men inventiveness is surprisingly extended. The above list of combinations, especially together, shows that looking positive might be judicious.

The *revolutions* have been quoted, many times, as divides leading to radically new life conditions. The *agricultural* revolution does not cease to surprise moving the anthropism (man' culture), spreading all over the lands (in Latin *agri*). Today, the *artificial* vegetation is so common, to make rather difficult singling out zones totally unaffected by anthropism. The *industrial* revolution, we have put as central topic of the study, is rather limited up-turn. The minimalism is, clearly, present in the term «industry» itself, *diligence*, well linked up with the *scientific* work-cycle reductionism. Now, the *industrialism* is already widened, to cover the *intelligent* work-organisation. The «robot age» uniformly incorporates *artificial* energy and intelligence, and the subsequent step towards *artificial* life is within the existing technologies reach, if we, hopefully, believe in *sustainable growth* horizons. The *cognitive* revolution, combining *artificial* energy, intelligence and life, is, at the moment, considered dramatic breakthrough by respect to the acknowledged ideas. It might result, only, a side falls-off of the old established *industrial* revolution, by widening the «robot age» spirit, along the up now sketched lines.

The «robot age» changeful knowledge, whether warning sign of revolution yet to be, or completion of in progress means, is here used to allow devising the *post-global* attitude suggesting stable and safe world progression opportunities. Further to the example five statements, recalled in the first paragraph of the section, more general propositions, such as the following, are noteworthy:

- the *communication/information* technologies are cross-boundary glues, so that the district castling is merely utopia, assuring the each other visibility of the economical and political rights and duties; the *cross-watch* practice is *benefit* (or *emergency*) affecting all future communities and individuals;
- the *world village* unified citizenship makes manifest the common destiny, good or bad fortune; each inhabitant behaviour invasively affects the *only* earth, and the surroundings fragmentation does not establish independent zones, where to vault riches or to segregate poisons and contamination;
- the *cautious headway* run-through needs to be up-dated and understood as the transparency practice, giving view on the in progress transformations, recording all falls-off and related origin/responsibility, and fully granting their *controlled evolution*, under the «robot age» watchful overseeing;
- the *hyper-market* spreading-out has to be brought back to the *internet* (or *world-*

wide-web) technicalities, and needs to be understood as regulated trade of the *four capital assets*, where the *natural resources bookkeeping* is compulsory *world village* authorities commitment and privilege;
- the *hyper-democracy* administration means accepting the *altruism* scheme, as the only way out to mankind survival and fair life quality; the political control is feature with unknown solution, because the citizens' vote is not fit to safeguard the future generations, face to today rigour and sacrifices.

The above listed are example propositions, showing mix of technology driven and socio-political options. The economy/ecology interplay remains the prevalent conditioning factor: if the *artificial* energy/intelligence/life transformations permit yielding plentiful (and safe) spendable riches, even the quizzical *hyper-democracy* administration of the (actually, real) *world village* will find solution. The *altruism*, basically, is equivalent to enacting news ownership rights, radically turned on the heritage protection, and new tax systems, drastically fostering the parsimony. The «robot age», as the initial paragraph five example statements show, details the few technical prerequisites of the hoped breakthrough. The conclusion, here, collects five example propositions, to put forward the civil society changes feasibility, on the condition that the technical capital innovation could support generating riches at the world populations' necessities range.

I like to join at this conclusion a couple of observations, the first quite trivial, quoting the saying: <*fortunæ suiquisque suæ faber est*> [3]; even the *opportunistic* economy does not exist without pickers and hunters, acting in their interest, and disturbing the universe. The second more personal. When working at the SWAN project, I happened to know about the «seaweed cultivation» venture, based on proliferating rightly modified plankton, to make effective the mass fish breeding, in open oceanic areas. The technical connect to the political problems (unless the *global village* is ruled with *hyper-democracy* administration); should the venture work, however, the over-population could (for a while !) become small concern, and the robot watching would grant from over-pollution. In my work, I happened, nonetheless, to find awkward reactions: for instance, the ideal experiment leading to genetically modified trees, directly yielding hydrocarbons, is feared; much less the sand-eater robots, as if the *artificial* silicon-based life is safe, not the *artificial* carbon-based life. In a few words, the «robot age» expands the way the people is perturbing the universe, provides spendable riches even from hostile surroundings (inhabited deserts, steep hills, ocean surfaces, etc.), allows the careful watching of every in progress transformations, and, thus, is worthwhile breakthrough to the *cognitive* revolution, should it be the case, replacing the «industry» minimalism, with the «knowledge» paradigms, explicitly putting *artificial energy*, *intelligence* and *life* together, to obtain resources for the mankind sustainable growth.

The <*fortunæ suiquisque suæ faber est*> saying is reasonable. No issue comes from the giraffe's attitude towards the storms, when these are not temporary gusts, but everlasting bad whether marks. Therefore, the book organises, to distinguish: the world socio-economical growth trends, through the *industry* to the *robot* age; from the personal experience, gradually built up in the engineer's problem solving doings. The general analyses are based on factual remarks:

- the «industry age» is captivating period of the mankind history, because it made possible the creation of spendable riches through *artificial* processes only affected, for a long while, by the men *application* trustfulness;
- the industrialism effectiveness was the result of pretty peculiar conditions, in which

---

[3] *Each one is author of his chance.*

the *western world* style plays the main role, corresponding with a mix of cultural, economic, political, social, etc. characterising traits;
- these peculiar conditions do not exist any more, being unrealistic drawing from the earth finite stocks, and giving back waste and pollution, hence a new mix of traits has to build on, replacing the disproved one;
- the «robot age» provides the new mix of traits, readily offering prop to the «to dematerialise» axiom, and allowing the use of the «to re-materialise» innovations, under the watch and control contrivances transparency;
- this new mix of traits brings to the *knowledge* paradigms (*vs.* the *industry* ones), showing the assessed *industrial* revolution limits, as compared with the (yet to be) *cognitive* breakthrough, granting alternative stocks.

The above short remarks follow, for the most part, technical prerequisites, not sufficient to explain the mankind industrialism parenthesis, and, even less, its end, should this need to fully stop, or to turn towards modified paradigms, with novel market staples to come. The analyses, accordingly, cover additional requirements, coming from the underlying cultural, social and political conditioning backstage. The collected considerations provide, nonetheless, an *engineering* viewpoint, and the search of viable solutions is main goal, perhaps, with anxious commitment.

6.4. AN OUTLOOK AT THE CONTENT

Then, the «robot age» becomes the short form of the hoped breakthrough, yet to come, but with smart advanced premonitions, when we look, more consciously, at the mankind civilisation progression. Following these lines, the book organises, to describe the world socio-economical growth trends, through the *industry*, to the *robot* age, in view to offer plausible account of the *cognitive* up-turn. The intents stimulate to sum up the underlying socio-political features in the build-up of the human civilisation. The three initial chapters involve background concepts, giving non-standard reading of the industrial revolution, from its initiating drivers, along current advances, to freshly recognised up-turns; the next two review professional developments, to exemplify case-studies; the last summarises alternative views, to orient towards more confident outcomes. The analysis of the facts, bringing from the «industry», to the «robot age», might show continuity in the technical capital spirit, using «*artificial intelligence*» added to «*artificial energy*», and opening the way to «*artificial life*». However, understanding the technology trends means to look necessarily at the supporting socio-economics and politico-legal motivations, and this is the non-standard reading of the developed study.

6.4.1. The Industrial Deployment Parable

The background concepts provide an interpretation of the recent civilisation deployment, based on reviewing:

- the «industry» paradigms, as they promoted the western world innovation, to obtain goods, by materials transformation efficiency;
- the industrialism, as manufacture economic system, changing into product-process supply chains, allowing the eco-footprint visibility;

- the «knowledge» paradigms, as «robot age» new characteristics, desirably giving the way to some future up-turn.

The listed concepts explore series of (on the whole) technical requisites, not sufficient to explain the mankind industrialism parenthesis. The analyses, hence, expand on additional facts, broadly, linked to the underlying cultural and social conditioning backstage. The digressions are essential, once acknowledged that the known manufacture economy faces imminent limits, unless the growth path turns towards modified paradigms, with novel market staples to come. The technologies are supposed, in the analyses, to be instrumental aids, requiring, even so, driving sustenance in the socio-political back-motivations.

The chapter 1 locates the «industry» paradigms, as they begun and developed in the worldwide contexts. The considered three aspects, in essence, outline:

- the strong bond with the structured *capitalism*, with, nonetheless the weird omission of the «natural» assets, when the manufacturing business happen to be described; the oversight needs to be acknowledged, to recognize why the common descriptions are lame; definitely, the industrial transformation efficiency is huge drawback, when looking at the «natural capital» stocks withdrawals and at waste/pollution increasing;
- the factually forced origin of the economical breakthroughs, when the old lifestyles were not providing sufficient riches, and alternative ways had to be found, repeating the *agricultural* revolution step, by the *industrial* one, until the step grants sound growth; the mankind history teaches us that the «industry» patterns are rather specialised opportunities, enjoying reliable and efficient outcomes into steady outer frames;
- the cultural peculiarity of the basic «industry» paradigms, making them to be the characterising feature of the socio-political western world lifestyle, notably, when the «diligence» attribute *minimalism* is retained, to specify the *scientific* work-organisation *reductionism*; the amazing features aims at side effects on the people life, nicely described by the demand/supply *law*, enabled in competitive *free* markets.

The *industrial* revolution impressive falls-out has marked the recent history, and the industrialism develops into the, today, reference *economical systems*. The chapter 2 offers the bird eye view of how the spreading is conditioning our way of life, according, fundamentally, to the reasons sketched in three sections:

- the work organisation changes, market (and technology) driven, off-setting the original «industry» *minimalism*, to exploit robot complexity, aiming at the economy of *scope* (in lieu of *scale*) and to the lifecycle product-service supply chains; the producers' lifestyle responsibility is significant issue, to include the *service engineering* and the *reverse logistics* as entrepreneurial requirements, to be incorporated;
- the entrepreneurial adjustment, to operate by the *product-service* delivery, on lifecycle span and end-of-life *take-back*, based on the *knowledge* value added parallel provision flow, making resort to the networked organisation efficiency; the function/facility market, joined to the networking options, allows setting apt infra-structures, up to the knowledge entrepreneurship of fully integrated *extended* enterprises;
- the *ecology* new demands, to establish bounds on the earth stocks looting and environment pollution, requiring, for all supply chains, compulsory recovery targets and eco-impact charges, which shall affect the all earth *global village*; the *resource*

manager is paradigmatic engineering function, required by the eco-sound *product-service* delivering, showing the already devised *industrialism* adaptivity.

The industrialism, hence, happens to modify from the old «industry», to the new «knowledge» paradigms, making clear, on one side, that the spendable riches obtained by transforming the earth resources into waste and pollution do not have future, and, the other side, that the «robot age» can provide value added, by the «to de-materialise» axiom, with intangible staples, and might prospect the «to re-materialise» axiom breakthrough, made possible by the *cognitive* innovation. The chapter 3, thereafter, provides an outlook all over the «knowledge» conditioning frame, discussing its three characterising prospects:

- for the technical frame, the net-concerns, with function/facility up-dating, and productive *flexibility* setting/fitting are emerging organisations, having potentials linked to the new «knowledge» paradigms;
- for the socio-economical frame, the capitalism trends, from city-states, to nation-states (up to the sub-continental size) is already gone through, to shape out local and global (citizens' consistent) quality of life;
- for the trade setting frame, the industrialism (we know) unsuitableness has to go through along the *global*, or *no-global*, or *post-global* track, opening questioning prospects, not without severe consequences on us.

The lessons, already taught by the «knowledge» paradigms, are well identified by the above three prospects, making clear some embedded corollaries:

- the *flexibility* of the *intelligent* work organisation allows experimenting the «complexity» of the on-process decision keeping, within (previously) fully structured surroundings; the resort to robots, here, only means to replace the *scientific* organisation front-operators, at the low instrumental level;
- the parallel *political* parable shows the tight relation of market efficiency and parliamentary democracy, up to the *ecology-economy* dilemma, in which the *global village* administration compels to look at protecting the *future* citizens, exploring the *hyper-democracy*, as option of «government through discussion», fully praising the *altruism*;
- the socio-economical track, similarly, finds the *global/no-global* conflict, and linked *hyper-market/autarchy* unfair outcomes, unless the *post-global* opportunities build eco-conscious frames, granting tangibles bookkeeping and knowledge entrepreneurship, through *extended* enterprise's synergic options, consistent with *solidarity* issue options.

The prospected broad analyses show viewpoints, up now, only in small parts acknowledged. The division in *industrial* vs. *developing* countries/societies is due to modern paradigms of the mankind growth, occurred after the recent *industrial* revolution. The «industry» paradigms, at least in their original form, lead to rather questionable growth, being based on the *transformation effectiveness*, through which the «natural capital» is withdrawn, to be reshaped in worthy consumables, quickly disposed into waste and pollution. Since the earth is, basically, finite and closed system, the wealth growth along such ideas will not last for long.

The today over-population, over-consumption and over-pollution bring facing the lane end, and the *global village* riches are useless bet, with no issue. The book suggests alternatives, in which the «knowledge» paradigms might support further wealth build-up, based on the value chains moved in intangibles, and, possibly, on the *cognitive* breakthrough, where the *transformation efficiency*, this time, applies not only on material inanimate resources, but also on *artificial* life deployments. Such breakthrough is tackled by the *post-global* opportunities, once removed the *global/no-global* opposed unfairness.

6.4.2. The Robot Age Helpful Advances

The «robot age», out of the viewpoints summarised into the recalled general analyses, presents with quite peculiar technicalities, already well established. The book continues with two chapters, based on engineering considerations, directly connected to the personal investigations, at the industrial robot design research group at the University of Genova, Italy. In the average, the activity gives hints on how the robot technologies help merging the information, to enhance the material value chain, adding enhanced shine to the manufacture economy we know.

The engagement, accordingly, is maximally turned to the *instrumental* robotic field, where all conceived devices are task-driven, and operate into domains, with soundly specified conditioning elements. The chapter 4 summarises some typical traits in manufacturing robotics, gathered into three sections:

- the *intelligent* work organisation outlines, with the linked on-line decision patterns and promotion aids; the flexibility and on-process choice tools are winning up-turn, rather than the un-manned factory, in actual fact, already working, also, with the fixed automation;
- the *flexible manufacturing* functions and facilities, with the related shop-floor logistics and structured duty surroundings; the shop-floor logistics is nice example, to replace the optimal *materials flow*, by timely apt choices, steered through the *information flow*, by the on-process decision logic;
- the *robot technologies* integration, with the linked design methodologies and duty-setting information aids; the robot is complex hardware/software mix, by which the on-duty task specification and job allotment are design accomplishments, asking the lifelong developers' awareness.

The *industrial* robotics is powerful way, to orient the engineering activity in view of *integrated design* paradigms, where the equipment on-duty performance is characterising constraint, to reach the supply chain effectiveness. The *flexibility* of the *intelligent* work organisation allows the pace-wise incorporation of the on-process decision-keeping style, based on well structured knowledge, going back at *scientific* work organisation reductionism. The appropriate rig and method design, in such frame, provides evidence to the *product-process simultaneous* engineering advantages, first step towards the broad *product-process-environment-enterprise* developments, leading to the *extended* enterprise concept.

The industrial robotics is exemplary instance, to assure the smoothest way to the «knowledge» paradigms, in the engineering activity. Its surroundings continue the reductionism of the western world cultural background and organisation style. The context, chiefly, explains the industrialism coherence, preserving its narrow timing and localisation issues in the mankind history. It might be used, even so, to explore alternative outlooks on how restructuring the manufacture business, to include «complexity». The industrial robotics minimalism, then, simply means to exploit the task-driven specialisation, as *design trick*, to take care of complexity through planned, timely enabling, features. The service robots development, then, takes benefit of the procedural patterns, suitably adapted to the specific problem solving requirements.

The short comments on the *service* robotics, accomplished in chapter 5, thus, follow as example explanatory review of case-engineering outcomes, confirming sound problem-solving helpfulness. The three sections embed:

- the method aids, to obtain the robot task-bend, exploiting helpful tools; the operation setting is the key challenge to make the equipment winning; the side implements are in the interface friendliness and in the job monitoring easiness;
- brief survey of example remote-operated robot equipment; the list gathers: 2MPA, *mixed-mode postal automation*; 3MD-VAS, *machine monitoring-maintenance by vibro-acoustic signatures*; ROBOCLIMBER; SBC, jacket *sub-bottom cutting*; MIDRA, *micro-tunnelling drainage*; RAMS, *robot aided mine sweeping*; CRHA, surgery *co-robotic handling appliance*; the issue are, always, motivated by existing engineering specifications;
- bird eye view on characteristic autonomous task-driven apparatuses; the list covers: GRA, *greenhouse robotic automation*; AMIR, *ambient intelligence ruling*, GECKO-COLLIE; SARA, *system for automotive refuel automation*; ROBO-RANGER; SWAN, *sub-sea wobble-free autonomous navigator*; RL-ELV, *reverse logistics for end-of-life vehicles*; always, moving from the embedded-stimuli of the explanatory cases.

The example achievements, very shortly sketched, provide preliminary hints on how the «robot age» mind helps merging tangible and intangible value added. In the sketched context, the robot deployments are instrumental organisation and cultural ingredients, supporting the «knowledge» paradigms alternatives, once the «industry» ones minimalism does not, any more, assure profit with reductionism. The changeful «knowledge» paradigms consent to look at the technically coherent «robot age» issues, on condition that the socio-political frames to come are nicely turned towards *active post-global* views. The engineering know-how gives visions over the instrumental robotic aids, from manufacture, to the service surroundings; these are deemed worthy spurs, to self-consistent policies, properly centred on the transparent bookkeeping of the managed «natural» capital.

The last chapter summarises the prospected topics and concepts, with especial attention on the realistic scenarios that the *global, no-global, post-global* views, most likely, engage the in progress mankind. The three devised scenarios are:

- the *hyper-market* arrangement, presided over by cross-border corporations, whose efficiency enhancement is sternly linked to the nation-state power crash, opening unknown set-ups, mostly, devoid of democratic control;
- the *cautious headway* road, based on strict *ecologic* concern, with frugality in mind, and focus on the autarchic self-sufficiency, to re-create thrifty protected zones, banning all (suspicious) technologies, as potentially risky;
- the *post-global* approach, looking at the «robot age» sustainable growth, aiming at *cognitive* breakthrough, to join the «to de-materialise» and «to re-materialise» axioms, with *altruism*-driven political world organisations.

The considerations in the book do not provide certainties. The industrialism absence of future is only statement given for sure. The *cognitive* revolution is a desirable hypothesis, and, in the mean time, the attention on the value chains with crucial intangible content is technology-driven back-up. The «robot age» is prised on its engineering gist. The durable growth is defy to accept. The «knowledge» driven opportunities cannot be neglected, if the demands challenge is faced, with resort to options, which need dramatically to modify the current paradigms of the industrial revolution.

The anthropic guesses, openly alleged in the book, lead to accept new cultural divides, each time the deployment of «technical» capital innovation is necessary for the man's survival. The *agricultural* revolution, in the pre-history ages and the *industrial* one, in recent times, are, definitely, multi-farious phenomena, including mixes of alternatives, to lead at more effective settings. The *industrial* revolution is somehow limited by the original minimalism (*diligence/industry* equivalence) of the *scientific* work organisation reductionism. Today, the innovative «robot age» vision provides the sound changeful «knowledge» paradigms, moving the old patterns, into «complexity» alternatives, more suited, to face the newly eco-protection requirements, enacted to approach sustainable growth. Unfortunately, technically coherent prospects are (necessary, but) not sufficient, to promote the crucial breakthrough, which will start new socio-political structures.

The fact is carefully acknowledged along the book, mixing the economic and political motivations, to the technological remarks. The all organises as «survey», providing the broad treatment of topics, which would be accessible to all people, entailed in the sketched facts. The eco-protection is expanding into numberless regulations, whose technical backdrop has strong political and economic flavour, making uneasy their parting, within the enacted measures. Likely, the «industry» patterns of the affluent society wealth build-up are hazardous market-driven steps, because the growth technically clashes against the practice to turn raw materials into waste and pollution. As noted, the industrial revolution is recent and sectional achievement, characteristic of the western world lifestyle, due to specific cultural peculiarities, not to stand for universal options. Now, its total up-turn is forced by the ecology demands; all contrary viewpoints are inconsistent, requiring careful up-grading, before providing helpful outcomes.

The background analyses are the book core part, leaving merely, explanatory function, to the case applications survey. The interwoven reliance of economics on technologies corresponds to some presumed cultural prominence all along the human progress. Concepts, such as *capitalism* or *democracy*, are strictly linked with the mankind evolution, and, if we believe in some anthropic principle, we need to find exacting cultural correlations and proper appraisals, justifying all the socio-political frames: the *capitalism* changes or *democracy* trends. The *industrial* revolution sectional achievements cannot escape similar imperatives; then, the *cognitive* breakthrough, should it be effective way out to the industrialism lack of future, will develop, when and where the cultural prerequisites establish on suited and receptive surroundings.

These surroundings are built by the world citizens. They might establish with sectional characteristics (e.g. the «industry» cultural connotation of Chap. 1); can evolve, incorporating new requests (e.g., the *industrialism* adaptation to lifestyle supply chains of Chap. 2); will profitably orient to safer eco-protection (e.g., the «knowledge» society prospects of Chap. 3). The technologies cultural interplay has rather multifaceted progression. The *reductionism* of the early «industry age» has, already, been dropped aside, with the economy of scope, permitting to cover the lifestyle supply chain demands. Looking at the changes, on merely technical options, the «robot age» brings to achievements, such as the ones outlined in the Chap. 4 and 5. Looking at the changes, on basically cultural premonitions, the anthropocentric understanding of the human progression suggests exploring the prospected new opportunities. These, summing up, match up with the *artificial energy, intelligence* and *life* exploitation, aiming at growth through «complexity». Indeed, the *ecology* means the end of the affluent society we know. The Chap. 6 sketches out the coming scenarios, to help choosing.

REFERENCES

CHAPTER 1. THE «INDUSTRY» PARADIGMS

Section 1.1.

The analysis on the on the industrial evolution is based on sets of ideas, widely available in the existing literature. The first section of the chapter addresses the central conditions, necessary for the manufacture effectiveness, discerning four capital assets: technology, finance, workforce and raw materials. This is original hypothesis, however, coherently derived from current approaches. The following references are suggestions for further reading.

Anderson J.C. & Narus J.A. (1991). *"Partnering as focused market strategy"*, California Management Review, April.
Antonelli C. (1995). *"The economics of localised technological change and industrial dynamics"*, Kluwer Acad. Pub., Boston.
Atwood M. (2008). *"Playback: debt and the shadow side of the wealth"*, House of Anansi Press, New York, p. 280.
Bacevich A.J. (2008). *"The limits of power: the end of American exceptionalism"*, Metropolitan Books, p. 206.
Best M.H. (2001). *"The competitive advantage"*, Oxford Uni. Press, Oxford.
Clement A. (1994). *"Computing at work: empowering actions by low levels users"*, ACM Communications, vol. 37, n° 1.
Collis B., Carliss Y.C. & Kim B. (1997). *"Managing in an age of modularity"*, Harvard Business Review, Sept. 1.
Corafas D.N. (2001). *"Enterprise architecture and new generation information systems: enterprise architecture and new generation information systems"*, CRC-St.Lucie.
Davenport T.H. & Short J.E. (1990). *"New industrial engineering: IT and business process redesign"*, Sloan Management Review, vol. 31, n° 4.
David P. (1994). *"Positive feedbacks and research productivity in science: re-opening an another black-box"*, O. Granstrand, (Ed.) Economics and Technology, Springer, Berlin.
Davidson W.H. (1993). *"Beyond engineering: the three phases of business transforms"*, IBM Systems J., vol. 32, n° 1.
Diamond L. (2008). *"The struggle to build free societies throughout the world"*, Times

Books, New York.

Fleischmann M. (2001). *"Quantitative models for reverse logistics"*, Springer, Berlin, p. 323.

Forey D. (2004). *"The economics of knowledge"*, MIT Press, Cambridge.

Frederix F. (2001). *"Enterprise co-operation leads to extended products and efficiency"*, Production and Operations Management Conf., Orlando, USA, Mar. 30-Apr. 2.

Granstrand O., Patel P. & Pavitt K. (1997). *"Multi-technology corporations: why they have 'distributed' rather than 'distinctive core' competencies"*, California Management Review, July 1.

Hall B., Grilliches Z. & Hausman J. (1986). *"Patents and R&D: is there a lag?"*, Intl. Economic Review, pp. 265-283.

Hazletine B. (2001). *"Field guide to appropriate technology"*, Harcourt Pub.

Johnston R. & Lawrence P.R. (1988). *"Beyond vertical integration: the rise of the value-adding partnership"*, Harvard Business Review, July 1.

Kaebernick H., Anityasari M. & Kara S. (2002). *"Technical and economic model of end of life options of industrial products"*, J. Environment Sustainable Development, vol 1-2, pp. 171-183.

Laudon K.C., Laudon J.P. (1996). *"Management information systems: organisation and technology"*, Prentice Hall Intl., Upper Saddle River.

Lindfors L.G., Christiansen K. Hoffman L. et al. (1995). *"The nordic guidelines on life-cycle assessment"*, Nord, Nordic Council of Ministers, n° 20, Copenhagen.

McDonald R. (2008). *"The death of the critics"*, Continuum, New York, p. 160.

Mandelbaum M. (2008). *"Democracy's good name: the rise and risks of the world's most popular form of government"*, Public Affairs, New York.

Mehli-Qaissi J., Coulibaly A. & Mutel B. (1999). *"Product data model for production management and logistics"*, J. Computers and Industrial Engineering, vol. 37, n. 1-2.

McGahan A. (1999). *"Competition, strategy and business performance"*, California Management Review, April 1.

Michelini R.C. (2008). *"Knowledge entrepreneurship and sustainable growth"*, Nova Sci. Pub., New York, p. XX-326.

Mosovosky J., Dickinson D. & Morabito J. (2000). *"Creating competitive advantage by resource productivity, eco-efficiency and sustainability in the supply chain"*, IEEE Intl. Sym. Electronics and Environment, San Francisco, CA, May 8-10, pp. 230-237.

Nelson R.R. & Winter S.G. (1993). *"An evolutionary theory of economic change"*, Harvard University Press.

Papanek V. (1971). *"Design for the real world: human, ecology and social changes"*, Pantheon Book, New York.

Ryan C. (1998). *"Designing for Factor 20 Improvements"*, J. of Industrial Ecology, MIT Press, Vol. 2, No. 2.

Schoensleben P. (2004)."*Integral logistics management: planning and control of comprehensive supply chains*", 2 ed., CRC Press, Boca Raton, p. 992.

Simpson G.G. (1964). *"This view of life: the world of an evolutionist"*, Harcourt, Brace & World, New York, p. 213.

Sinnott E.W. (1955). *"The biology of the spirit"*, Viking Press, New York.

Strijbos S. & Basden A. (2006). *"In search of an integrative vision for technology: interdisciplinary studies in information systems"*, Springer, London, p. 310.

Vernadat, F.B. (1996). *"Enterprise modelling & integration: principles & applications"*,

Chapman & Hall, p. 513.

Wingand R., Picot A. & Reichwald R. (1997). *"Information, organisation & managing: expanding markets and corporate boundaries"*, John Wiley, Chichester.

Wise R. & Baumgartner P. (1999). *"Go downstream: the new profit imperative in manufacturing"*, Harvard Business Review, Sept. 1.

Womack, J.P., Ross, D. & Jones, D.T. (1990). *"The machine that changed the world"*, Rawson Ass. New York.

Section 1.2.

The second section provides a short outlook on the wealth accumulation and past market staples; the growth sustainability, however, shows that the experimented paradigms face, in the short future, abrupt limitations, and the «knowledge» market option is devised. The sketched outlines are inferred by current investigations, and the next literature is notable example domain reference.

Abe J.M., Dempsey P.E. & Basset D.A. (1998). *"Business ecology: giving your organisation the natural edge"*, Butterworth-Heinemann, London.

Amin S. (1990).*"Maldevelopment: anatomy of a global failure"*, Zed Book, London, pp. 244.

Braungart M. & Engelfried J. (1993). *"The intelligent product system"*, Bulletin EPEA, Hamburg, p. 36.

Cahn E., Rowe J. (1992). *"Time dollars"*, Rodale Press, Emmaus, PA, pp. 272.

Camarinha-Matos L.M., Afsarmanesh H. & Rabelo R.J. (2000) *"E-business and virtual enterprises: managing business-to-business co-operation"*, (Eds.) Kluwer Pub.

Casti J.L. (1991). *"Searching for certainty: what scientists can know about the future"*, Morrow, New York.

Charbonneau S. (2006). *"Droit communautaire de l'environnement"*, L'Harmattan, Paris, pp. 295.

Charter M. & Tischner U. (2001). *"Sustainable solutions: developing products and services for the future"*, Greenleaf Pub., Feb.

Cunha M.M., Putnik G.D. (2006). *"Agile virtual enterprises: implementation and management support"*, IDEA Group Pub., p. 382.

Dembinski P.H. (2008). *"Finance servante ou finance trompeuse ?"*, Desclée de Brouwer, p. 206.

Ferguson N. (2008). *"The ascent of money: a financial history of the world"*, Penguin Press, p. 442.

Gilpin R. (1981). *"War and change in world politics"*, Cambridge Uni. Press, Cambridge.

Hout T.M., Porter M.E. & Rudden E. (1982). *"How global companies win out"*, Harvard Business Review, Sept. 1.

Jansson K. & Thoben K.D. (2002). *"The extended products paradigm: an introduction"*, Intl. Conf. on Design of Information Infrastructure Systems for Manufacturing. Osaka, Japan, pp. 39-48.

Jha P.S. (2006). *"The twilights of the nation states"*, Pluto Press, London, p. 659.

Jolly V.K. (1997). *"Commercialising new technologies"*, HBS Press Book, Sept. 26.

Kinsley M. & Lovins L.H. (1995). *"Paying for growth, prospering from development"*,

Rocky Mountain Inst. Pub., Denver.

Latouche S. (2004). *"Survivre au développement: de la décolonisation de l'imaginaire économique à la construction d'une société alternative"*, Mille et une Nuits (Edit), Paris, pp. 126.

Matsumoto S. & Yagi J. (1999). *"Study on post mass production paradigm, PMPP, in the case of construction industry"*, First Intl. Symp. on Environmentally Conscious Design and Inverse Manufacturing. Eco-design '99, Tokyo, Japan.

Meijkamp R. (1997). *"Changing consumer needs by eco-efficient services: an empirical study on car sharing"*, Conf. towards Sustainable Product Design, The Centre for Sustainable Design, The Surrey Institute of Art and Design. 17 July.

Michelini R.C. (1976). *"XIV BIAS foreword: automation and resources utilisation"*, Intl. Conf. Automation & Instrumentation, FAST, Milano, 23-24 Nov., pp. 1-6.

Morris C.R. (2008). *"The trillion dollars meltdown: easy money, high rollers and the great credit crash"*, Public Affairs, London, pp. 224.

Nevens T.M., Summe G.L. & Uttal, B (1990). *"Commercializing technology: what the best companies do"*, Harvard Business Review, May 1.

Normann R., Ramirez R. (1993) *"From value chain to value constellation: designing interactive strategy"*, Harvard Business Review, July 1.

Orbinski J. (2008). *"An imperfect offering: humanitarian action for the twenty-first century"*, Walker & Co., London, p. 448.

Ottman J.A. (1998). *"Five strategies for business reinvention: development of sustainable products and services"*, J. Emerging Opportunities, Vol. 5, No. 5, pp. 81-89.

Partant F. (1997). *"La fin du développement"*, F. Maspero, Paris, pp. 186.

Pezzey J. (1989). *"Economic analysis of sustainable growth and sustainable development"*, World Bank, WP 15, New York, pp. 88.

Redclift M., Woodgate G., eds. (1997). *"The international handbook of environmental sociology"*, Edward Elgar, Cheltenham, pp. 512.

Sachs W. (1999). *"Planet dialectics"*, Zed Book, London, pp. 226.

Samuelson P.A., Nordhaus W.D. (1985). *"Economics"*, McGraw Hill, New York.

Sen, A.K. (1981). *"Poverty and famine: an essay on entitlement and deprivation"*, Oxford Univ. Press, Oxford.

Vachon R., ed. (1988). *"Alternatives au développement: approches interculturelles à la bonne vie et à la coopération internationale"*, Centre interculturel Monchanin, Montréal, pp. 372.

Volterra V. (1926). *"Variations and fluctuations of the numbers of individuals in animal species living together"*, in R.M. Chapman, Ed., 'Animal ecology', McGraw Hill, New York.

Weizsäcker E.V., Lovins A.B. & Lovins L.H. (1997). *"Factor four: doubling wealth, halving resource use"*, The new Report to the Club of Rome. London: Earthscan Publications Ltd.

Woods D.D., Hollnagel E. (2006). *"Joint cognitive systems: patterns in cognitive systems engineering"*, CRC Taylor & Francis, Boca Raton, p. 232.

Section 1.3.

The third section presents a short analysis to explain the original industrial revolution as outcome of the western world cultural background and political lifestyle. These issues, too, have well-established traditions, but are prospected as premonition of current trends revision, if «complexity» will replace the «reductionism». The listed further reading will help explaining the suggested point of view.

Barmé G.R. (2007). *"The forbidden city"*, Harvard Uni. Press, Boston, p. 251.
Bass G.J. (2008). *"Freedom's battle: the origin of humanitarian interventions"*, Knopf, New York, p. 509.
Becker J. (2007). *"City of heavenly tranquillity: Beijing in the history of China"*, Oxford Uni. Press, London, pp. 371.
Bhagwati N. (2004). *"In defence of globalisation"*, Oxford Univ. Press, New York, p. 298.
Bowles S., Gordon, D.M., Weisskopf T.E. (1983). *"Beyond the waste land: a democratic alternative to economic decline"*, Anchor Press Doubleday, Garden City.
Burgelman R.A. & Tzuo T. (1997). *"Asymmetric digital subscriber line: prospects in 1997"*, Stanford University, Aug. 1.
Dorf R.C. (2001). *"Technology, humans and society: toward a sustainable world"*, Harcourt Pub., 2001.
Dosi G. (1982). *"Technological paradigms and technological trajectories: a suggested interpretation of the determinants and directions of technological change"*, Research Policy, vol. 11, pp. 147-162.
Doz Y.L. & Hamel G. (1998). *"Alliance advantage"*, HBS Press Book, July 1.
Eldredge N. (2002). *"Life on earth: an encyclopaedia of biodiversity, ecology and evolution"*, ABC-CLIO, Inc. Santa Barbara.
Friedman T.L. (2005). *"The world is flat: a brief history of the twenty-first century"*, Farrar Straus & Giroux, New York, p. 488.
Godehardt N., (2008). *"The Chinese meaning of just war and its impact on the foreign policy of the People's Republic of China"*, Giga Working Papers, ssrn.com/abstract= 1287161.
Gottschall J. (2008). *"Literature, science and the new humanities"*, Palgrave MacMillan, New York, p. 240.
Harrison A.W. & Reiner R.K., Jr. (1993). *"The influence of individual differences on skill in end-user computing"*, J Management Information Systems, vol. 9, n° 1.
Hillary R. (2000). *"Small and medium sized enterprises and the environment business imperatives"*, Greenleaf Pub., Feb.
Hitchings H. (2008). *"The secret life of words: how English became English"*, John Murray, London.
Huang Y. (2008). *"Capitalism with Chinese characteristics: entrepreneurship and the state"*, Cambridge Uni. Press, New York, p. 366.
Jacquart R., Ed. (2004). *"Building the information society"*, Kluwer, Acad. Pub., Boston.
Kagan R. (2003). *"Of paradise and power: America and Europe in the new world order"*, Alfred Knopf, New York.
Kanter R.M. (1994). *"Collaborative advantage: the art of alliances"*, Harvard Business Review, July 1.
Kherdjemil B., Panhuys H., Zaoual H. (1998). *"Territoires et dynamiques économiques"*,

L'Harmattan, Paris, pp. 228.
Kothary R. (1989). "*Rethinking development: in search of human alternatives*", New Horizons Press, NJ, pp. 233.
Malherbe M. (1995). "*Les langages de l'humanité*", Robert Laffont, Paris, pp. 1734.
Meadows D.H. (2004)."*Limit to growth: the 30 years update*", Chelsea Green, Boston, pp. 368.
Mendes C., Castoriadis C. (1977). "*Le mythe du développement*", Seuil, Paris, pp. 277.
Menzies G. (2004)."*1421: the year China discivered America*", Harper Perennial, New York.
Michelini R.C., Capello A. (1985) "*Misure e strumentazione industriali: segnali e strumenti di misura*", UTET, Torino, p. 415.
Michelini R.C. (1992). "*Decision anthropocentric manifold for flexible-specialization manufacturing*", 4th ASME-ISCIE Intl. Symp. Flexible Automation, SanFrancisco, July 12-15, vol. 1, pp. 467-474.
Michelini R.C., Razzoli R.P. (2000). "*Affidabilità e sicurezza del manufatto industriale: la progettazione integrata per lo sviluppo sostenibile*", Tecniche Nuove, Milano, pp. 278.
Naess A. (1989). "*Ecology, community and lifestyle: an eco-sophy outline*", Cambridge Univ. Press, Cambridge.
Nandy A. (1987). "*The intimate enemy*", Oxford Univ. Press, Bombay, pp. 194.
Nisbett R.E. & Ross L. (1980). "*Human inference: strategies and shortcoming of social judgements*", Englewood Cliffs, Prentice Hall.
Nisbett R.E. (2003). "*The geography of thought*", Free Press, New York, p. 263.
Norenzayan A. (1999). "*Rule- & experience-based thinking: cognitive consequences of intellectual traditions*", Univ. Michigan Rpt., Ann Arbor.
Partant F. (1997). "*La fin du développement*", F. Maspero, Paris, pp. 186.
Pin M.-X. (2006). "*China's trapped transition: the limits of developmental autocracy*", Harvard Unii. Press, Cambridge.
Prestowitz C. (2005). "*Three billion new capitalists: the great shift of wealth and power to the east*", Basic Book, New York.
Polanyi K. (1991). "*The great transformation: the political and economic origin of our time*", Beacon Press, Boston, pp. 315.
Putman R.J. (1994). "*Community ecology*", Chapman and Hall, London.
Riegel K.F. (1975). "*Dialectical operations: the final period of cognitive development*", Human Development, vol. 18, p. 430-443.
Roarch S.S. (1984). "*Industrialisation of information economy*", Morgan Stanley, New York.
Robertson J. (1980). "*The same alternative: a choice of futures*", (2nd Ed.) River Basin Pub., St. Paul.
Royce J. (1908). "*The world and the individual*", Macmillan, New York.
Rumbaugh J. & Blaha M. (1991). "*Object oriented modelling and design*", Prentice Hall, Englewood Cliffs.
Sachs W. (1999). "*Planet dialectics*", Zed Book, London, pp. 226.
Salter W.E.G. (1996). "*Productivity & technical change, 2nd ed.*"; Univ. Cambridge, Dept. Appl. Economics, Monograph 6; Cambridge Univ. Press. London.
Schelling T. (1971). "*Dynamic models of segregation*", J. Mathematical Sociology, vol. 1, pp. 143-186.
de Shalit A. (1998). "*Why posterity matters*", Rutledge, London.
Smith RL. (1996). "*Ecology and field biology*", 3 ed., HarperCollins, New York.

Stiglitz J.E., ed. (2001). *"Attacking poverty: world development report"*, Oxford Univ. Press, New York.

Stiglitz J.E., ed. (2002). *"Can anyone hear us? - Crying out for change - From many lands"*, World Bank Reports 1/2/3, Washington.

Stitch S. (1990). *"The fragmentation of reason"*, MIT Press, Cambridge.

Sussman R. (1997). *"The biological basis for human behaviour"*, Simon and Schuster Custom Pub., New York.

Teune H. (1988). *"Growth"*, Sage Publication, London, pp. 141.

Thaler, R. (1994). *"The winner's course"*, Princeton Univ. Press, Princeton.

Tomiyama T. (1999). *"The post mass production paradigm"*, Symp. Environmentally Conscious Design and Inverse Manufacturing. Eco-design '99. Tokyo, Japan.

Toussaint S. (2008). "Humanismes de Ficin à Heidegger", Les Belles Lettres, Paris, p. 334.

Triantaphyllou R.E. (1999). *"Procedures for the evaluation of conflicts in rankings of alternatives"*, J. Computers and Industrial Engineering, vol. 36, n. 1.

Tweed R.G., Lehman D. (2002). *"Learning considered within a cultural context: Confucian and Socratic approaches"*, American Psychologist, vol. 57, p. 89-99.

Vachon R., ed. (1988). *"Alternatives au développement: approches interculturelles à la bonne vie et à la coopération internationale"*, Centre interculturel Monchanin, Montréal, pp. 372.

Wang D.J. (1979). *"The history of Chinese logical thought"*, People's Press, Shangai.

Whorf B.L. (1956). *"Language, thought and reality"*, J. Wiley, New York.

Williamson M. *"High-tech training"*, Byte, Dec. 1994.

Winchester S. (2008). *"The man who loved China: the fantastic story of the eccentric scientist who unlocked the mysteries of the middle kingdom"*, Harper Collins, p. 336.

Witkin H.A., Berry J.W. (1975). *"Psychological differentiation in cross-cultural perspectives"*, J. Cross-Cultural Psychology, vol. 6, p. 4-87.

Worster D. (1988). *"The end of the earth: perspectives on modern environment history"*, (Ed) Cambridge Univ. Press, New York.

Yates J.F., Lee J., Bush J. (1997). *"General knowledge overconfidence: cross-national variations"*, Organisational Behaviour & Human Decision Processes, vol. 63, p. 138-147.

Zakaria F. (2008). *"The post-American world"*, W. W. Norton & Company, New York.

CHAPTER 2. INDUSTRIALISM TWILIGHTS

Section 2.1.

The first section is devoted to discuss the progress of the industrialism organisation, from the economy of scale, to economy of scope, up to the very important new achievements of the *extended* enterprises, assuring lifestyle product-service delivery. The analysis aims at fixing the evolution critical nature, and the suggested further reading permits to make clear the characterising features.

Alting L., Legarth J.B. (1995). *"Life-cycle engineering and design"*, CIRP Annals, vol. 44, n° 2, pp. 569-580.

Alting L. & Jorgensen J. (1993). *"The lifecycle concept as basis for sustainable industrial production"*, CIRP Annals, vol. 42, n° 1, pp. 163-167.

Bras B. (1997) *"Incorporating environmental issues in product design and realization"*, UNEP Industry and Environment, January-June.

Cairns J. (1996). *"Determining the balance between technological and ecosystem services"*, in P.C. Jr Schultze, (Ed.) 'Engineering with ecological constraints,' Ntl. Academy Press, Washington, pp. 12-30.

Crenna F., Michelini R.C. et al. (1999). *"Product/process data management for engineer competitiveness"*, 11th Intl. Conf. Design Tools & Methods in Industrial Engineering, Palermo 8-12 Dec., vol. D, pp. 17-24.

Deming W. (1982). *"Quality productivity and competitive position"*, Cambridge Univ. Press, Boston.

Dyckhoff H., Lackes R. & Reese J. (2004). *"Supply chain management and reverse logistics"*, Springer, p. 426.

Eagan P.D. & Joeres E. (2002). *"The utility of environmental impact information: a manufacturing case study"*, J. of Cleaner Production, Vol. 10, n° 1, pp. 75-83.

Fiksel J. (1996). *"Design for environment: creating eco-efficient products and processes"*, (Ed.) New York, N.Y. McGraw-Hill, p. 513.

Giudice F., LaRosa G. & Risitano A. (2006). *"Product design for the environment: a lifecycle approach"*, CRC Taylor & Francis, Boca Raton, p. 482.

Hanssen O.J. (1998). *"Environmental impacts of product systems in a life cycle perspective; a survey of five product types based on life cycle assessments studies"*, J. of Cleaner Production, Elsevier, Vol. 6, n° 3-4, Sept., pp. 299-311.

James P. (1997). *"The sustainability cycle: a new tool for product development and design"*, J. of Sustainable Product Design, July, pp.52-57.

Leontieff W.W. (1966). *"Input-output economics"*, Oxford Uni. Press, New York.

Lovelock J.E. (1969). *"Gaia: a new look at life on the earth"*, Oxford Univ. Press, Oxford.

Luttropp C. & Lagerstedt J. (1999). *"Customer benefits in the context of life cycle design"*, Eco-Design '99: 1st Intl. Symp. on Environmental Conscious Design and Inverse Manufacturing, Tokyo, Feb.

Mayne R.W. (2005). *"Introduction to Windows and graphic programming by Visual C++.Net"*, World Sci., Singapore, pp. 340.

Michelini R.C. & Kovacs G.L. (1999). *"Knowledge organisation and govern-for-flexibility in lean manufacturing"*, A.B. Baskin, G.L. Kovacs, G. Jacucci, Eds: 'Co-operative Knowledge Processing for Engineering Design', Kluwer Academic Pub., Norwell Mass., pp. 61-82.

Michelini R.C. & Kovacs G.L. (2001). *"Integrated design for sustainability: intelligence for eco-consistent extended-artefacts"*, Invited Lecture, Intl. IFIP Conf. Digital enterprise: new challenges: lifecycle approach to management and production, Budapest, 7-9 Nov., pp. 5-35.

Nevins J.L., Whitney D.E. & DeFazio T.L. (1989). *"Concurrent design of products and processes"*, McGraw Hill, New York.

Pneuli Y., Zussman E., Kriwet A., et al. (1994). *"Evaluating product end-of-life value and improving it by redesign"*, IIE Trans. Design and Manufacturing.

Portney PR. (1993). *"The price is right: making use of lifecycle analyses"*, Issues in Science and Technology, vol. 10, Winter, pp. 69-75.

Prudhomme G., Zwolinski P., Brissaud D. (2003). "*Integrating into the design process the needs of those involved in the product lifecycle*", J. Engineering Design, vol. 14, n. 3, p. 333-353.

Schmidt M., Joao E., Albrecht E., Eds. (2005). "*Implementing strategic environment assessment*", Springer, London, p. 742.

Steingart G. (2008). "*The war for wealth: why globalisation is bleeding the west of its prosperity*", McGraw-Hill, New York.

Stevels A. (1999)."*Integration of eco-design into business, a new challenge*", First Intl. Symp. on Environmentally Conscious Design and Inverse Manufacturing. *Eco-design '99* Tokyo, Japan, pp. 27-32.

Suh N.P. (1990). "*The principles of design*", Oxford University Press, New York.

Subramanian A. (2008). "*India's turn: understanding the economic transformation*", Oxford Uni. Press, Oxford, p. 320.

Tang D., Zheng L., Li Z., et al. (2000). "*Re-engineering of the design process for concurrent engineering*", J. Computers and Industrial Engineering, vol. 38, n. 4.

Thierry M.C., Salomon M., vanNunen J., et al. (1995). "*Strategic issues in product recovery management*", California Management Review, vol. 37, n 2, pp. 114-135.

Vogtländer J.G., Bijma A. & Brezet H.C. (2002). "*Communicating the eco-efficiency of products and services by means of the eco-costs/value model*", J. Cleaner Production, Vol. 10, n 1, pp. 57-67.

Wackernagel M., Rees W. (1995). "*Our ecological footprint: reducing human impact on the earth*", New Society Pub., Gabriola Island.

Westkaemper E., Alting L. & Arndt G. (2000). "*Lifecycle management and assessment: approaches and visions towards sustainable manufacturing*", CIRP Annals, vol. 49, n. 2, pp. 501-522.

Section 2.2.

The «knowledge» entrepreneurship is winning prospect to expand the intangibles market and to lower the «natural» capital dependence. The analysis brings in new concepts, with light and shadow, when the growth sustainability needs to emerge, through account of the resources bookkeeping. The literature in the domain is quickly expanding, and the listed items are worthy reference..

Alting L., Hauschild M. & Wenzel H. (1989). "*Elements in a sustainable industrial culture: environmental assessment in product development*", Robotics and Computer-Integrated Manufacturing, vol. 14, pp. 429-439.

Benyus J.M. (1997). "*Bio-mimicry: innovation inspired by nature*", William Morrow, New York.

Cafasso R. "*Re-thinking re-engineering*", Computerworld, March 15, 1993.

Camarinha-Matos L.M., Afsarmanesh H. & Erbe H.H. (2000). "*Advances in networked enterprises: virtual organisations balanced automation and systems integrations*", (Eds.) Kluwer Academic Pub.

Chan C.K. & Lee H.W.J. (2005). "*Successful strategies in supply chain management*", IDEA Group Inc., IRM Press, p. 300.

Carr G. (2001). "*The digital enterprise: how to reshape your business for a connected world*", HBS Press Book, May 23.

Durlauf S. & Blume L.E. eds. (2007). "*The new Palgrave dictionary of economics*", Macmillan.

Eagan P.D. & Streckewald K.E. (1997). "*Striving to improve business success through increased environment awareness and design for the environment education; case study: AMP incorporated*", J. of Cleaner Production, Vol. 5, n° 3, pp. 219-223.

Ellis C.D. (2008). "*The partnership: the making of Goldman Sachs*", Penguin Press, New York, p. 729.

Etzion O., Scheuermann P., Eds. (2001). "*Co-operative information systems*", Springer, Berlin, p. 322.

Fischer M.M., Frolich J., Eds. (2001). "*Knowledge, complexity and innovation systems*", Springer, p.322.

Forey D. (2004). "*The economics of knowledge*", MIT Press, Cambridge.

Fredendall L.D. (2000). "*Basic of supply chain management*", CRC Press, Boca Raton, p. 288.

Hagel J. III & Singer M. (1999). "*Net worth*", HBS Press Book, Jan. 8

Hanssen O.J. (1995). "*Preventive environmental strategies for product systems*", J. of Cleaner Production, Elsevier, Vol. 3, n° 4, Dec., pp. 181-187.

Hayes R.H. & Pisano G.P. (1994). "*Beyond world-class: the new manufacturing strategy*", Harvard Business Review, Jan. 1.

Hinterberger F. & Luks F. (1999). "*Demateralization, employment and competitiveness in a globalized economy*", Wuppertal Inst. for Climate, Environment and Energy, Conf. Intel. Society for Ecological Economics (ISEE): Beyond Growth: Policies and Institutions for Sustainability, p. 9.

Griffith D.A. & Palmer,J.W. (1999). "*Leveraging the web for corporate success*", Business Horizons/Indiana University, Jan. 15.

Hirsch B.E. & Eschenbacher J. (2000). "*Extended products in dynamic enterprises*", in B. Stanford-Smith, P.T. Kidd, (Eds.) The e-business: key issue, applications and technologies, IOS Press.

Kapp K.M., Latham W.F., Ford-Latham H. (2001). "*Integrated learning for ERP success: a learning requirements planning approach*", CRC Press, Boca Raton, p. 322.

Kasai J. (1999). "*Lifecycle assessment: evaluation method for sustainable growth*", J SAE Review, Elsevier, vol 20, pp. 387-393.

King J. & Cafasso R. (1994). "*Client/server trimming*", Computerworld, Dec. 19.

Klein S. & Poulymenakou A. (2006). "*Managing dynamic networks: organisational perspectives of technology enabled inter-firm collaboration*", Springer, London, p. 308.

Landes D. (1999). "*The wealth and poverty of nations*", W.W. Norton, New York.

Linn R.J., Zhang W. & Li Z.Y. (2000). "*Intelligent management system for technology management*", J. Computers and Industrial Engineering, Vol. 38, n. 3, Oct.

Mahadevan B. (2000). "*Business models for internet-based e-commerce: an anatomy*", California Management Review, July 1.

Manzini E. (2000). "*Different scenarios of products and services combinations*", Conf. Challenges of Sustainable Development, INES. Industry and Sustainability: Pioneer Industries on Sustainable

Meijkamp R. (1994). "*Service-products, a sustainable approach? A case study on call-a-car in the Netherlands*", The Eco-Efficient Services seminar at the Wuppertal Inst. Wuppertal: September 18-19.

Mertins K., Heisig P. & Vorbeck J. (2001). "*Knowledge management: best practice in Europe*", Springer, Berlin, p. 347.

Michelini R.C. (2008). "*Knowledge entrepreneurship and sustainable growth*", Nova Sci. Pub., New York, p. XX-326.

Mont O. (1999). "*Strategic alliance between products and services*", World Service Congress '99, Atlanta, Georgia.

Myerson J.M. (2001). "*Enterprise systems integration*", 2nd ed, CRC Press, Boca Raton.

Olsthoorn X., Wieczorek A.J., Eds. (2005). "*Understanding industrial transformation: views from different disciplines*", Springer, London, p. 226.

Parker S. (2006). "*The lifecycle of entrepreneurial ventures*", Springer, London.

Polanyi, M. (1958). "*Personal knowledge*", Univ. Chicago Press, Chicago.

Powell W.W. (1998). "*Learning from collaboration: knowledge and networks in biotechnology and pharmaceutical industries*", California Management Review, April 1.

Putnik G., Cunha M.M., Eds. (2005). "*Virtual enterprises integration: technological and organisational perspectives*", IGI Pub. Hershey, p. 454.

Rasmussen L.B., Beardon C. & Munari S. (2000). "*Computers and networks in the age of globalisation*", (Eds.) Kluwer Academic Pub., ISBN 0-7923-7253-0.

Shin M., Holden T. & Schmidt R.A. (2000). "*From knowledge theory to management practice: towards an integrated approach*", J. Information Processing & Management, vol. 37, n. 2, Mar.

Simon H.A. (1981). "*The science of the artificial*", 2nd Ed., MIT Press, Cambridge.

Stahel W.R. (1997). "*The functional economy: cultural and organisational change*", The Implications for Environmental Design and Management, National Academy Press, Washington, D.C. pp. 91-100.

Stalk G. Jr., Evans P. & Shulman L.E (1992). "*Competing on capabilities: the new rules of corporate strategy*", Harvard Business Review, March 1.

Stoughton M., Shapiro K., Feng L., et al. (1998). "*The business case for extended producer responsibility: a feasibility study for developing a decision-support tool*", Boston, MA: Tellus Institute.

Tambe M., Lewis-Johnson M., Jones R.M. et al. (1995). "*Intelligent agents for interactive simulation environnements*", AI Magazine, vol. 16, n. 1, pp. 15-39.

Teece D.J. (1998). "*Capturing value from knowledge assets: the new economy, markets for know-how and intangible assets*", California Management Review, April 1.

Weaver P., Jansen L., Von Grootveld G., et al. "*Sustainable technology development*", Greenleaf Pub., Apr. 2000.

Walton J. & Whicker L. (1996). "*Virtual enterprise: myth and reality*", J. Control, Oct.

Walzer, M. (1995). "*Towards a global civil society*", (Ed) Berghahn Books, Providence.

Welford R.J. (1996). "*Corporate environmental management: systems and strategies*", Earthscan, London.

Wu J. (1999)."*Distributed systems design*", CRC Press, Boca Raton.

Xue D., Yadav,S. & Norrie D.H. (1999). "*Knowledge base and database representation for intelligent concurrent design*", J. Computer-Aided Design, Vol. 31, n 3.

Yusuf Y.Y. Sarhadi M. & Gunasekaran A. (1999). "*Agile manufacture: drivers, concepts and attributes*", Intl. J. Production Economics, n 62, pp. 33-43.

Zack M. (1999). *"Developing a knowledge strategy"*, California Management Review, Apr. 1.

Section 2.3.

The life-quality granted through the industrialism is conceived as non releasable standard, and the socio-political falls-off, of the end the affluent society habits, face the *global, no-global, post-global* issues, still, difficult to interpret. The prospected analysis looks at the «knowledge» entrepreneurship as useful aid, and the listed literature are basic references to help understanding the trends.

Amoretti M.C., Vassallo N.. eds. (2008). *"Knowledge language and interpretation: on the philosophy of Donald Davidson"*, Ontos Verlag, Frankfurt, p. 224.
Barbera A.J. & McConnell V.D. (1990) *"The impact of environmental regulations on industry productivity: direct and indirect effects"*, J. Environmental Economics and Management, vol. 18, Jan., pp. 50-65.
Bendell J. *"Terms of endearment: business, NGOs and sustainable development"*, Greenleaf Pub., July 2000.
Bhide A. (2008). *"The venturesome economy: how innovation sustains prosperity in a more connected world"*, Princeton Uni. Press, Princeton, p. 520.
Boldizzoni F. (2008). *"L'idea del capitale in occidente"*, Marsilio, Venezia, p. 258.
Brower D. (1995) *"Let the mountains talk, let the rivers run"*, Harper Collins, New York.
Charter M. & Polonsky M.J. (1999). *"Greener marketing: a global perspective on greening marketing practice"*, Greenleaf Pub., July.
Coyne K.P. & Dye R. (1998). *"The competitive dynamics of network-based businesses"*, Harvard Business Review, Jan. 1.
Crane D., Kawashima N., Kawasaki K., Eds. (2002). *"Global culture: media, arts, policy and globalisation"*, Rutledge, London.
Deaglio M. (2004). *"Postglobal"*, Laterza, Bari, p. 160.
Doumeingts G., Ducq Y., Clave F., et al. (1995). *"From CIM to global manufacturing"*, Proceedings CAPE'95, Computer Applications in Production Engineering, Chapman & Hall.
Ferguson N. (2004). *"The rise and demise of the British world order and the lessons for global power"*, Basic Books, New York.
Flynn S. (2007). *"The edge of disaster: rebuilding a resilient nation"*, Random House, New York.
Graedel T.E. & Allenby B.R. (1997). *"Design for environment"*, Prentice Hall.
Greenspan A. (2007). *"The age of turbulence"*, Pinguin Press, New York, p. 580.
James L. (1996). *"The rise and fall of the British empire"*, St. Martin's Press, New York.
Michelini R.C. (2002). *"Information infrastructures and eco-labelling"*, IFIPWG 5.5, COVE Newsletter 4, pp. 2-8, Oct.
von Mises L. (1948). *"Human action: a treatise on economics"*, Yale Uni. Press, New Haven.
Mo J.P.T., Nemes L. (2001). *"Global engineering, manufacturing and enterprise networks"*, (Eds.) Kluwer Academic Pub.
Molles C.M., Jr. (1999). *"Ecology: concepts and applications"*, WCB McGraw Hill, New York.

Morris J. (1980). *"Pax Britannica: climax of an empire"*, Harcourt Brace, New York.

Pomeranz K. (2000). *"The great divergence: China, Europe and the making of the modern world economy"*, Princeton Uni. Press, Princeton.

Paolucci,M., Sacile R. (2005). *"Agent-based manufacturing and control engineering"*, CRC Press, Boca Raton, p. 288.

Rodriguez F. & Rodrik D. (2001). *"Trade policy and economic growth: a sceptic's guide to the cross-national evidence"*, B. Bernanke, K.S. Rogoff, (Ed.): Macroeconomics Annuals 2000, MIT Press, Cambridge.

Rowledge L.R., Barton R.S. & Brady K.S. (with the co-operation of: J.A. Fava, C.L. Figge, K. Saur, S.B. Young) (1999). *"Mapping the journey: case studies in strategy an action toward sustainable development"*, Greenleaf Pub., Dec..

Schama S. (2008). *"The American future: a history"*, The Bodley Head, London, p. 392.

Scruton R. (2006). *"A political philosophy"*, Continuum, London, p. 256.

Shughaert W.F. & Tollison R. (2006). *"Policy changes and political responses: public choice perspectives on the post 09.11 world"*, Springer, London, p. 250.

Schumpeter J.A. (1981).*"Business cycles"*, McGraw Hill, New York.

Stiglitz J. (2001). *"Globalisation and its discontents"*, W.W. Norton, New York, p. 282.

Sunstein C.R. (2007). *"Worst-case scenarios"*, Harvard Uni. Press, Cambridge.

Tremonti G. (2008). *"La paura e la speranza"*, Mondadori, Milano, p. 260.

Trcek D. (2006). *"Managing information systems security and privacy"*, Springer, London, p. 236.

Wang P. (2005). *"The economics of foreign exchange and global finance"*, Springer, London, p. 351.

Wise S. (2008). *"The blackest streets: the life and death of Victorian slums"*, Bodely Head, London, p. 240.

Woodward B. (1994). *"Maestro: Greenspan's FED and American boom"*, Simon & Schuster, New York.

CHAPTER 3. THE «KNOWLEDGE» PARADIGMS

Section 3.1.

The «knowledge» paradigms characterise the new entrepreneurship trends of the «robot» age, based on the net-concerns «complexity», having on-process decision keeping aids to manage the operation flexibility. The references that follow are further reading proposals.

Bernstein W.J. (2008). *"A splendid exchange: how trade shaped the world"*, Atlantic Books, London, p. 467.

Buchanan M. (2002). *"Nexus, small worlds and the groundbreaking science of networks"*, Norton & Co, New York.

Camarinha-Matos L.M., Afsarmanesh H. (1998). *"Virtual enterprises: lifecycle tools and technologies"*, A. Molina, J. Sanchez, A. Kusiak Eds. Handbook of Life Cycle Engineering: Concepts, Tools and Techniques, Chapman and Hall.

Chabrow E.R. (1995). *"On-line employment"*, Information Week, Jan. 23.

Collis D.J. & Montgomery C.A. "*Competing on resources: strategy in the 1990s*", Harvard Business Review, July 1, 1995.

Davis S.A. & Bostrum R.P. (1993). "*Training end-users: an investigation on the role of interfaces and training methods*", MIS Quarterly, vol. 17, n° 1.

El-Elran M. (2008). "*When markets collide: investments strategies for the age of global economic change*", McGraw Hill, New York, p. 304.

George R. (2008). "*The big necessity: the unmentionable world of human waste and why it matters*, Metropolitan Books, New York, p. 304.

Gonzales-Torre P.L., Adenso-Diaz B. & Artiba H. (2004). "*Environmental and reverse logistics policies in European bottling and packaging firms*", J. Production Economics, Elsevier, vol 88, pp. 95-104.

Hall J. (2000). "*Environmental supply chain dynamics*", J. of Cleaner Production, Vol. 8, n° 6, Dec., pp. 455-471.

Harford T. (2008). "*The logic of life: the rational economics of an irrational world*", Random House, New York, p. 272.

Holmes R. (2008). "*The age of wonder: how romantic generations discovered the beauty and the terror of science*", Harper Press, London, p. 380.

Huang G.Q., Lee S.W. & Mak K.L. (1999). "*Web-based product and process data modelling in concurrent design for X*", J. Robotics and Computer-Integrated Manufacturing, vol. 15, n. 1, July.

Huang G.Q. & Mak K.L. (2000). "*We bid: a web-based framework to support early supplier involvement in new product development*", J. Robotics and Computer-Integrated Manufacturing, vol. 16, n. 2-3, Apr.

Huang G.Q., Yee W.Y. & Mak K.L. (2001). "*Development of a web-based system for engineering change management*", J. Robotics and Computer-Integrated Manufacturing, vol. 17, n. 3, June.

Kim L. (1997). "*Imitation to innovation*", HBS Press Book, Jan. 28.

Koschatzky K., Kulicke M., Zenker A., Eds. (2001). "*Innovative networks: concepts and challenges in the European perspective*", Springer, Berlin, p. 212.

Kropf P., Babin G., Plaice J., et al. (2001). "*Distributed communities in the web*", (Eds.) Springer, Berlin, p. 335.

Lacity M.C., Willcocks L.P. & Feeny D.E. (1995). "*Information technology outsourcing maximise flexibility and control*", Harvard Business Review, May-June.

Newman M.E.J. (2003). "*Mixing patterns in networks*", Phys. Rev. S. E, vol. 67, art. n° 026126.

Newman M.E.J. (2003). "*The structure and function of complex networks*", SIAM Rev., vol. 45, n° 2, pp. 167-256.

Porter, M.E. (1998). "*Clusters and the new economics of competition*", Harvard Business Review, Nov. 1.

Purba S. (2001). "*New directions in Internet management*", (Ed). Auerbach Pub.

Rampersad H.K. (2001). "*Total quality management: an executive guide to continuous improvement*", Springer, Berlin.

Savolainen R.(1999). "*The role of the internet in information seeking: putting networked services in context*", J. Information Processing & Management, vol. 35, n. 6.

Sherman H.D., Zhu J. (2006). "*Service productivity management: improving service performance by data envelopment analysis*", Springer, London, p.344.

Sirkin H., Hemmerling J., Bhattacharya A. (2008). *"Globality: competing with everyone from everywhere for everything"*, Business Plus, London, p.304.

Sudoh O. (2005). *"Digital economy and social design"*, Springer, London, p. 236.

Suhas H.K. (2001). *"From quality to virtual enterprise: an integrated approach"*, CRC Press, Boca Raton.

Swanson K., McComb D., Smith J., et al. (1991). *"The application of software factory: applying total quality techniques to systems developments"*, MIS Quarterly, vol. 15, n° 4.

Tapscott D. (2008). *"Grown up digital: how the net generation is changing your world"*, McGraw Hill, New York, p. 384.

Teisberg E.O., Porter M.E. & Brown G.B. (1994). *"Making competition in health care work"*, Harvard Business Review, July.

Vollmann, T.E. (1996). *"Transformation imperative"*, HBS Press Book, May 20.

Wang P., Hawk W.B. & Tenopir C. (2000). *"Users' interaction with World Wide Web resources: an exploratory study using a holistic approach"*, J. Information Processing and Management, vol. 36, n. 2, Feb.

Watson H.J., Reiner R.K. Jr. & Chang E.K. (1991). *"Executive information systems: a frame for development"*, MIS Quarterly, vol. 15, n° 1.

Watts D.J. (1999). *"Small worlds"*, Princeton Univ. Press, Princeton.

Watts D.J. (2003). *"Six degrees: the science of a connected age"*, Norton, New York.

Watts D.J. & Strogatz S.H. (1998). *"Collective dynamics of small-world networks"*, Nature, vol. 393, pp. 440-442.

Wengel J., Warnke P., Lindbom J. (2000). *"Case study: automotive industry personal cars"*, Future of Manufacture in Europe 2015-2020: The Sustainability Challenge, Fraunhofer Inst. Systems and Innovation Research, Karlsruhe, JCR-IPTS.

Wright P. (1998). *"An encroachment too far"*, in A. Barnet, R. Scruton, eds., 'Town and country', Jonathan Cape, London.

Zachman J.A. (1987). *"A framework for information systems architectures"*, IBM Systems J., vol. 26, n 3, pp. 276-292.

Zanette D.H. (2001). *"Critical behaviour of propagation on small-world networks"*, Phys. Rev. S. E, vol. 64, art. n 056115.

Section 3.2.

The socio-political surroundings happen to be conditioning driver not less relevant of the technological revolutions. The prospected outlook summarises the capitalism phases, to devise how the wealth build-up might make up innovative tracks. The listed literature is example reference for further deepening.

Bergen S.D., Bolton S.M. & Fridley J.L. (2001). *"Design principles for ecological engineering"*, Ecological Engineering, Elsevier, vol. 18, pp. 201-210.

Bergsma G. & Kroese M. (1994). *"Entrepreneurial approaches to energy efficiency in a service economy"*, The Eco-Efficient Services Workshop at the Wuppertal Institute, Wuppertal. Sept. 18-20.

Borg J.C., Farrugia P.J. & Camilleri K.P. (2004). *"Knowledge intensive design technology"*, (Eds.) Kluwer Ac. Pub., Boston.

Boorstin D.J. (1985). *"The discoverers"*, Vintage Books, New York.
Cerveni R.P., Garrity E.J. & Sanders G.L. (1986). *"The application of prototyping to systems development"*, J Management Information Systems, vol. 2, n° 3.
Chatterjee S. (1998). *"Delivering desired outcomes efficiently: the creative key to competitive strategy"*, California Management Review, Jan. 1.
Funazaki A., Taneda K., Tahara K., et al. (2003). *"Automobile assessment issues at end of life and recycling"*, JSAE Review, Elsevier, vol. 24, pp. 381-386.
Haeckel S.H. (1999). *"Adaptive enterprise: creating and leading sense-and-respond organisations"*, HBS Press Book, June 15.
Hammer M. & Stanton S.A. (1995). *"The re-engineering revolution"*, Harper Collins, New York.
Howe J. (2008). *"Crowdsourcing: why the power of the crowd is driving the future business"*, Random House, New York, p.320.
Kennedy P. (1987). *"The rise an fall of the great powers: economic changes and military conflicts from 1500 to 2000"*, Random House, New York.
Landes D.S. (1983). *"Revolution in time: clocks and the making of the modern world"*, Harvard Uni. Press, Cambridge.
Machlup F. (1962). *"The production and distribution of knowledge in the US"*, Princeton Univ. Press, Princeton.
Meijkamp R. (1998). *"Changing consumer behaviour through eco-efficient services: an empirical study on car sharing in the Netherlands"*, Business Strategy and the Environment, No. 7, pp. 234-244.
Michelini R.C. (2008). *"Knowledge entrepreneurship and sustainable growth"*, Nova Sci. Pub., New York, p. XX-326.
Olson J.M. & Judith S. (1991). *"User-centred design of collaboration technology"*, J. Organisational Computing, vol. 1, n° 1.
Orlikowski W.J. (1992). *"The duality of technology: rethinking the role of technology in organisations"*, Organisation Science, vol. 3, n° 2.
Parsaei H., Usher J., Roy U. (1998). *"Integrated product and process development: methods, tools, and technologies"*, New York: John Wiley, p. 412.
Petrovic D. (1998). *"Modelling and simulation of a supply chain in an uncertain environment"*, Europ. J. Operational Research, n. 109, pp. 299-309.
Phadke M.S. (1989). *"Quality engineering using robust design"*, Prentice Hill, New York.
Rayport J.F. & Sviokla J.J. (1995). *"Exploiting the virtual value chain"*, Harvard Business Review, Nov. 1.
Rosenblatt, M., Roll, Y. & Zyse, V. (1993). *"A combined optimization and simulation approach for designing automated storage-retrieval systems"*, IIE Transactions, vol. 25, n 1, pp. 40-50.
Schläler A. (2008). *"Die Kraft der schöpferischen Zerstörung"*, Campus Verlag, Frankfurt, p. 286.
Scheer A.W. (1994). *"Business process engineering: reference models for industrial enterprises"*, Springer, Berlin.
Shostack G.L. (1977). *"Breaking free from product marketing"*, J. of Marketing, Vol. 41, No. 2, April, pp. 73-80.
Sorensen R. (2008). *"Seeing dark things: the philosophy of shadows"*, Oxford Uni. Press, Oxford, p. XIII-310.

Thaler R.H., Sunstein C.R. (2008). *"Nudge: improving decisions about health, wealth and happiness"*, Yale Uni. Press, New York, p. 304.

Thoben K.D., Jagdev H. & Eschenbacher J. (2001). *"Extended products: evolving the traditional product concepts"*, 7th Intl. Conf. Concurrent Enterprising: Engineering the Knowledge Economy through Co-operation, Bremen, 27-27 June.

Wagner B., Enzier S., Eds. (2006). *"Material flow management: improving cost efficiency and environmental performance"*, Springer, London, p. 206.

Waldo J. (2002). *"Virtual organisations, pervasive computing and infrastructures for networking at the edge"*, J. Information Systems Frontiers, vol. 4, n° 1.

Westcott R. (1995)."*Client satisfaction: the yardstick for measuring the MIS success"*, J Information Systems Management, Oct.

Williamson O. (1985). *"The economic institutions of capitalism"*, (Ed) The Free Press, New York.

Yourdon E. (1989). *"Modern structured analysis"*, Prentice Hall, Englewood Cliffs.

Zuest R. & Caduff G. (1997). *"Lifecycle modelling as an instrument for lifecycle engineering"*, CIRP Annals, vol. 46, n. 1, pp. 351-354.

Section 3.3.

Three scenarios are sketched, to describe the times to come: the *global* «hyper-market», the *no-global* «autarchy» and the *post-global* «altruism». The last is deemed utopia, still, no coherent alternative is easily acknowledged, if the next generations future has to be preserved. The recalled literature allows to recognise fundamental bets to face.

Alavi M., & Carlson P. (1992). *"A review on management information systems researches and disciplinary developments"*, J. Management Information Systems, vol. 8, n° 4.

Albrecht S.M. (2001). *"Forging new directions in science and environmental politics and policy: how can co-operation, decision and deliberation be brought together?"*, Environment, Development and Sustainability, vol. 3, pp. 323-341.

Attali J. (2006). *"Une brève histoire de l'avenir"*, Fayard, Paris, p.320.

Ayres R.U. (1998). *"Technological progress: a proposed measure"*, Technological Forecasting and Social Change, No. 59, pp. 213-233.

Bennet M. & James P. (1998)."*The green bottom line. Environmental accounting for management: current practice and future trends"*, Greenleaf Pub., Sept.

Bennet M. & James P. (1999). *"Sustainable measures: evaluation and reporting of social and environmental performance"*, Greenleaf Pub., June.

Benson A.P., Hinn D.M. & Lloyd C. (2001). *"Visions of quality: how evaluators define, understand and represent program quality"*, Elsevier Sci., Amsterdam.

Camarinha-Matos L.M. &. Afsarmanesh H. (1999). *"Infrastructures for virtual enterprises: networking industrial enterprises"*, Proc. PRO-VE'99, Kluwer Academic Pub.

Davenport T.H. & Prusak L. (1997). *"Working knowledge,"* HBS Press Book, Oct. 29.

Evans T.P. & Wurster T.S. (1997)."*Strategy and the new economics of information"*, Harvard Business Review, Sept. 1.

Fisher I. (1911). *"The purchaser power of money"*, Macmillan, New York.

Forey D. (2001). *"The economics of the knowledge and of the knowledge-based economy: a*

changing discipline in an evolving society", Kluwer Academic Pub.
Friedman M. & Friedman R. (1980). "*Free to chose*", Harcourt Brace Jovanovich, New York.
Friend G. (1994). "*The end of ownership? Leasing, licensing, and environmental quality*", The New Bottom Line, Vol. 3, No. 11.
Fuentes X. (2002). "*International law-making in the field of sustainable development: unequal competition between development and environment*", Intl. Environmental Agreements: Politics, Law and Economics, Kluwer, vol. 2, pp. 109-133.
Garten J.E. (1999). "*World view: global strategies for the new economy*", HBS Press Book, Dec. 9.
Hagel J. III, Singer M. (1999) "*Unbundling the corporation*", Harvard Business Review, March 1.
Hardt M., Negri A. (2000). "*Empire*", Harvard Uni. Press, Boston, p. 512.
Hawken P., Lovins A.B. & Hunter Lovins L. (2001). "*Natural capitalism: creating the next industrial revolution*", Little, Brown & Co., Boston, p. 340.
Hayes R.H. (1985). "*Strategic planning: forward in reverse?*", Harvard Business Review, Nov. 1.
Hope J. & Hope T. (1997). "*Competing in the third wave*", HBS Press Book, Aug. 15.
Keyser T.K., Davis R.P. (2000). "*Statistically assessing distributed computing approaches*", J. Computers and Industrial Engineering, vol. 38, n. 2, July.
Kleinert J. (2004). "*The role of multinational enterprises in globalisation*", Springer, Kieler Studien, Berlin, p. 211.
Lee S. & Leifer R.P. (1992). "*A framework for linking the structure of information systems with organisational requirements for information saving*", J Management Information Systems, vol. 8, n° 4.
Leonard D. "*Wellsprings of knowledge*", HBS Press Book, March 25, 1998.
Liebovitz J. (2001). "*Knowledge management: learning from knowledge engineering*", CRC Press, Boca Raton, p. 233.
Matheson J. & Matheson D. (1997). "*Smart organisation*", HBS Press Book, Oct. 24.
Meadows D.H., Meadows D.L. & Randers J.J. (1992). "*Beyond the limits: global collapse or a sustainable future*", Earthscan, London, p.131.
Miller D. & Whitney J.O. (1999). "*Beyond strategy: configuration as a pillar of competitive advantage*", Business Horizons/Indiana University, May 15.
Morin P.J. (1999). "*Community ecology*", Blackwell Scientific, New York.
Muller E. (2001). "*Innovation interactions between knowledge-intensive business services and small-and-medium sized enterprises: an analysis in terms of evolution, knowledge and territories*", Springer, Berlin.
Nagel C. & Meyer P. (1999). "*Caught between ecology and economy: the end-of-life aspects of environmentally conscious manufacturing*", J. Computers & Industrial Engineering, vol. 36, n. 4, Dec
O'Neill J. (2001). "*Ecology, policy and politics*", Cambridge Univ. Press, London, 2001.
Oesterle H., Fleisch E. & Alt R (2001). "*Business networking: shaping collaboration between enterprises*", Springer, Berlin, p. 343.
Peterson P. (1999). "*How the coming age wave will transform America and the world*", Times Books, New York.
Phillips F.Y. (2001). "*Market-oriented technology management: innovating for profit in entrepreneurial time*", Springer, Berlin.

Quadrio-Curzio A., Fortis M. (2005). *"Research and technological innovation: the challenge for a new Europe"*, Springer, London, p. 290.

Rapoport A. & Horvath W. (1961). *"A study of large sociogram"*, J. Behavioural Sci., vol. 6, pp. 279-291.

Porter M.E. (2001). *"Competition in global industries"*, (Ed.) HBS Press Book, Jan.

Press F., Siever R. (2000). *"Understanding earth"*, W.H. Freeman & Co., 3 ed., New York.

Rittemberger V. (1994). *"Beyond anarchy: international co-operation and regimes"*, (Ed.) Oxford Uni. Press, Oxford, 1994.

Roll E. (1942). *"A history of economic thought"*, Prentice Hall, New York.

Rosenau, J.N. (1990). *"Turbulence in world politics: a theory of change and continuity"*, Princeton Univ. Press.

Russo N.L., Fitzgerald B., DeGross J.I., eds. (2001). *"Re-aligning research and practice in information systems development: the social and organizational perspective"*, Kluwer Academic Pub..

Sapir A., ed. (2008). *"Fragmented power: Europe and the global economy"*, Bruegel Books, p. 332.

Sarkis J. (1999). *"A methodological framework for evaluating environmentally conscious manufacturing programs"*, J. Computers & Industrial Engineering, vol. 36, n. 4.

Sarvary M. (1999). *"Knowledge management & competition in the consulting industry"*, California Management Review, Jan. 1.

Sassen S. (1988). *"The mobility of labour and capital: a study in international investment and labour flow"*, Cambridge Uni. Press, 2 vol., Cambridge.

Scherer F.M. (1999). *"New perspectives on economic growth and technological innovation"*, Brookings Institution Press, Washington,

Shapiro C. & Varian H.R. *"The art of standards wars"*, California Management Review, Jan. 1, 1999.

Stebbins G.L. (1969). *"The basis of progressive evolution"*, Univ. North Carolina Press., Chapel Hill.

Stephanidis, C. *"Ambient intelligence in the context of universal access"*, ERCIM News, n° 47, Oct. 2001, pp.10-11.

Stephanson,A. (1995). *"Manifest destiny: America expansionism and the empire of right"*, Hill & Wang, New York.

Sulzmaier S. (2001).*"Consumer-oriented business design: the case of airport management"*, Springer, Berlin.

Summers L.H. & Summers V.P. (1989). *"When financial markets work too well: a cautious case for a security transaction tax"*, J. Financial Services Research, vol. 3, pp. 161-186.

Tobin J. (1978). *"A proposal for international monetary reform"*, Eastern Economic J., vol. 4, pp. 153-159.

Turner B.S. (1990). *"Theories of modernity and post-modernity"*, Sage, London.

Usher J.M., Roy U., Parsaei H.R. (1998). *"Integration product and process development methods, tools and technologies"*, John Wiley & Son, New York.

Vickers J. & Yarrow G. (1988). *"Privatisation: an economic analysis"*, MIT Press, Cambridge.

Wallerstein I. (1979). *"The capitalist world-economy"*, Cambridge Uni. Press, Cambridge.

Walrand J., Bagchi K., Zubrist G. (1998). *"Network performance: modelling and simulation"*, G&B Sci, Harwood Pub.

van Weenen H. (1996). *"Discovery: from collision to co-operation"*, Third European Roundtable on Cleaner Production, Kalundborg and Copenhagen. 31 Oct.-4 Nov.
Wassermann O. (2001). *"The intelligent organisation: winning the global competition with supply chain idea"*, Springer, Berlin.
Wheeler J.A. (1979). *"Frontier of the time"*, North-Holland, Amsterdam.
Williamson O. (1975). *"Markets and hierarchies: analysis and antitrust implications"*, The Free Press, New York.
Yudice G. (1995). *"Civil society, consumption and govern-mentality in an age of global restructuring"*, Social Text, n. 45, pp. 1-25.
Yunas M. *"The Grameen bank"*, Scientific American, n. 281, Nov. 1999, pp. 114-119.
Zarnekow R., Brenner W., Pilgram U. (2006). *"Integrated information management: applying successful industrial concepts in information technology"*, Springer, London, p. 158.

CHAPTER 4. ROBOTS IN MANUFACTURE JOBS

Section 4.1.

The chapter is based on the results of research projects accomplished at the University of Genova by the «Industrial robot design research group», established Dec. 1978, within the Engineering School, University of Genova, within the «Laboratory on computation methods for the design of dynamical systems», operating, under my direction, in the areas of space engineering and artificial spacecrafts control. This initial section collects series of general purpose developments, aiming at specifying the surrounding infrastructures, typically conditioning the robotic equipment. Indeed, the successfulness of any specific solution highly depends on how the innovation is made working, to fully exploit the economy of scope requirements. The main issues are object of explanatory comments.

Michelini R.C., Molfino R.M., Acaccia G.M. (1983). *"The development of modular simulation procedures for the design of task-dependent industrial robots"* Intl. Symp. Simulation '83, Lugano, pp. 135-140.
Michelini R.C. (1986). *"Current developments of robotics researches in Italy"*, P. Scott, Ed, The World Yearbook of Robotics Research and Developments, Kogan Page, London, pp. 118-120.
Acaccia G.M., Michelini R.C., Molfino R.M., Rossi G.B.(1987). *"Knowledge-based simulators as off-line programming facilities in production engineering"*, 2nd Intl. Conf. Computer-Aided Production Engineering, Edinburg, 13-15 Apr., pp. 71-78.
Acaccia G.M., Michelini R.C., Molfino R.M., Piaggio P.A. (1987). *"Information data-based structures for flexible manufacturing simulators"*, 3rd IFIP Conf. Advances in Production Management Systems, Winnipeg, Aug. 11-14, pp. 649-662.
Acaccia G.M., Michelini R.C., Molfino R.M., Rossi G.B. (1987). *"Expert-simulation for evaluating the dynamical behavior of robotized manufacturing plants"*, European Simulation Multiconference, Wien, July 7-10, pp. 59-65.
Acaccia, G.M., Michelini, R.C., Molfino, R.M. (1988). *"Information data-based systems in

flexible manufacturing", A. Kusiak, Ed.: 'Advanced Production Management Systems', Elsevier, Amsterdam, pp. 649-662.

Acaccia G.M., Michelini R.C., Molfino R.M., Ragonese A. (1988). *"Failure analysis and maintenance model for a flexible manufacturing cell"*, 7th Intl. Conf. Modelling, Identification & Control, Grindelwald, Feb. 16-18, pp. 80-86.

Michelini R.C. (1989). *"Application of artificial intelligence tools in the programming facilities for factory automation"*, EERP Survey Report, DEC Woods Meeting, Stresa, 7-9 June, pp. V.1-67.

Acaccia G.M., Michelini R.C., Molfino R.M., Rossi G.B. (1990). *"Computer-intelligence options in flexible manufacturing"*, IFIP Intl. Conf. Artificial Intelligence: Industrial Applications, Leningrad, 15-19, Apr., pp. 127-132.

Michelini R.C., Pampagnin F., Rossi G.B. (1990). *"Computer-integrated manufacturing trends vs. work-organization trends"*, IFIP Intl. Conf. Men in Flexible Automated Production, Praha, June 19-21, pp. 88-95.

Michelini R.C., Kovacs G. (1994). *"Knowledge organization & govern-for-flexibility in manufacturing"*, 3rd. Intl. Workshop on Cooperative Knowledge Processing for Engineering Problem Solving, Rovereto, may 29- june 1, pp. 48-57.

Michelini R.C., Pampagnin F., Razzoli R. (1994). *"Lean engineering and performace assessment in flexible manufacturing"*, 27th. ISATA Lean/Agile Manufacturing, Aachen, 31 Oct.-4 Nov., pp. 279-286.

Michelini R.C., Acaccia G.M., Callegari M., Molfino R.M., Razzoli R.P. (2001). *"Robots design with application to manufacturing"*, C.T. Leondes, Ed.: Artificial Intelligence and Robotics in Manufacturing, Vol. VII, CRC Press LLC, pp. 7.01-7.62.

Michelini R.C., Kovàcs G.L. (2002). *"Integrated design for sustainability: intelligence for eco-consistent products-and-services"*, The Estonian Business School Review, Tallin, Winter 2002-3 issue, n° 15, Dec., pp. 81-95.

Michelini R.C., Razzoli R.P. (2004). *"Product-service eco-design: knowledge-based infrastructures"*, Intl. J. Cleaner Production, Elsevier, vol. 12, n° 4, pp. 415-428.

Michelini R.C., Razzoli R.P. (2005). *"Collaborative networked organisations for eco-consistent supply-chains"*, in G.D.Putnik, M.M.Cunha, Eds., 'Virtual Enterprise Integration', IGI Press, Hershey, pp. 45-77.

Acaccia, G.M., Kopàcsi, S., Kovàcs, G., Michelini R.C., Razzoli R.P. (2007). *"Service engineering and extended artefact delivery"*, G.D. Putnik, M.M. Cunha, Eds., 'Knowledge and Technology Management in Virtual Organisations', IGI Press, Hershey, PA, pp. 45-77.

Section 4.2.

The second section specifically addresses the shop-floor operations, and how these need to be modified to enable robotic efficiency. Referring to the Genova «Industrial robot design research group» activity, the development of «expert controllers», with embedded *artificial intelligence* to operate to on-process decision-keeping tasks, is the most notable outcome. The efforts are, today, with the existing hardware/software instruments, out of use, but, at the time, appeared quite advanced implementations.

Acaccia G.M., Ferrari R., Michelini R.C., Molfino R.M. (1985). "*Computer-aids for the modelling and the performances evaluation of an FMS*", 13th Intl. Symp. Modelling & Simulation, Lugano, 24-26 June, pp. 42-46.

Acaccia G.M., Michelini R.C., Molfino R.M., Piaggio P.A. (1986). "*X-SIFIP: knowledge-based special-purpose simulator for developing flexible manufacturing cells*", IEEE Intl. Conf. Robotics & Automation, San Francisco, Apr., 7-10, pp. 645-653.

Acaccia G.M., Bovone M., Michelini R.C., Molfino R.M., Spinosa F. (1987). "*Rule-based dispatching-govern for flexible manufacturing*", IEEE Intl. Conf. Robotics & Automation, Raleigh, Mar. 30-Apr. 3, pp. 558-567.

Acaccia G.M., Michelini R.C., Molfino R.M. (1987). "*Knowledge-based simulators in manufacturing engineering*", D.Sriram, R.A.Adey Eds.: Computational Mechanics Pub., Unwin Brothers, Southampton, pp. 327-344.

Acaccia G.M., Michelini R.C., Molfino R.M. (1987). "*A knowledge-based computer-simulator for the functional-scaling and the perfomances evaluation of automated tool-handling services*", 9th Intl. Conf. Production Research, Cincinnati, Aug. 17-20, pp. 1201-1209.

Acaccia G.M., Callegari M., Michelini R.C., Molfino R.M., Piaggio P.A. (1988). "*X-ARS: a consultation program for selecting the industrial robot architectures*", J.S. Gero, Ed., Artificial Intelligence in Engineering: Robotics & Processes, pp. 35-58.

Michelini R.C., Acaccia G.M., Molfino R.M. (1989). "*Design of intelligent governors for the material-handling equipment of automated factories*", T.Sata, G.Olling, Eds.: 'Software for Factory Automation', North-Holland, Amsterdam, , pp. 297-312.

Acaccia G.M., Michelini R.C., Molfino R.M., Raffaelli G. (1989). "*An expert-scheduler for the tool-stock management in a CIM-environment*", Intl. J. Advan. Manufacturing Engineering, vol. 1, n°4, pp. 203-210.

Acaccia G.M., Michelini R.C., Molfino R.M., Rossi G.B. (1989). "*Shopfloor logistics for flexible manufacturing with distributed intelligence*", Intl. J. Advan. Manufacturing Technology, vol. 4, n° 3, pp. 231-242.

Acaccia G.M., Michelini R.C., Molfino R.M., Stolfo F., Tacchella A. (1989). "*Development of a decentralized control for flexible manufacturing facilities*", Intl. J. Computer Applications in Technology, vol. 2, n° 2., pp. 89-100.

Acaccia G.M., Michelini R.C., Molfino R.M., Stolfo F., Tacchella A. (1989). "*A distributed-interconnected control for computer integrated manufacturing*", Intl. J. Computer-Integrated Manufacturing Systems, vol. 2, n° 2, pp. 108-114.

Acaccia G.M., Campolonghi F., Michelini R.C., Molfino R.M. (1989). "*Expert-simulation of a tool-dispatcher for factory automation*", Intl. J. Computer Integrated Manufacturing, vol. 2, n° 3, pp. 131-139.

Michelini R.C., Acaccia G.M., Callegari M., Molfino R.M. (1990). "*Expert-simulational environment for robotized manufacturing*", 5th Intl. Conf. Systems Research, Informatics and Cybernetics, Baden-Baden, Aug. 6-12, pp. 101-106.

Acaccia G.M., Michelini R.C., Molfino R.M., Piaggio M. (1990). "*Concurrent management of shopfloor operations: an onprocess expert-scheduler*", ISCIE-ASME Symp. on Flexible Automation, Kyoto, Jul. 9-12, pp. 1165-1171.

Michelini R.C., Acaccia G.M., Callegari M., Molfino R.M. (1990), "*Integrated management of concurrent shopfloor operations*", Intl. J. Computer-Integrated Manufacturing Systems, vol. 3, n° 1, pp. 27-37.

Acaccia G.M., Michelini R.C., Molfino R.M. (1991). "*Govern-for-flexibility of tool-stock and delivery operations for factory automation*", G.J.Olling & Z.Deng Eds Information Technology for Advanced Manufacturing Systems, Elsevier, pp. 483-490.

Michelini R.C., Acaccia G.M., Callegari M., Molfino R.M. (1992). "*An expert simulator for factory automation*", G.J.Olling, F.Kimura Eds: Human Aspects in Computer-Integrated Factories, North-Holland, pp. 797-804.

Michelini R.C., Acaccia G.M., Callegari M., Molfino R.M. (1992). "*Case studies in modelling the control-and-recovery strategies of robotized manufacturing cells*", 1st. Worldwide Gensym Users Group Meeting, Santa Margherita, June 24-26, pp. 50-65.

Acaccia G.M., Michelini R.C., Molfino R.M. (1992). "*Computer-intelligence options in flexible manufacturing*", J.L.Alty & L.I. Mikulich Eds.: Industrial Applications of Artificial Intelligence, North-Holland, Amsterdam, pp. 344-349.

Michelini R.C., Acaccia G.M., Callegari M., Molfino R.M. (1992). "*Simulation facilities for the development of computer-integrated manufacturing*", Int. J. of Advanced Manufacturing Technology, Vol. 7, n°4, pp. 238-250.

Acaccia G.M., Callegari M., Michelini R.C. Molfino R.M. (1993)."*Control automation of a multi-operational section for the flexible manufacturing of highly-diversified products,*" I. Mezgar, P. Bertok, Eds. 'Knowledge Based Hybrid Systems', North-Holland, Amsterdam, pp. 193-202.

Acaccia G.M., Michelini R.C., Molfino R.M., Piaggio M. (1993). "*The govern-for-flexibility of manufacturing facilities: an explanatory example*", Intl. J. Computer-Integrated Manufacturing Systems, vol. 6, n° 3, pp. 149-160.

Michelini R.C., Acaccia G.M., Molfino R.M. (1994). "*Expert scheduler for adaptive planning of batch-varying highly-diversified products*", Intl. ASME-ISCIE Symp. on Flexible Automation, Kobe, July 11-18, pp. 752-756.

Acaccia G.M., Callegari M., Michelini R.C., Milanesio R.,Molfino R.M, Rossi A. (1994). "*Simulation assessment and effectiveness investigation of a pilot FMS prototype*", Intl. European Simulation Symp. ESS'94, Istanbul, Oct. 9-12, vol. 2° pp. 55-59.

Michelini R.C., Acaccia G.M., Callegari M., Molfino R.M. (1994). "*Flexible manufacturing with integrated control and management*", M.B. Zaremba, B. Prasad Eds., '*Modern Manufacturing: Control & Technology*', Springer, pp. 225-253.

Acaccia G.M, Firenze G.A, Firenze G.G, Michelini R.C, Molfino R.M (1994). "*Integrated control and management in the flexible manufacturing of pharmaceutical products*", 10th ISPE·IFAC Intl. Conf. CAD/CAM, Robotics and Factories of the Future, Ottawa, Aug. 21-24, pp. 342-347.

Michelini R.C., Acaccia G.M., Callegari M., Molfino R.M., Razzoli R.P. (1995). "*Knowledge-based emulation-simulation for flexible-automation manufacturing*", Eurosim Congress '95, TU Vienna, Sept. 11-15, pp. 1259-1264.

Acaccia G.M., Callegari M., Michelini R.C., Milanesio R.,Molfino R.M, Rossi A. (1995). "*Pilot CIM implementation for lean engineering experimentation*", Intl. J. Computer-Integrated Manufacturing Systems, vol. 8, n° 3, pp. 185-192.

Acaccia G.M., Michelini R.C., Callegari M., Firenze G.A., Firenze G.G., Molfino R.M., Razzoli R.P. (1995). "*Intelligent real-time scheduling and control of pharmaceutical plants*", 3rd. Worldwide Gensym Users Group Meeting, Milano, June 24-26, pp. 50-65.

Acaccia, G.M., Callegari, M., Michelini, R.C., Molfino R.M. (1995)."*Pilot CIM implementation for lean engineering experimentation*", Intl. J. Computer-Integrated

Manufacturing Systems, vol. 8, n° 3, pp. 185-192.

Acaccia G.M., Callegari M., Marzapani R., Michelini R.C., Molfino R.M. (1996). "*Development of modular assembly facilities with store-up and by-pass management*", 8th. Europ. Simulation Symposium ESS '96, Genova, Oct. 24-26, , pp. 289-293.

Michelini R.C., Acaccia G.M., Callegari M., Molfino R.M., Razzoli R.P. (1996). "*Integrated product-and-process govern of robotic assembly cells*", 27th. Intl. Symp. on Industrial Robots, Milano, 6-8 Oct. pp. 415-420.

Michelini R.C., Acaccia G.M., Callegari M., Molfino R.M., Razzoli R.P. (1997). "*Shop controller-and-manager for intelligent manufacturing*", S. Tzafestas Ed., Management and Control of Manufacturing Systems, Springer, 1997, pp. 219-254.

Acaccia G.M., Chiavacci A., Michelini R.C., Callegari M. (1999). "*Benchmarking the clothing industry effectiveness by computer simulation*", 11th European Simulation Symposium and Exhibition, ESS99, October 26-28, Castle, Friedrich-Alexander University, Erlangen-Nuremberg, Germany, pp. 519-524.

Michelini R.C., Molfino R.M., Piras D., Callegari M. (1999). "*Development and simulation assessment of a modular assembly facility for automotive derivation boxes*", 11th European Simulation Symposium and Exhibition, ESS99, October 26-28, 1999, Castle, Friedrich-Alexander University, Erlangen-Nuremberg, pp. 525-529.

Molfino R.M., Lacchini A., Maggiolo G., Michelini R.C., Razzoli R.P.(1999). "*Re-engineering issues in automatic assembly*", G. Jacucci, G.J. Olling, K. Preiss, M. Wozny, Eds., Globalisation of Manufacture in the Digital Communication Era', Kluwer, Boston, pp. 603-616.

Acaccia G.M., Conte M., Maina D., Michelini R.C., Molfino R.M. (1999). "*Integrated manufacture of high-standing dresses for customised satisfaction*", G. Jacucci, G.J. Olling, K. Preiss, M. Wozny, Eds., 'Globalisation of Manufacturing in the Digital Communication Era', Kluwer, Boston, pp. 511-523.

Michelini R.C., Acaccia G.M., Callegari M., Molfino R.M., Razzoli R. (1999). "*Artefact integration by concurrent enterprises and productive break-up*", G.Jacucci, G.J.Olling, K.Preiss, M.Wozny, Eds.: 'Globalisation of Manufacturing in the Digital Communication Era', Kluwer, Boston, 1999, pp. 221-234.

Acaccia G.M., Marelli A., Michelini R.C., Zuccotti A. (2001). "*The fabric feeding management for automatic clothing manufacturing*", G.L. Kovacs, P. Bertok, G. Haidegger, Eds.: Digital Enterprise, New Challenges, Kluver, Boston, pp. 416-427.

Michelini R.C., Acaccia G.M., Callegari M., Molfino R.M., Razzoli R.P. (2001). "*Computer integrated assembly for cost effective developments*", C.T. Leondes, Ed.: Computer Integrated Manufacturing, Vol. II, CRC Press, Boca Rato, pp. 2.01-2.68.

Acaccia G.M., Marelli A., Michelini R.C., Zuccotti A. (2003). "*Automatic fabric storing and feeding in quality clothing manufacture*", Intl. J. Intelligent and Robotic Systems, vol. 37, Aug., pp. 443-465.

Acaccia G.M., Conte M., Maina D., Michelini R.C. (2003). "*Computer simulation aids for the intelligent manufacture of quality cloths*", Computers in Industry, vol 50, n°1, pp. 71-84.

Michelini R.C., Acaccia G.M. (2006). "*Distributed intelligence in shop-floor organisation: quality clothes manufacture from fabric warehousing to sewn garments*", in J.X. Liu, Ed.: New Development in Robotic Research, Nova Science, New York, pp. 121-172.

Section 4.3.

This last chapter section addesses the development of particular robotic fixtures. The special contribution of the Genova «Industrial robot design research group» comes out from the integrated approach, where the robot equipment is viewed as component of the *intelligent automation*, not as standalone achievement. The special purpose design aids, accordingly, required hard labour, to standardise and simplify the duty-oriented choice of the front-end robots.

Michelini R.C., Polledro P.L., MarcantoniTaddei C. (1978). *"Position steering of industrial robots by nonlinear controllers"*, 8th Intl. Symp. on Industrial Robots and 4th Conf. on Industrial Robot Tecnology, Stuttgart, 31 may - 1 june.

Acaccia G.M., Michelini R.C., Molfino R.M. (1984). *"Computer-aided design procedures for industrial robots: online generation of the dynamics equations"*, Intl. Conf. CAD '84, Nice, 19-21 June, pp. 183-187.

Bonsignorio F., Michelini R.C., Molfino R.M., Piaggio P.A. (1985). *"Polynomial control for assembly robots"*, 7th IASTED Intl. Conf. Robotics & Automation, Lugano, June 24-26, pp. 98-101.

Acaccia G.M., Callegari M., Michelini R.C., Molfino R.M. (1986). *"Architectur analysis of robotic industrial manipulators"*, Intl. Conf. IASTED-AFCET: Robotics & Artificial Intelligence (Vol. II), Toulouse, June 18-20, pp. 579-598.

Acaccia G.M., Michelini R.C., Molfino R.M. (1987). *"Development of CAD-codes for the job-integration of industrial robots"*, Intl. J. Robotics, vol. 3, n° 3/4, pp. 371-388.

Acaccia G.M., Michelini R.C., Molfino R.M., Pampagnin F, Rossi GB. (1988). *"Knowledge-based programming instruments for robotised manufacturing facilities"*, Intl. J. Advanced Manufacturing Technology, vol.3, n°3, pp. 53-66.

Michelini R.C., Pampagnin F., Rossi GB. (1988). *"Developments in factory automation: assembly specifications and govern-for-flexibility attainments"*, 3rd Intl. Conf. Computer Aided Production Engineering, Ann Arbor, June 1-3, pp. 131-149.

Acaccia G.M., Michelini R.C., Molfino R.M., Recine M.A. (1990). *"Simulational programming environment for the development of industrial multirobot systems"*, ISCIE-ASME Symp. on Flexible Automation, Kyoto, July 9-12, pp. 849-855.

Acaccia G.M., Callegari M., Michelini R.C., Molfino R.M. (1990). *"A tactical tool-supplier for flexible manufacturing"*, 2nd Intl. Conf. on Advanced Manufacturing Systems and Technology (Vol. II), Trento, Italy, 19-21 Jun., pp. 454-460.

Acaccia G.M., Michelini R.C., Molfino R.M., Recine M.A. (1991). *"Information reference set-up for the development of industrial multirobot systems"*, Intl. J. Computer Applications in Technology, vol. 4, n° 3, pp. 137-148.

Acaccia G.M., Michelini R.C., Molfino R.M., Recine M.A. (1991). *"Modeling the coordination of multirobot equipment"*, 6th Intl. Conf. CAD/CAM, Robotics and Factories of the Future, London, 19-22, Aug., pp. 870-876.

Acaccia G.M., Michelini R.C., Molfino R.M., Recine M.A. (1991). *"Assessment of position/force dynamic control performances for advanced robotics"*, 5th Intl. Conf. on Advanced Robotics ICAR'91, Pisa, 20-22 June, pp. 1465-1469.

Acaccia G.M., Callegari M., Michelini R.C., Molfino R.M., Orlando P. (1992). *"Assessing the dynamics of robotic manipulators with articulated closed loops"*, Convegno

Nazionale AIMETA, Trento, 29 Sept.-2 Oct. 1992, vol. 3, pp. 9-14.

Acaccia G.M., Callegari M., Caracciolo R., Michelini R.C., Molfino R.M., Torbidoni M. (1992). "*Redundant position/force control for advanced robotic applications*", 6th Intl. Conf. Systems Research Informatics and Cybernetics, BadenBaden, Aug. 17-23, pp. 63-72.

Acaccia G.M., Aiachini C., Callegari M., Michelini R.C., Molfino R.M., (1993). "*Dynamic control for advanced robotic applications*", Intl. J. Systems Automation: Research & Applications - SARA J. Ablex Publ., n° 4, pp. 221-227.

Acaccia G.M., Callegari M., Michelini R.C., Molfino R.M. (1993). "*Virtual reality technique for the development and integration of robotic manipulators*", Intl. Conf. on CAD-CAM, Robotics and Factories of the Future, St. Petersburg, May 17-20, Vol. II, pp. 425-430.

Acaccia G.M., Cagetti P., Callegari M., Michelini R.C., Molfino R.M. (1994). "*Modeling the impact dynamics of robotic manipulators*", Intl. IFAC Symp. on Robot Control SYROCO '94, Capri, Sept. 19-21, pp. 559-564.

Cagetti P., Michelini R.C., Pampagnin F., Razzoli R. (1994). "*SIRIAT: an animation module for virtual reality simulation of robotic manipulators*", 27th. ISATA on Mechatronics, Aachen, 31th Oct.-4th Nov., pp. 609-616.

Acaccia G.M., Callegari M., Consano L., Michelini R.C., Molfino R.M., Pampagnin S., Razzoli R.P. (1995). "*Universal master for remote micro-manipulation*", Intl. Conf. Advanced Robotics and Intelligent Automation, Athens, 6-8 Sept., pp. 499-504.

Michelini R.C., Acaccia G.M., Callegari M., Molfino R.M., Razzoli R.P. (1996). "*Innovation in robotics: concurrency operation, co-operation and mobility and control redundancy*", 27th. Intl. Symp. on Industrial Robots, Milano, 6-8 Oct., pp. 37-42.

Acaccia G.M., Callegari M., Michelini R.C., Molfino R.M., Razzoli R.P. (1996). "*Assessing the dynamics of articulated manipulators with closed kinematic chains*", 27th. Intl. Symp. on Industrial Robots, Milano, 6-8 Oct., pp. 575-580.

Acaccia G.M., Callegari M., Michelini R.C., Molfino R.M. (1996). "*The impact dynamics of robotic arms*", Intl. Conf. Advanced Robotics and Intelligent Automation, Wien, Sept. 26-28, pp. 313-320.

Acaccia G.M., Callegari M., Michelini R.C., Molfino R.M., Razzoli R.P. (1997). "*Dynamics of a co-operating robotic fixture for supporting automatic deburring tasks*", Intl. Conf. Informatics and Control, St. Petersburg, June 9-13, pp. 1244-1254.

Michelini R.C., Kovàcs G.L. (1999). "*Knowledge organisation and govern-for-flexibility in lean manufacturing*", A.B. Baskin, G.L. Kovàcs, G. Jacucci, Eds: 'Co-operative Knowledge Processing for Engineering Design', Kluwer, , pp. 61-82.

Acaccia G.M., Callegari M., Michelini R.C., Molfino R.M., Razzoli R.P. (1998). "*Exploiting functional and command redundancy for the process attuning of instrumental robots*", 4th. ECPD Intl. Conf. Advanced Robotics, Intelligent Automation & Active Systems, Moscow, Aug. 24-26, pp. 373-383.

Acaccia G.M., Bruzzone L., Callegari M., Michelini R.C., Molfino R.M., Razzoli R.P. (1998). "*Functional assessment of the impedance controller of parallel actuated robotic six d.o.f. rig*", Proc. 6th IEEE Mediterranean Conference on Control and Systems MCCS, Tornambè A., Conte G. & Perdon A.M. Eds., June 9-11, Alghero, Italy, pp. 397-402.

Michelini R.C., Molfino R.M., Cattaneo D., Callegari M. (1999). "*The conceptual design of a parallel-kinematics manipulator for high speed assembly tasks*",. Intl. Workshop on Parallel Kinematics Machines (PKM99), Milano, Italy, Nov. 30th, pp. 220-223.

Acaccia G.M., Bruzzone L., Michelini R.C., Molfino R.M., Razzoli R.P. (1999). *"Parallel robot co-operating with a 3 D.O.F. machining center in deburrig tasks"*, Intl. Workshop on Parallel Kinematics Machines (PKM99), Milano, Italy, Nov. 30th, pp. 47-52.

Cavallo E., Michelini R.C., Molfino R.M., Razzoli R.P. (2001). *"Task-driven design of a reconfigurable gripper for the robotic picking and handling of limp sheets"*, Intl. CIRP Seminar: Design in the new economy, Stockholm, 6-8 June, pp. 79-82.

Bruzzone L.E., Michelini R.C., Molfino R.M., Zoppi M. (2003). *"Constraints singularities of force transmission in nonredundant parallel robots with less than six degrees-of-freedom"*, ASME Trans. J. of Mechanical Design, vol. 125, pp. 557-563.

Michelini R.C., Zoppi M. (2004). *"Under-actuated hands for rods balanced handling"*, Intl. Conf. Intelligent Manipulation & Grasping, Genova, July 1-2, pp. 312-317.

CHAPTER 5. ROBOTS IN SERVICE APPLICATIONS

Section 5.1.

This chapter, as well as the previous, provides the bird eye view of the personal activity. The previously existing research group on industrial robotics, extended its activity, and the previous «Laboratory on computation methods for the design of dynamical systems», transformed into the «Laboratory on integrated design for expert automation, robotics and measurements». This first section of the chapter gathers series of general studies aiming at the «action choices», preliminary and most important duty, when the design of service robot has to be accomplished.

Acaccia G.M., Callegari M., Hageman D., Michelini R.C., Molfino R.M., Pampagnin S., Razzoli R.P., Schwenke H. (1995). *"The design of a robotic head for active vision"*, Workshop on Active Vision Hardware, the European Computer Vision Network, Grenoble, 1-3 Feb., pp. 7·1-48.

Acaccia G.M., Callegari M., Hageman D., Michelini R.C., Molfino R.M., Pampagnin S., Razzoli R.P., Schwenke H. (1995). *"Robotic fixture to experiment anthropomorphic vision"*, ICAR '95 Intl. Conf. Advan. Robotics, Barcelona, Sept. 20-22, pp. 237-244.

Michelini R.C., Callegari M., Rossi GB. (1996). *"Robots with uncertainty and intelligent automation"*, Intl. Conf. Advanced Robotics and Intelligent Automation, Wien, Sept. 26-28, pp. 31-39.

Michelini R.C., Kovàcs G.L. (2003). *"Intelligent integrated design for sustainability: products-services"*, 5th Intl. Conf. Computer Science and Information Technologies, Ufa, Russia, Sept. 16-18, vol 1, pp. 31-38.

Michelini R.C., Kovàcs G.L. (2005). *"Information infrastructures and sustainability"*, in: L. Camarinha Matos, Ed., Emerging Solutions for Future Manufacturing Systems, Springer, pp.347-356.

Section 5.2.

This second chapter section collects example projects, basically, dealing with the master-slave architectures, where the front end robots are extension or specialisation of human opertors, permitting remote manipulation, overseeing, etc. tasks, for the many reasons that migth impose or suggest the elimination of on-process attendants.

Acaccia G.M., Callegari M., Michelini R.C., Molfino R.M., Razzoli R.P. (1995). *"Dynamics of a multi-powered platform for task-steered instrumental robots"*, IX IFToMM World Conf. on Theory of Machines and Mechanisms, Milano, Aug. 30-Sept. 2, pp. 1816-1821.

Michelini R.C., Molfino R.M., Razzoli R.P., Rio A., Truffelli F. (1999). *"A parallel kinematics robotic arm for deep sea operations"*, Intl. Conf. Parallel Kinematics Machines (PKM99), Milano, Nov. 30th, pp. 243-250.

Acaccia G.M., Cavallo E., Garofalo E., Michelini R.C.,.Molfino R.M, Callegari M. (1999). *"Remote manipulator for deep-sea operations: animation and virtual reality assessment"*, 2nd Conf. Harbour, Maritime & Logistics Modelling and Simulation, HMS99, 16-18 Sept., Genova, pp.57-62.

Bruzzone L.E., Molfino R.M., Acaccia G.M., Michelini R.C., Razzoli R.P. (2000). *"A tethered climbing robot for firming up high-steepness rocky walls"*, Intl. Conf. Intelligent Autonomous Robotic Systems, Venezia, July 25-27, pp 307-312.

Anthoine P., Armada M., Carosio S., Comacchio P., Cepolina F., Gonzales P., Klopf T., F.Martin, R.C.Michelini, R.M.Molfino, S.Nabulsa, R.P.Razzoli, E.Rizzi, L.Steiniche, Zannini R., Zoppi M. (2003). "The *roboclimber project*", Intl. Conf. Advances in Service Robotics AseR 03, Verona, March 13-15.

Cepolina F., Matteucci F., Michelini R.C. (2004). *"Self-adaptable clamping tools for multiple seizure"*, Intl. Conf. Intelligent Manipulation & Grasping, Genova, July 1-2, pp. 308-311.

Cavallo E., Michelini R.C., Molfino R. (2004). *"A remote-operated robotic platform for undewater decommissioning tasks"*, 35th Intl. Symposium on Robotics, ISR 2004, Paris, March 23-26, p. 99 (5) th 14-2.

Barbieri A., Michelini R.C., Zoppi M. (2004). *"An underground robotic equipment for leachate draining and landfills remediation"*, 35th Intl. Symposium on Robotics, ISR 2004, Paris, March 23-26, 2004, p. 101 (5) th 14-5.

Cepolina F., Michelini R.C. (2004). *"Robots in medicine: a survey of in-body nursing aids. Introductory overview and concept design hints"*, 35th Intl. Symposium on Robotics, ISR 2004, Paris, March 23-26, p. 67 (5) we 23-5.

Anthoine P., Armada M., Carosio S., Comacchio P., Cepolina F., Gonzales P., Klopf T., F.Martin, R.C.Michelini, R.M.Molfino, S.Nabulsa, R.P.Razzoli, E.Rizzi, L.Steiniche, Zannini R., Zoppi M. (2004). *"A four-legged climbing robot for rocky slope consolidation and monitoring"*, World Automation Congress, WAC04, Seville, June 28- July 1.

Cepolina F., Michelini R.C. (2004). *"Review of robotic fixtures for minimal invasive surgery"*, Intl. J. Medical Robotics & Computer-Assisted Surgery, vol. 1, n° 1, pp. 43-63.

Cavallo E., Michelini R.C., Molfino R. (2004). *"A robotic system for off-shore plants decommissioning"*, IFAC Intl. Conf. Control Applications in Marine Systems, CAMS 04, Ancona, 7-9 July, pp. 131-137.

Cavallo E., Michelini R.C., Molfino R. (2004). *"The decommissioning of submerged structures: prototype equipment design and assessment"*, Intl. Symp. Offshore and Polar Engineering, ISOPE, S. 6, Toulon, May 22-28, vol. I, pp. 509-514.

Belotti V., Crenna F., Michelini R.C., Rossi G.B. (2005). *"Remote sensing and control with client-server architecture, with application to a robot for landfill definitive remediation"*, Intl. IMEKO TC1-TC7 Symp., Ilmenau, Sept. 21-24, 2005.

Belotti V., Michelini R.C., Zoppi M. (2005). *"Remote control and monitoring of an underground robotic drilling equipment for landfills remediation"*, Intl. Symp. Automation and Robotics in Construction, ISARC, Ferrara, Sep. 11-14, pp. 58-68.

Cepolina F., Michelini R.C. (2005). *"Trends in robotic surgery"*, Intl. J. Recent Advances in Urology, vol. 19, n° 8, pp. 940-951.

Frumento S., Michelini R.C., Konietschke R., Hagn U., Ortmaier T., Hirzinger G. (2005). *"A co-robotic positioning device for carrying surgical end-effectors"*, Intl. Conf. ASME-ESDA Torino, July 4-7, paper 95308 pp. 1-8.

Belotti V., Crenna F., Michelini R.C., Rossi GB. (2006). *"Wheel-flat diagnostic tool, using the wavelet transform"*, Intl. J. Mechanical Systems and Signal Processing, vol. 20, pp. 1953-1966.

Belotti V., Michelini R.C., Zoppi M. (2006). *"Remote-controlled underground robot for landfill drainage"*, Intl. Conf. ASME-ESDA Torino, July 4-7, paper 95465 pp. 1-8.

Belotti V., Crenna F., Michelini R.C., Rossi G.B. (2007). *"Remote sensing and control with client-server architecture, with application to a robot for landfill definitive remediation"*, Intl. J. Measurement, vol. 40, pp. 109-122.

Belotti V., Michelini R.C. (2007). *"On-board virtual instrument to ethernet control of an underground drilling machine for drainage piping lay-dawn"*, Intl. Conf. Remote Engineering and Virtual Instrumentation REV07, Porto, June 24-27.

Belotti V., Hemapala M.U., Michelini R.C., Razzoli R.P. (2008). *"Remote robotic control and mine clearing"*, Intl. Conf ASME-ESDA 08., Haifa, July 5-7, paper n° 59397.

Michelini R.C., Razzoli R.P., Hemapala M.U. (2008) *"Humanitarian demining: efficiency by intelligent planning and low-cost robotics"*, Intl. EUROSIS Industrial Simulation Conf. ISC, Lyon, June 9-11.

Michelini R.C., Razzoli R.P. (2008). *"Co-operative minimal invasive robotic surgery"*, Industrial Robot J., vol. 35, n. 4, p. 347-369.

Belotti V., Hemapala M.U., Michelini R.C., (2008). *"Humanitarian demining: path planning and remote robotic sweeping"*, Industrial Robot J., vol. 35, n. 3, ID IR-07-572.

Section 5.3.

The last chapter section assemble explanatory projects requesting autonomic operation robots, at different levels of the local intelligence (due to on-board sensors) and decisio-keeping capabilities. The shortly commented approaches are chosen provide hints on the large veriety of situations that the «robot age» spirit permit to solve.

Amodeo G., Guglielmino E., Messina M., Michelini R.C. (1996). *"A robotic fixture for orange harvesting"*, 27th. Intl. Symp. on Industrial Robots, Milano, 6-8 Oct., pp. 173-176.

Acaccia G.M., Callegari M., Garbato M., Grosso S., Michelini R.C., Molfino R.M., Razzoli

R.P. (1997). "*The roboranger project: fuzzy controller for four powered wheels*", 2nd. IARP Intl. Conf. on Service and Personal Robots: Technol. & Appl., Genova, Oct. 23-24, pp. 8.01-8.16.

Callegari M., Khan A., Michelini R.C., Molfino R.M., Razzoli R.P. (1997). "*The robo-ranger project: styling and concept design*", 10th. ADM Intl. Conf. Design Tools and Methods in Industrial Engineering, Firenze, Sept. 17-19, pp. 1-6.

Michelini R.C., Acaccia G.M., Callegari M., Molfino R.M., Razzoli R.P. (1997). "*Robot harvesting of citrus fruits*", Intl. Conf. on Advanced Robotics, Intelligent Automation and Active Systems, Bremen, Germany, Sept. 15-17, pp. 447-452.

Acaccia G.M., Callegari M., Michelini R.C., Molfino R.M., Razzoli R.P. (1998). "*The Roboranger: development of the city car independent powered wheels controller*", 2nd Intl. ATA Conf. Control and Diagnostics in Automotive Applications, Genova, Oct. 29-30, pp. 155-164.

Acaccia G.M., Callegari M., Michelini R.C., Molfino R.M., Razzoli R.P. (1998). "*Greenhouse's automation: produce tillage and phytopatologies treatments*", 4th. ECPD Intl. Conf. Advanced Robotics, Intelligent Automation & Active Systems, Moscow, Aug. 24-26, pp. 339-343.

Acaccia G.M., Callegari M., Michelini R.C., Molfino R.M., Razzoli R.P. (1998). "*Underwater robotics: example survey and suggestions for effective devices*", 4th. ECPD Intl. Conf. Advanced Robotics, Intelligent Automation & Active Systems, Moscow, Aug. 24-26, pp. 409-416.

Cepolina F., Michelini R.C., Molfino R.M., Razzoli R.P. (1999). "*Domestic-chores automation: multi-media analysis and assessment study*", XI ADM Intl. Conf. Design Tools and Methods in Industrial Engineering, Palermo 8-12 Dec., pp. 139-146.

Callegari M., Cavallo E., Garofalo E., Michelini R.C., Molfino R.M., Razzoli R.P. (1999). "*The design of diving robots: set-up assessment by virtual mock-ups*", ADM Intl. Conf. on Design Tools and Methods in Industrial Engineering, Palermo 8-12 Dic., pp. 147-154.

Cepolina F., Michelini R.C., Molfino R.M., Razzoli R.P. (2000). "*Gecko-Collie: home-cleaning automation of floors, walls and cupboards*", 3rd Intl. Conf. CLAWAR Climbing and Walking Robots, Madrid, Oct. 2-4, pp. 803- 812.

Bruzzone L.E., Cavallo E., Michelini R.C., Molfino R.M., Razzoli R.P. (2000). "*The design of a robotic equipment for deep-sea maintenance operations*", 5th ASME Intl. Conf. Engineering Systems Design and Analysis, Montreux, July 10-13, pp. 113-120.

Cepolina F., Michelini R.C., Molfino R.M., Razzoli R.P. (2001). "*Char-robot: the design of a co-operative equipment for kitchen cleaning and sanitising tasks*", European Workshop on Service Robots, Santorini, Greece, June 25-27, pp. 111-118.

Acaccia G.M., Corvi M., Ferebauer W., Michelini R.C. (2001). "*Assisted mail sorting and forwarding stands: performance analysis and ergonomic assessment*", G.L. Kovacs, P. Bertok, G. Haidegger, Eds.: Digital Enterprise, New Challenges, Kluver, Boston, pp. 543-554.

Cavallo E., Michelini R.C., Molfino R.M., Razzoli R.P. (2001). "*Robotic equipment for deep-sea operation: digital mock-up and assessment*", G.L. Kovacs, P. Bertok, G. Haidegger, Eds.: Digital Enterprise, New Challenges, Kluver, Boston, 2001, pp. 533-542.

Becchi F., Michelini R.C., Molfino R.M., Razzoli R.P. (2001). "*SARA: a robotic system for car fuelling*", Intl. J. Intelligent and Robotic Systems, n° 32, pp. 37-54.

Cepolina F., Michelini R.C., Molfino R.M., Razzoli R.P. (2001). "*Char-Robot: design of a co-operative equipment for kitchen cleaning and sanitising tasks*", Systems Science J., special issue on Virtual reality and mobile/service robots, May, pp. 109-115.

Fukuda T., Michelini R.C., Potkoniak V., Tzafestas S., Valavanis K., Vukobratovic M. (2001). "*How far away is 'Artificial Man'*", IEEE Robotics & Automation Magazine, vol. 7, n° 1, pp. 66-73.

Cepolina F., Michelini R.C. (2002). "*Gecko: the walls cleaner*", Intl. J. Industrial Robots, vol. 29, n° 6, pp. 538-543.

Acaccia G.M., Michelini R.C., Molfino R.M., Razzoli R.P. (2002). "*Mobile robots in greenhouse cultivation: inspection and treatment of plants*", CLAWAR News, n° 9, Nov. 2002, pp.14-16.

Cepolina F., Michelini R.C., Molfino R.M., Razzoli R.P. (2003). "*Gecko project*", Intl. Conf. Advances in Service Robotics, AseR 03, Verona, March 13-15.

Acaccia G.M., Michelini R.C., Molfino R.M., Razzoli R.P. (2003). "*Robots in greenhouses farming: plants inspection and treatment*", Intl. Workshop on Advances in Service Robotics, AseR 03, Verona, March 13-15.

Cavallo E., Michelini R.C. (2004). "*A robotic equipment for the guidance of a vectored thrustor AUV*", 35th Intl. Symposium on Robotics, ISR 2004, Paris, March 23-26, p. 88 (5) we 43-1.

Cavallo E., Michelini R.C., Molfino R.M. (2004). "*The unrigging of submerged structures: prototype design and assessment*", 7th ASME Conf. Engineering Systems and Design Analysis, ESDA 2004, Manchester, 19-22 July.

Acaccia G.M., Michelini R.C., Qualich N. (2005). "*End-of-life vehicles collection and disassembly: modelling and simulation*", Joint ESM-MESM Conf., EUROSIS 05, Porto, Oct. 24-26, pp. 34-40.

Cavallo E., Michelini R.C., Filaretov V.F., Ukimets D.A. (2004). "*Path guidance and attitude control of a vectored thrustor AUV*", 7th ASME Conf. Engineering Systems and Design Analysis, ESDA 2004, Manchester, 19-22 July.

Cavallo E., Michelini R.C. (2004). "*Operation assessment of selfcompensating vectored thrustor AUV*", Intl. Symp. Offshore and Polar Engineering, ISOPE 2004, Toulon, May 22-28, vol. II, pp. 244-249.

Cavallo E., Michelini R.C., Filaretov V.F. (2004). "*Conceptual design of an AUV equipped with a 3dof vectored thrustor*", Intl. J. Intelligent and Robotic Systems, vol. 39, n. 4, pp. 365-391.

Michelini R.C., Razzoli R.P. (2004). "*Product-service for environmental safeguard: a metric to sustainability*", Intl. J. Resources, Conservation and Recycling, vol. 42, n° 1, pp. 83-98.

Acaccia G.M., Michelini R.C., Qualich N. (2005). "*Mixed automation of a sorting postal facility: design and performance assessment*", Joint ESM-MESM Conf., EUROSIS 05, Porto, Oct. 24-26, pp. 41-49.

Michelini R.C., Filaretov V.F. (2005). "*The ecoIMP, eco-impact marine patroller, project: development of reference set-ups and operation tools for the protected exploitability of the marine environment*", 6th. EU Framework Programme, RUSERA Brokerage Event, Project № 50, Moscow 27-28 January.

Michelini R.C., Filaretov V.F. (2005). "*The SWAN, sub-marine wobble-free autonomous navigator, project: development of an AUV new generation control, for under-water eco-system survey and reclamation/safety missions*", 6th. EU Framework Programme, RUSERA

Brokerage Event, Project № 48, Moscow 27-28 January.

Cavallo E., Michelini R.C., Filaretov V.F., Ukimets D.A. (2005). "*The design features and control system of autonomous underwater vehicle with a thruster for its spatial movement*", Problemy Mashinostroeniya i Nadezhnosti Mashin (J. of Machine Manufacture and Reliability), vol. 6, 2005, pp. 98-107, *in Russian*.

Cavallo E., Michelini R.C., Filaretov V.F., Ukimets D.A. (2005). "*Control features of a vectored thruster underwater vehicle*", 16th IFAC World Congress, Prague, July 4-8.

Cavallo E., Michelini R.C., Filaretov V.F., Ukimets D.A. (2005). "*The designing and control features of autonomous underwater vehicles using a single thruster*", J. Mechatronics, Automation and Control, n° 11, pp.21-36, *in Russian*.

Acaccia, G.M., Michelini, R.C., Qualich, N. (2007). "*Sustainable engineering management: end-of-life vehicles with recycling in mind*", World Review of Science, Technology and Sustainable Development, WRSTSD, vol. 4, n. 2/3, pp. 105-125.

Belotti V., Michelini R.C., Razzoli R.P. (2007). "*Remote overseeing of car brake pedal for reuse/recycle and pre-collision history investigation*", Intl. Conf. Remote Engineering and Virtual Instrumentation REV2007, Porto, June 24-27.

Kovàcs G., Kopàcsi S., Haidegger G., Michelini R.C. (2006). "*Ambient intelligence in product lifecycle design*", J. Engineering Applications of Artificial Intelligence, Vol. 19, n. 8, pp. 953-965.

Acaccia G.M., Michelini R.C., Penzo L., Qualich N. (2006). "*Automotive systems: end-of-life vehicles dismantling facilities*", Intl. Conf. ASME-ESDA Torino, July 4-7, paper 95332 pp.1-8.

Acaccia G.M., Michelini R.C., Qualich N. (2006). "*End-of-life vehicles models with recycling in mind*", ECEC & FUBUTEC Intl. Conf., Athens, Apr. 17-19, pp. 76-82.

Acaccia G.M., Michelini R.C., Penzo L., Qualich N. (2006). "*Modelling and simulation of car dismantling facilities*", Joint ECEC & FUBUTEC Intl. Conf., Athens, Apr. 17-19, pp. 70-75.

Acaccia G.M., Kopàcsi S., Kovàcs G., Michelini R.C., Razzoli R.P. (2007). "*Service engineering and extended artefact delivery*", G.D. Putnik, M.M. Cunha, Eds., Knowledge and Technology Management in Virtual Organisations, IGI Press, Hershey, pp. 45-77.

Acaccia G.M., Michelini R.C., Penzo L., Qualich N. (2007). "*Reverse logistics and resources recovery: modelling car dismantling facilities*", World Review of Science, Technology and Sustainable Development, WRSTSD, vol. 4, n. 4/5, pp. 127-136.

A.Anufriev, S.Kopàcsi, G.Kovàcs, R.C.Michelini: "*Ambient intelligence as enabling technology for modern business paradigms*", J. Robotics and Computer Integrated Manufacturing, Elsevier, Vol. 23, n. 2, April 2007, pp. 242-256.

Michelini R.C., Razzoli R.P. (2007). "*Take-back of end-of-use vehicles, recovery of parts/material and landfill management*", XVI ADM Intl. Conf. Tools and Methods Evolution in Engineering Design, 6-9 June, Perugia, MSP1 18 pp. 017.

Michelini R.C., Razzoli R.P. (2007). "*Ubiqutous computing & communication for product monitoring*", M.Khosrow-Pour, Ed.: Encyclopaedia of Information Science and Technology, 2nd Ed, IDEA Group Inc., 2007.

Belotti V., Michelini R.C., Razzoli R.P. (2008). "Lifecycle monitoring for automotive eco-sustainability", ASME-ESDA 08 Conf., Haifa, July 5-7, paper n° 59395.

Michelini R.C., Razzoli R.P. (2008) "Modelling and simulation features for products-services", Intl. Conf. EUROSIS Industrial Simulation ISC, Lyon, June 9-11, 2008, pp.

256-261.

CHAPTER 6. REMARKS, PROSPECTS AND CONCLUSION

Section 6.1.

The ideas of the chapter have been object, in recent years, of discussions with colleagues, being, however, quite personal viewpoints. The «robot age» is announced technological turmoil, which, still, might enter within standard engineering trends, whether the outer economical constraints do not interfere with the known industrialism patterns. In the first chapter section, the further reading suggestions privilege such continuity, imagining the multiple industrial revolution issues, as it happened with the agricultural revolution.

Bernstein W.J. (2008). *"A splendid exchange: how trade shaped the world"*, Atlantic Books, London, p. 467.
Bhagwati N. (2004). *"In defence of globalisation"*, Oxford Univ. Press, New York, p. 320.
Chambers R. (1995). *"Poverty and livelihoods"*, Environment and Urbanization, Vol. 7, No. 1, 173-204.
Chang H-J. (2008). *"Bad Samaritans: rich nations, poor policies & the threat of developing world"*, Random House, London, p. 276.
Diamond J. (2005). *"Collapse: how societies choose to fail or succeed"*, Viking, New York, p. 575.
Dreher A., Gaston N., Martens P. (2008). *"Measuring globalisation, and gauging its consequences"*, Springer, London, p. 224.
Easterly W. (2001).*"The elusive quest for growth: the economists' adventures and misadventures"*, MIT Press, Cambridge, pp. 342.
Freeland C. (2000). *"Sale of the century: inside story of the second Russian revolution"*, Little, Brown & Co., London, pp. 370.
Hoogendijk W. (1993).*"The economic revolution: towards a sustainable future by freeing the economy from money-making"*, International Books, Utrecht, pp. 208.
Humphreys M., Sachs J.D., Stiglitz J.E. (2008). *"Escaping the resource curse"*, Columbia Uni. Press, New York, p. 408.
Jones E. (2003). *"The European miracle: environments, economies and geopolitics in the history of Europe and Asia"*, 3rd ed., Cambridge Uni. Press, Cambridge.
Kachru B.B. (1983). *"The Indianization of English: the English language in India"*, Oxford Uni. Press, Oxford.
Keynes, J.M. (1936). *"The general theory of employment, interest and money,"* Harcourt Brace, New York.
Malherbe M. (1995). *"Les langages de l'humanité"*, Robert Laffont, Paris, pp. 1734.
Meyer M. (2007). *"The last days of old Beijing: life in the vanishing backstreets of a city"*, Bloomsbury Co., London, p. 353.
Minc A. (2008). *"Une sorte de diable: les vies de John-Maynard Keynes"*, Grasset, Paris.
Ndione E. (1994). *"Réinventer le présent"*, Enda-Graf Sahel, Dakar, pp. 131.
Needham J. (1969). *"Within the four seas: the dialogue of east and west"*, Allen & Unwin,

London.

Pasinetti L. (2008). *"Keynes and Cambridge Keynesians: a revolution in economics to be accomplished"*, Cambridge Uni. Press, London, p. 368.

Patten C. (2005). *"Not quite the diplomat: home truths about world affairs"*, Allen Lane, London.

Perna T. (1998). *"Fair trade: la sfida etica al mercato mondiale"*, Bollati Boringhieri, Torino, pp. 192.

Pezzey J. (1989). *"Economic analysis of sustainable growth and development"*, World Bank, WP 15, New York, pp. 88.

Polanyi K. (1977). *"The livelihood of man"*, Academic Press, New York, pp. 335.

Putnam R.D. (2000). *"Bowling alone: the collapse and revival of American community"*, Simon & Shuster, New York, pp. 544.

Reich R.B. (2007). *"Supercapitalism: the transformation of business, democracy and everyday life"*, Borzoi Book, New York, p. 272.

Roberts J.M. (1993). *"History of the world"*, Oxford Uni. Press, Oxford.

Sen A. (2001). *"Development as freedom"*, Oxford Univ. Press., New York, pp. 366.

Soros G. (2008). *"The new paradigms for financial markets: the credit crunch of 2008 and what it means"*, Public Affairs, New York, p. 208.

Steckel R., Floud R., eds. (1997). *"Health and welfare during industrialisation"*, Univ. of Chicago Press., Chicago, pp. 465.

Stiglitz J.E. (2007). *"Making globalisation work"*, W.W. Norton, New York, pp. 320.

Stillman D. (2008). *"Mustang: the saga of the wild horse in the American west"*, Houghton Mifflin, p. 348.

Sussman R. (1998). *"The biological basis for human behaviour"*, Prentice & Hall, NJ, pp. 448.

Targetti F., Fracasso A. (2008). *"Le sfide della globalizzazione: storia, politica e istituzioni"*, F. Brioschi, Milano, p. 604.

Teune H. (1988). *"Growth"*, Sage Publication, London, pp. 141.

Tremonti G. (2008). *"La paura e la speranza: Europa crisi globale che si avvicina e la via per superarla"*, Mondadori, Milano, p. 114.

Turner G. (2008). *"The credit crunch: housing bubbles, globalisation and the worldwide economic crisis"*, Pluto Press, London, p. 268.

Villar F. (1996). *"Los indoeuropeos y los origines de europa: language y historia"*, Editorial Gredos, Madrid.

Wang D.J. (1979). *"The history of Chinese logical thought"*, People's Press, Shanghais.

Whorf B.L. (1956). *"Language, thought and reality"*, MIT Press, MA, pp. 278.

Winchester S. (2008). *"The man who loved China: the fantastic story of the eccentric scientist who unlocked the mysteries of the middle kingdom"*, Harper Collins, New York, pp. 336.

Womack, J.P., Ross, D. & Jones, D.T. (1990). *"The machine that changed the world,"* Rawson Ass. New York, pp. 323.

Section 6.2.

The second chapter section deals with the mainly *no-global* concepts, providing basic overviews on why the ecologic concerns are fundamental requirements.

Allenby B.R., Graedel T.E. (1995). *"Industrial ecology"*, Prentice Hall.
Amin S. (1990)."*Maldevelopment: anatomy of a global failure*", Zed Book, London, pp. 244.
Cahn E., Rowe J. (1992). *"Time dollars"*, Rodale Press, Emmaus, PA, pp. 272.
Charbonneau S. (2006). *"Droit communautaire de l'environnement"*, L'Harmattan, Paris, pp. 295.
Connelly M. (2008). *"Fatal misconception: the struggle to control the world population"*, Harvard Uni. Press, Boston, p. 544.
Diamond J. (2005). *"Guns, germs and steel: the fates of human societies"*, W. W. Norton, New York.
Ferguson N. (1988). *"The pity of war"*, Penguin Books, New York.
Friedrich R., Bickel P. (2001). *"Environmental external costs of transport"*, Springer Berlin, p. 321.
Ghani A., Lockhart C. (2008). *"Fixing failed states: a framework for rebuilding a fractured world"*, Oxford Unii. Press, Oxford, p. 254.
Goldsmith E., Piélat T. (2002)."*Le Tao de l'écologie*", Edition du Rocher, Monaco, pp. 499.
Eldredge N. (2002). *"Life on earth: an encyclopaedia of biodiversity, ecology and evolution"*, ABC-CLIO, Inc. Santa Barbara, pp. 793.
Hillary R. (2000). *"Small and medium sized enterprises and the environment business imperatives"*, Greenleaf Pub., UK, pp. 391.
Kempf H. (2007). *"Comment les riches détruisent la planète"*, Editions du Seuil, Paris, pp. 147.
Kherdjemil B., Panhuys H., Zaoual H. (1998). *"Territoires et dynamiques économiques"*, L'Harmattan, Paris, pp. 228.
Kothary R. (1989). *"Rethinking development: in search of human alternatives"*, New Horizons Press, NJ, pp. 233.
Latouche S. (2004). *"Survivre au développement: de la décolonisation de l'imaginaire économique à la construction d'une société alternative"*, Mille et une Nuits (Edit), Paris, pp. 126.
Latouche S. (2007). *"Petit traité de la décroissance sereine"*, Mille et une Nuits (Edit), Paris, pp. 176.
Lévy B-H. (2008). *"Left dark times: a stand against the new barbarism"*, Random House, New York, p. 256.
Meadows D.H. (2004)."*Limit to growth: the 30 years update*", Chelsea Green, Boston, pp. 368.
Mendes C., Castoriadis C. (1977). *"Le mythe du développement"*, Seuil, Paris, pp. 277.
Michelini R.C. (2008). *"Knowledge entrepreneurship and sustainable growth"*, Nova Sci. Pub., New York, pp. 325.
Myrdal G. (1968). *"Asian drama: an inquiry into the poverty of nations"*, Pantheon, New York, pp. 228.
Naess A. (1989). *"Ecology, community and lifestyle: an eco-sophy outline"*, Cambridge Univ. Press, Cambridge, pp. 223.
Nandy A. (1987). *"The intimate enemy"*, Oxford Univ. Press, Bombay, pp. 194.
Nebel B.J., Wright R.T. (1993). *"Environmental science: the way the world works"*, 4 ed., Prentice Hall, Englewood Cliffs.
Ostrom E., Ahn T.K., eds. (2005). *"Foundation of social capital"*, Edward Elgar Pub., Cheltenham, pp. 590.

Partant F. (1997). "*La fin du développement*", F. Maspero, Paris, pp. 186.

Polanyi K. (1991). "*The great transformation: the political and economic origin of our time*", Beacon Press, Boston, pp. 315.

Pörsken U. (1989). "*Plastikwörter: die Sprache einer internationalen Diktatur*", Klett-Cotta, Stuttgart, pp. 128.

Prado de las Escosura L. (2005). "*Growth, inequality and poverty in Latin America: historical evidence and controlled conjectures*", Working Paper n° 05-41, Univ. Carlos III, Madrid.

Putnam R.D., Leonardi R., Nanetti R.Y. (1993). "*Making democracy work: civic tradition in modern Italy*", Princeton Univ. Press, Priceton, pp. 280.

Rawis J. (1999). "*A theory of justice*", Oxford Univ. Press, Oxford, pp. 538.

Redclift M., Woodgate G., eds. (1997). "*The international handbook of environmental sociology*", Edward Elgar, Cheltenham, pp. 512.

Ricklefs R.E., Miller G.L. (2000). "*Ecology*", 4 ed., W.H. Freeman, New York.

Rist G. (1996). "*Le développement: histoire d'une croyance occidentale*", Presse de Sciences Po, Paris, p. 426.

Sachs J.D. (2008). "*Common wealth: economics for a crowded planet*", Penguin Press, London, p. 374.

Sachs W., ed. (1992). "*The development dictionary*", Zed Book, London, pp. 306.

Sachs W. (1999). "*Planet dialectics*", Zed Book, London, pp. 226.

Said E.W. (1994). "*Culture and imperialism*", Vintage Books, New York, pp. 380.

Sahlins M. (1972). "*Stone age economics*", Aldine-Atherton, Chicago, pp. 348.

Seabrook J. (1993). "*Victims of development*", Verso, London, pp. 250.

de Shalit A. (1998). "*Why posterity matters*", Rutledge, London, pp. 192.

Shiva V. (1989). "*Staying alive: woman, ecology and development*", Zed Book, London, pp. 240.

Schumacher F. (1999). "*Small is beautiful*", Hartley & Marks Pub., pp. 286.

Simmons I.G. (1993). "*Environmental history: a concise introduction*", Blackwell, Oxford.

Stiglitz J.E., ed. (2001). "*Attacking poverty: world development report*", Oxford Univ. Press, New York.

Subramanian A. (2008). "*India's turn: understanding the economic transformation*", Oxford Uni. Press, Oxford, p. 237.

Terrasson F. (2002). "*En finir avec la nature*", Editions du Rocher, Monaco, pp. 309.

Thurow L.C. (1985). "*The zero-sum solution: building a world-class American economy*", Simon & Shuster, New York, pp. 414.

Uchitelle L. (2006). "*The disposable America: layoffs and their consequences*", Knopf, New York, pp. 283.

Vachon R., ed. (1988). "*Alternatives au développement: approches interculturelles à la bonne vie et à la coopération internationale*", Centre interculturel Monchanin, Montréal, pp. 372.

Veleva V., Ellenbecker M. (1996). "*Ecological design*", Inland Press, Washington.

Walker G., King D. (2008). "*The hot topic: what we can do about global warming*", Harcourt, New York, p. 276.

Worster D. (1988). "*The end of the earth: perspectives on modern environment history*", (Ed) Cambridge Univ. Press, New York, pp. 352.

Weizsäcker E.U., Young O.R., Finger M., eds. (2005). "*Limit to privatisation: how to avoid too much a good thing*", Earthscan Pub., London, pp. 414.

Wenzel H., Hauschild M., Alting L. (1997). "*Environmental assessment of products: methodology, tools and case studies in product development*", Chapman and Hall, London, p. 550.

Watkins K. (2002). "*Cultivating poverty: the impact of US cotton subsidies on Africa*", Oxfam Briefing paper n° 30, pp. 1-37.

Zanotelli A. (2006). "*Avec ceux qui n'ont rien*", Flammarion, Paris, pp. 303.

Section 6.3.

The last chapter section deals with the mainly *post-global* concepts, providing basic overviews on why the economic/ecologic concerns are fundamental requirements. The listed references aim at providing "further reading" hints, just to exemplify the topics intricacy, without, anyway, claim at all to be exhaustive, or, simply, to have singled out the most relevant references.

Abe J.M., Dempsey P.E., Basset D.A. (1998) "*Business ecology: giving your organisation the natural edge*", Butterworth-Heinemann, London.

Brown A. (2008). "*Fishing in utopia: Sweden and the future that disappeared*", Granta, London, p. 261.

Carr G. (2001) "*The digital enterprise: how to reshape your business for a connected world*", HBS Press Book.

Chan C-K., Lee H.W.J. (2005) "*Successful strategies in supply chain management*", IDEA Group Inc., IRM Press, p. 300.

Dean P., Cole W.A. (1969). " *British economic growth*", 2nd ed., Cambridge Univ. Press, Cambridge.

Dyson F. (1979). "*Disturbing the universe*", Harper & Row, New York.

Elliott L., Atkinson D. (2008). "*The gods that failed: how blind faith in markets has cost us our future*", Bodley Head, New York, p. 336.

El-Erian M. (2008). "*When markets collide: investment strategies for the age of global economic changes*", Mc-Grow Hill, New York, p. 344.

Fiksel J., ed. (1996) "*Design for environment: creating eco-efficient products and processes*", McGraw-Hill, New York, p. 513.

Friedman T.L. (2005). "*The world is flat: a brief history of the twenty-first century*", Farrar Straus and Giroux, New York, pp. 488.

Gandolfi A. (2008). "*Formicai, imperi, cervelli:introduzione alla scienza della complessità*", Boringhieri, Torino , p. 304.

Hacker P.M.S. (2008). "*Human nature: the categorical framework*", Blackwell, Oxford, p. 326.

Hammer M., Stanton S.A. (1995) "*The re-engineering revolution*", Harper Collins, New York.

Hobsbbawm E. (1996). "*The age of revolution*", Vintage, New York.

Kagan R. (2008). "*The return of history and the end of dreams*", Knopf, New York, p. 112.

Karl T.L. (1997). "*The paradox of plenty: oil booms and petrol-states*", Univ. of California Press, Berkeley, pp. 380.

Khanna P. (2008). *"The second world: empires, and influence in the new world order"*, Random House, New York, p. 496.

Kleinert J. (2004) *"The role of multinational enterprises in globalisation"*, Springer, Kieler Studien, Berlin, p. 211.

Korten D. (1999). *"The post-corporate world: life after capitalism"*, Bartlett-Koehler, San Francisco.

Krugman P. (2007). *"The conscience of a liberal"*, Norton, New York, p. 296.

Kruta V. (2000). *"Les Celtes: histoire et dictionnaire"*, Ed. Robert Laffont, Paris.

Marraffa M., DeCaro M., Ferretti F. eds. (2008). *"Cartographies of the mind"*, Springer, London, p. 374.

Mc Person N. (1994). *"Machines and economic growth"*, Greenwood Press, Westport.

Michelini R.C., Capello A. (1985) *"Misure e strumentazione industriali: segnali e strumenti di misura"*, UTET, Torino, p. 415.

Michelini R.C., Razzoli R.P. (2000). *"Affidabilità e sicurezza del manufatto industriale: la progettazione integrata per lo sviluppo sostenibile"*, Tecniche Nuove, Milano, pp. 278.

Michelini R.C. (2008). *"Knowledge entrepreneurship and sustainable growth"*, Nova Sci. Pub., New York.

Mo J.P.T., Nemes L., eds. (2001) *"Global engineering, manufacturing and enterprise networks"*, Kluwer Academic Pub., New York.

Muller E. (2001) *"Innovation interactions between knowledge-intensive business services and small-and-medium sized enterprises: an analysis in terms of evolution, knowledge and territories"*, Springer, Berlin.

Muther A. (2001). *"Electronic customer care: the supplier-consumer relationship in the information age"*, Springer, Berlin.

Nelson R.R., Winter S.G. (1993) *"An evolutionary theory of economic change"*, Harvard Uni. Press, Boston.

von Neumann J. (in: Burks A.W., 1966). *"Theory of self-reproducing automata"*, Univ. Illinois Press, Urbana.

Newman P., Kenworthy J., Robinson L. (1992). *"Winning back the cities"*, Pluto Press Australia, Leichhardt.

Oesterle H., Fleisch E., Alt R. (2001) *"Business networking: shaping collaboration between enterprises"*, Springer, Berlin, p. 343.

O'Rourke K.H., Williamson J. (1999). *"Globalisation and history: the evolution of the nineteenth century Atlantic economy"*, MIT Press, Cambridge, p. 155.

Penzias A. (1995). *"Harmony: business, technology and life after paperwork"*, HarperCollins, New York.

Peterson P. (1999). *"How the coming age wave will transform Americas and the world"*, Times Books, New York.

Posner R.A. (2008). *"How judges think"*, Harvard Uni. Press, p. 380.

Popov F., DeSimone F.D. (1997) *"Eco-efficiency: the business link to sustainable development"*, The MIT Press, Cambridge.

Putnam H. (2003). *"The collapse of the fact: value dichotomy and other essays"*, Harvard Uni. Press, Cambridge.

Putnik G.D., Cunha M.M. eds. (2005). *"Virtual enterprise integration"*, IDEA Group Pub., Hersey Pa.

Putnik G.D., Cunha M.M. (2007). *"Knowledge and technology management in virtual organisations*", IDEA Group Pub., Hersey Pa.

Quadrio-Curzio A., Fortis M. (2005) "*Research and technological innovation: the challenge for a new Europe*", Springer, London, p. 290.

Rasmussen L.B., Beardon C., Munari S., eds. (2000) "*Computers and networks in the age of globalisation*", Kluwer Academic Pub., New York.

Rathje W., Murphy C. (1992). "*Rubbish!: the archaeology of garbage*", HarperCollins, New York.

Reisner M. (1993). "*Cadillac desert: the American west and its disappearing water*", Penguin Books, New York.

Rifkin J. (2000). "*The age of access: the new culture of hypercapitalism, where all the life is a paid-for experience*", Penguin Putman, New York, p. 312.

Roarch S.S. (1984). "*Industrialisation of the information economy*", Morgan Stanley, New York.

Russo N.L., Fitzgerald B., DeGross J.I., eds. (2001). "*Re-aligning research and practice in information systems development: the social and organizational perspective*", Kluwer Academic, New York.

Sadurski W. (2008). "*Equality and legitimacy*", Oxford Uni. Press, Oxford, p. 260.

Sassen S. (1988). "*The mobility of labour and capital: a study in international investment and labour flow*", Cambridge Uni. Press, 2 vol., Cambridge.

Scarpa R., Alberini A.A. (2005) "*Application of simulation in environment and resource economics*", Springer, London, p. 410.

Schor J. (1991). "*The overworked American: the unexpected decline of leisure*", Basic Books, New York.

Schumpeter J.A. (1939). "*Business cycles*", McGraw Hill, New York.

Scott P., ed. (1986). "*The word yearbook of robotics research and development*", 2nd ed., Kogan Page, London.

de Shalit A. (1998). "*Why posterity matters*", Rutledge, London.

Shille R.J. (2008). "*The sub-prime solution: how today's global financial crisis happened*", Princeton Uni. Press, Princeton, p. 208.

Simon H.A. (1981). "*The science of the artificial*", 2nd Ed., MIT Press, Cambridge.

Stahel W.R., Børlin M. (1987). "*Stratégie économique de la durabilité*", Soc. de Banque Suisse, Genève.

Stahel W.R. (1989). "*The limits to certainty: facing risks in the new service economy*", Kluwer Acad., Dordrecht.

Stark J. (2005). "*Product life-cycle management*", Springer, London, p. 441.

Stebbins G.L. (1969). "*The basis of progressive evolution*", Univ. North Carolina Press., Chapel Hill.

Stein R.E. (2003). "*Re-engineering the manufacturing system applying the theory of constraints*", 2nd ed., CRC Press, Boca Raton, p. 384.

van Stel, A. (2006). "*Empirical analysis of entrepreneurship and economic growth*", Springer, London, p. 235.

Stiglitz J.E., Charlton A. (2005). "*Fair trade for all: how trade promotes development*", Oxford Univ. Press, New York, pp. 315.

Sudoh O. (2005). "*Digital economy and social design*", Springer, London, p. 236.

Suhas H.K. (2001). "*From quality to virtual enterprise: an integrated approach*", CRC Press, Boca Raton.

Thaler R.H., Sustein C.H. (2008). *"Nudge: improving decisions about health, wealth and happiness"*, Yale Uni. Press, New York, p. 294.

Turner B.S. (1990). *"Theories of modernity and post-modernity"*, Sage, London.

Vernadat F.B. (1996) *"Enterprise modelling and integration: principles and applications"*, Chapman & Hall, New York, p. 513.

Wackernagel M, Rees W. (1995). *"Our ecological footprint: reducing human impact on the earth"*, New Society Pub., Gabriola Inland BC.

Wall S. (2008). *"A stranger in Europe: Britain and the EU from Thatcher to Blair"*, Oxford Uni. Press, Oxford, p. 230.

Walzer M., ed. (1995). *"Towards a global civil society"*, Berghahn Books, Providence.

Wann D. (1990). *"Biology: environmental protection by design"*, Johnson Books, Boulder.

Warren D.M., Sikkerveer L.J., Brokensha D. eds. (1991). *"The cultural dimension of development: indigenous knowledge systems"*, Intermediate Technology, London.

Wassermann O. (2001). *"The intelligent organisation: winning the global competition with supply chain idea"*, Springer, Berlin.

Weber M.K., Hemmelskamp J. (2004). *"Toward environmental innovation systems"*, Springer, Berlin, p. 351.

Weiner J. (1990). *"The next one hundred years"*, Bantam, New York.

Wilczek F. (2008). *"The lightness of being: mass, ether and the unification of forces"*, Basic Books, New York, p. 288.

Wiley E.O. (1981). *"Phylogenetics: the theory and practice the cladist"*, J. Wiley, New York.

Williamson O., ed. (1985). *"The economic institutions of capitalism"*, The Free Press, New York.

Wingand R., Picot A., Reichwald R. (1997). *"Information, organisation and management: expanding markets and corporate boundaries"*, John Wiley, Chichester.

Wolf M. (2004). *"Why globalisation works"*, Yale Univ. Press, New York, pp. 398.

Worster D. ed. (1988). *"The ends of the earth: perspectives on modern environmental history"*, Cambridge Univ. Press, New York.

Wrigley E.A. (1969). *"Populations and history"*, McGraw-Hill, New York.

Yegin D. (1991). *"The prize, the epic: quest for oil, money and power"*, Simon and Shuster, London.

Zakaria F. (2008). *"The post-American world"*, Norton & Co., New York, p. 292.

Zepezauer N.A. (1996). *"Tacking the rich off welfare"*, Odonian Press, Tucson Az.

Section 6.4.

The final section is added, simply, to provide an overview of the all book, and no further reference is purposely addressed.

INDEX

A

ABC, 277, 307
academic, ix, 98, 175, 240
accessibility, 97, 165, 168, 182
accidental, 40
accidents, 41, 192, 193, 199
accountability, 25, 77, 189
accounting, 19, 25, 56, 68, 95, 140, 259, 289
accreditation, 75, 79
accuracy, vii, 142, 167, 168, 169, 170, 171, 173, 174, 175, 176, 179, 180, 182, 183, 186, 191, 206, 208, 255
achievement, xiii, 6, 8, 11, 14, 20, 38, 39, 40, 42, 44, 47, 50, 52, 61, 71, 72, 74, 84, 97, 115, 116, 131, 150, 171, 183, 215, 250, 258, 260, 261, 272, 297
ACI, 228
acid, 211
ACM, 273
acoustic, 192, 225, 271
acoustic emission, 192
acoustic waves, 192
activation, 177
actuation, 81, 147, 167, 168, 169, 170, 171, 174, 175, 182, 195, 199, 205, 207, 214, 219, 220, 221, 223, 224, 225
actuators, 168, 181, 196, 201, 205, 206, 217, 223
adaptation, 70, 77, 90, 108, 115, 128, 143, 155, 174, 218, 256, 272
adaptive control, 105
addiction, 47
additives, 217
adhesives, 168, 171
adjustment, 52, 65, 169, 202, 268
administration, 25, 36, 50, 81, 211, 238, 239, 240, 252, 260, 266, 269

administrative, 62, 64, 72, 75, 76, 78, 90, 93, 96, 122, 132, 139, 187, 228, 229, 230, 232, 237, 238, 252, 261, 262
advertising, 137, 215
Africa, 233, 309
age, viii, ix, x, xi, xii, xiii, xvi, 9, 12, 14, 22, 23, 31, 43, 45, 48, 50, 52, 53, 54, 55, 56, 63, 93, 97, 98, 102, 117, 122, 130, 135, 144, 145, 147, 148, 151, 152, 164, 167, 179, 180, 182, 183, 184, 185, 186, 187, 188, 192, 193, 205, 214, 227, 228, 229, 235, 245, 246, 248, 250, 251, 253, 254, 255, 257, 262, 263, 264, 265, 266, 267, 268, 269, 270, 271, 272, 273, 283, 284, 285, 286, 287, 290, 292, 301, 305, 308, 309, 310, 311
agent, 60, 103, 106, 177
agents, 103, 109, 123, 143, 166, 185, 210, 211, 283
aggregation, 22, 27, 73, 76, 95
aggression, 132
aging, viii
agrarian, 16, 21, 22, 23, 24, 26, 43, 199, 251
agricultural, x, xiii, xv, 1, 2, 13, 20, 21, 22, 23, 28, 30, 31, 37, 200, 201, 202, 222, 226, 227, 231, 232, 233, 234, 235, 242, 244, 247, 249, 250, 251, 255, 256, 257, 258, 259, 265, 268, 272, 305
agriculture, 24, 25, 48, 171, 181, 209, 211, 226, 234
aid, 29, 32, 69, 73, 77, 84, 93, 99, 110, 112, 119, 126, 128, 129, 131, 136, 142, 148, 159, 182, 184, 197, 210, 211, 212, 213, 244, 284
air, 30, 210, 211, 256
airports, 22, 220
alcohol, 256
Alps, 41

alternative, ix, x, xi, 1, 2, 3, 4, 7, 8, 14, 19, 20, 24, 31, 39, 40, 42, 44, 48, 49, 50, 51, 54, 58, 75, 79, 81, 83, 95, 97, 103, 105, 110, 112, 116, 117, 118, 123, 124, 126, 129, 132, 136, 137, 148, 152, 159, 160, 163, 170, 184, 188, 194, 195, 199, 201, 209, 214, 230, 231, 240, 241, 246, 249, 250, 261, 264, 267, 268, 270, 276, 277, 278, 289, 307
alternatives, xi, xiii, xvi, 20, 23, 26, 34, 39, 67, 75, 80, 82, 87, 106, 113, 114, 117, 118, 127, 147, 151, 156, 159, 160, 163, 167, 173, 187, 189, 209, 219, 244, 246, 249, 256, 261, 265, 269, 271, 272, 278, 279, 307
altruism, xi, xiii, xvi, 18, 19, 36, 48, 50, 85, 116, 117, 123, 124, 127, 129, 132, 135, 136, 137, 142, 144, 145, 231, 232, 239, 240, 248, 253, 260, 261, 262, 264, 266, 269, 271, 289
ambiguity, 75, 100, 122, 129, 235, 258, 261, 263
amendments, 1, 2
American colonies, 12
Amsterdam, 6, 7, 8, 12, 120, 289, 292, 293, 294, 295
anatomy, 204, 206, 275, 282, 307
Andes, 233
anger, 245
angular velocity, 221
animals, 24, 31, 117, 128, 233, 247, 250
anomalous, 59
antagonism, 85, 124, 238, 242
antagonistic, vii, 135
antecedents, 68
anthropic, xiii, 24, 26, 46, 48, 118, 119, 124, 127, 136, 137, 204, 220, 231, 238, 249, 254, 256, 257, 258, 272
anthropic principle, xiii, 136, 254, 257, 258, 272
anti-scientific, 37, 245
antitrust, 292
ants, 129
anxiety, 118
apparel, 195, 196, 197
application, 9, 26, 28, 53, 71, 112, 155, 163, 166, 167, 171, 176, 179, 180, 181, 182, 183, 184, 192, 194, 220, 221, 225, 230, 242, 253, 266, 287, 288, 293, 301
appraisals, xiii, 42, 239, 272
appropriate technology, 3, 274
aptitude, 167
aquifers, 6, 7, 24
arbitration, 80, 86
argument, 227, 261
Aristotle, 31, 39, 40, 41, 44
armed forces, 14, 199
ARS, 294

articulation, 180
artificial intelligence, viii, 9, 23, 26, 30, 31, 32, 36, 37, 43, 51, 81, 90, 103, 112, 126, 127, 128, 129, 244, 250, 254, 267, 293
artificial life, xi, xvi, 23, 26, 30, 31, 32, 36, 43, 128, 129, 179, 180, 226, 232, 250, 254, 265, 267, 270
Asia, 39, 40, 41, 43, 233, 234, 305
Asian, viii, 49, 121, 307
Asian countries, viii, 121
aspiration, 14, 21, 34
assault, 22
assessment, 26, 38, 47, 57, 58, 59, 65, 66, 68, 69, 70, 71, 74, 77, 78, 79, 101, 105, 111, 112, 113, 115, 117, 137, 138, 140, 144, 149, 155, 161, 163, 166, 172, 173, 174, 182, 189, 203, 215, 228, 242, 252, 253, 254, 274, 281, 282, 288, 293, 295, 296, 298, 300, 301, 302, 303, 309
assessment procedures, 71
assets, x, xv, 1, 2, 3, 4, 5, 6, 7, 8, 11, 14, 17, 18, 19, 20, 21, 25, 26, 27, 29, 62, 63, 69, 70, 83, 85, 89, 99, 107, 109, 119, 120, 124, 130, 132, 133, 139, 140, 141, 142, 149, 229, 240, 244, 248, 251, 252, 253, 257, 258, 259, 262, 264, 266, 268, 273, 283
assignment, 53, 205
assimilation, 114, 116
assumptions, 39, 40, 45, 47, 48, 49
asymptotic, 47
ATF, 92, 229
Athens, 298, 304
Atlantic, 169, 285, 305, 310
atmosphere, 248
atoms, 44
attacks, 22, 95
attitudes, 127, 234
attribution, 48, 50, 117
Australia, 12, 310
authority, 4, 11, 14, 33, 77, 119, 135, 139, 200, 232, 237, 261, 264
automata, 310
automation, 55, 56, 57, 63, 81, 98, 99, 101, 104, 108, 110, 112, 141, 148, 152, 154, 155, 157, 159, 160, 161, 162, 165, 167, 181, 183, 184, 188, 189, 191, 201, 202, 209, 210, 211, 212, 214, 215, 216, 217, 229, 270, 271, 276, 281, 293, 294, 295, 297, 299, 302, 303
autonomous communities, 242
autonomous navigation, 210
Autonomous Robotic Systems, 300
autonomy, 6, 8, 42, 43, 49, 71, 72, 97, 98, 99, 105, 128, 142, 144, 180, 188, 191, 195, 197,

215, 216, 219, 220, 221, 223, 224, 226, 227, 237, 238, 242, 250, 259, 261
availability, 3, 5, 7, 9, 23, 26, 32, 59, 97, 102, 104, 109, 113, 120, 126, 136, 153, 167, 180, 181, 184, 192, 222, 225, 238, 245
avoidance, 63, 141, 176
awareness, 19, 33, 54, 85, 92, 103, 119, 127, 132, 133, 185, 228, 234, 242, 262, 263, 270, 282

B

background noise, 192
backlash, 167, 169, 174, 175
bacteria, 211
baking, 97
balance sheet, 69
balanced budget, 186
ballast, 224
bankers, 6, 7, 8, 123
banks, 15
bargaining, 119
barley, 233
barrier, 13, 201, 260
barriers, 5, 84, 91, 123, 130, 134, 139, 216, 230, 231, 232, 237, 245, 247
barter, 6, 7, 11
basic needs, 117
batteries, 220
beams, viii, 194, 211
behavior, 292
behaviours, 17, 19, 31, 33, 49, 54, 56, 58, 61, 70, 78, 80, 86, 90, 91, 92, 94, 124, 200, 218, 228, 241, 242, 247, 264
Beijing, 277, 305
benchmark, 259
benchmarking, 77, 149
benchmarks, 103, 241
bending, 224
benefits, 13, 14, 17, 20, 21, 24, 30, 32, 35, 36, 43, 68, 82, 84, 90, 91, 96, 98, 102, 116, 117, 122, 123, 124, 126, 132, 133, 141, 168, 186, 193, 198, 200, 202, 219, 222, 226, 231, 235, 237, 239, 254, 257, 260, 280
beverages, 117
Bhagwati, 277, 305
bias, 2, 3, 15, 16, 68, 86, 91, 122, 130, 153, 167, 169, 174, 228, 235
binding, 19, 25, 33, 44, 48, 56, 79, 85, 90, 93, 124, 234
biodiversity, 256, 277, 307
biological processes, 31, 222, 255, 256
biosphere, 257
biotic, 210, 211, 212

biotic factor, 211, 212
birds, 24, 72
birth, 37, 48, 97
black-box, 273
bleeding, 281
blindness, 264
blocks, 71, 103, 104, 105, 111, 143, 153, 154, 156, 158, 173, 176, 182, 183, 199, 207, 216, 220, 221, 224
blood, 14, 180
blood vessels, 180
blurring, 53, 64, 70
bonds, 21, 49, 68, 76, 95, 155, 187, 259
booms, 18, 309
borderline, 134
Boston, 273, 277, 278, 280, 283, 287, 290, 296, 302, 307, 308, 310
bottleneck, 102
bottlenecks, 154
bottom-up, 4, 5, 12, 22, 63, 66, 95, 111, 119, 120, 137, 149, 151, 152, 153, 156, 157, 162, 163, 201, 203, 234, 238, 239, 245, 261
bounds, 59, 126, 141, 268
brain, 22, 32, 41, 68, 128
branching, 152, 159
Brazil, 130, 232
breakdown, 59
breeding, 20, 21, 23, 26, 43, 209, 222, 223, 226, 227, 232, 236, 250, 266
Britain, 312
brokerage, 77
bubbles, 259, 306
buffer, 161, 194, 197, 198
buildings, 35, 213
burning, 227
business model, 115
business organisation, 28, 87, 136
business policy, 155
buyer, 16, 53, 56, 74, 82, 88, 149, 243
by-products, 136

C

C++, 280
cabinets, 216
cables, 194
CAD, 55, 56, 295, 297, 298
calibration, vii, 176
CAM, 55, 56, 295, 297, 298
campaigns, 172, 210
Canada, 12
CAP, 55, 56
capital consumption, 94

capital intensive, 138
capital productivity, 129, 149, 254
capitalism, x, xi, xiii, 1, 2, 4, 5, 6, 7, 8, 11, 12, 13, 14, 17, 18, 20, 22, 24, 26, 27, 32, 34, 36, 97, 99, 118, 119, 120, 121, 122, 127, 130, 131, 132, 134, 237, 238, 242, 244, 251, 260, 263, 268, 269, 272, 287, 289, 290, 310, 312
capitalist, 33, 291
caps, 197, 217
CAR, 299
carbon, 248, 251, 266
carbon dioxide, 251
caretaker, 219
carmakers, 92, 93, 217, 228, 229, 230
carrier, 47, 168, 174, 190, 191, 201, 205, 206, 218, 222
case study, 280, 282
case-studies, 267
cast, 151, 216
CAT, 55, 56, 165
catheter, 180
Catholic, 38
cattle, 226
causal attribution, 38, 48
causal inference, 42, 48, 111
causal model, 154
causal reasoning, 39
causality, 9, 155
cell, 293
cellular phone, 185
cellulose, 256
central bank, 15
central planning, 14
cereals, 233
CERES, 84
certificate, 228
certification, 36, 66, 68, 71, 81, 85, 86, 92, 144, 186, 192, 193, 212, 229, 243
chain transfer, 243
channels, 80, 143, 224, 225
chaos, 241
charm, 242
cheating, 249
chemicals, 166, 210, 211, 236
children, 41
China, 12, 14, 22, 28, 46, 48, 121, 130, 232, 234, 277, 278, 279, 285, 306
chlorophyll, 30, 36, 128
Christianity, 38
Cincinnati, 294
circulation, 207, 211, 228, 236
citizens, xv, 4, 5, 10, 14, 15, 16, 17, 18, 22, 25, 28, 32, 33, 34, 35, 36, 38, 41, 49, 50, 61, 68, 78, 85, 91, 93, 94, 116, 118, 119, 120, 121, 122, 124, 125, 132, 135, 136, 137, 139, 142, 144, 229, 231, 232, 234, 235, 237, 240, 242, 247, 252, 253, 254, 259, 260, 261, 262, 265, 266, 269, 272
citizenship, 118, 231, 265
citrus, 209, 233, 302
civil servant, 11, 12, 25, 237, 240, 252
civil servants, 25, 237, 240, 252
civil society, 266, 283, 312
class struggle, 4, 13, 14, 26, 34, 53, 252
classes, 14, 25, 34, 49, 51, 61, 100, 131, 175, 179, 188, 216, 218
classical, vii
classification, 69, 104, 168, 177, 179, 209
cleaning, 214, 215, 216, 302, 303
clients, 19, 28, 58, 63, 64, 70, 75, 79, 80, 85, 86, 87, 94, 96, 126, 131, 133, 138, 144, 149, 192, 213, 219, 220, 230, 235, 237
climate change, 16, 91
clone, 256
closed-loop, 25, 26, 126, 160, 225
closure, 49, 122, 129, 134, 162, 168, 197, 217, 241, 243, 247
clothing industry, 296
clustering, 64, 73, 74, 81, 261
clusters, 21, 64, 95, 106, 234, 264
Co, 84, 176, 203, 276, 280, 282, 285, 289, 290, 291, 298, 301, 305, 312
CO2, 248
coaches, 217
coal, 13
coal mine, 13
Coalition for Environmentally Responsible Economies, 84
coalitions, 238, 245
codes, 30, 103, 104, 128, 136, 191, 199, 297
coding, 40, 47, 103, 104, 190, 191
co-existence, 251
cognition, 30
cognitive abilities, 27, 250
cognitive development, 278
cognitive domains, 179
cognitive process, 251
cognitive system, 276
coherence, 7, 8, 25, 42, 249, 257, 270
cohesion, 245
collaboration, 38, 70, 74, 145, 150, 151, 177, 186, 187, 213, 215, 216, 251, 282, 283, 288, 290, 310
collision avoidance, 176
collisions, 172
collusion, 228

Columbia, 305
combined effect, 245
combustion, 248, 251
commerce, 117
communication, 6, 8, 11, 18, 21, 27, 38, 41, 43, 68, 69, 71, 74, 80, 96, 103, 106, 115, 123, 129, 130, 143, 144, 149, 177, 184, 185, 186, 189, 191, 198, 199, 207, 210, 212, 213, 219, 224, 225, 237, 239, 245, 263, 265, 304
communication technologies, 11, 27, 129, 130, 189, 239
communism, 14, 34
communities, 6, 8, 11, 18, 21, 22, 25, 28, 35, 37, 38, 42, 49, 73, 83, 95, 117, 121, 130, 135, 139, 228, 231, 234, 237, 242, 245, 249, 259, 265, 286
community, xv, 9, 13, 22, 25, 29, 33, 35, 37, 40, 44, 48, 50, 53, 89, 91, 96, 117, 137, 139, 141, 200, 216, 227, 236, 240, 245, 247, 248, 278, 306, 307
community service, 137
compatibility, 71, 74, 184, 248
compensation, 10, 22, 137, 167, 168, 169, 171, 173, 176, 182, 199, 205, 221, 223
competence, 27, 64, 92, 261
competency, 54, 72, 74, 111, 180
competition, 2, 3, 14, 27, 28, 32, 33, 34, 35, 38, 39, 42, 43, 49, 50, 52, 53, 54, 58, 60, 62, 63, 71, 81, 83, 84, 87, 88, 91, 92, 97, 117, 118, 124, 125, 126, 130, 131, 132, 161, 187, 189, 193, 229, 235, 238, 247, 252, 286, 287, 290, 291
competitive advantage, 1, 2, 84, 87, 99, 123, 124, 131, 226, 273, 274, 290
competitive markets, 93
competitiveness, 15, 16, 53, 80, 81, 84, 86, 91, 122, 144, 149, 188, 189, 193, 215, 237, 244, 252, 280, 282
competitor, 78, 100
complement, 12, 40, 41, 166, 182, 235, 246
complementarity, 208
complex systems, 28, 66, 95
complexity, xvi, 28, 29, 37, 38, 39, 40, 42, 43, 44, 45, 48, 49, 50, 51, 57, 66, 67, 72, 74, 81, 94, 95, 99, 101, 103, 104, 106, 107, 109, 110, 111, 112, 113, 114, 115, 116, 128, 133, 142, 150, 152, 155, 156, 157, 161, 162, 163, 164, 166, 167, 172, 177, 184, 206, 223, 235, 253, 254, 262, 268, 269, 270, 272, 277, 282, 285
compliance, vii, 66, 78, 88, 93, 114, 167, 169, 173, 174, 175, 176, 208, 219
complications, 203

components, xi, 63, 100, 105, 106, 151, 158, 159, 169, 170, 181, 185, 186, 196, 205, 217, 224
composition, 217
comprehension, 245
computation, 47, 73, 292, 299
computer simulation, 296
computer technology, 148
computing, 28, 102, 143, 144, 184, 185, 277, 304
concealment, 134
concentrates, 207
concentration, 149, 211, 247
conception, 49, 58, 76, 153, 155, 203, 206
concrete, 197, 198
concurrency, 109, 298
concurrent engineering, 281
conditioning, 16, 26, 40, 45, 46, 49, 75, 98, 101, 105, 111, 115, 125, 145, 151, 153, 159, 160, 161, 162, 164, 167, 170, 180, 199, 258, 266, 267, 268, 269, 270, 287, 292
confidence, viii, 37, 83, 160, 208, 258
configuration, 73, 74, 100, 108, 143, 158, 206, 207, 239, 290
conflict, 5, 14, 31, 33, 35, 68, 84, 105, 124, 245, 247, 257, 258, 262, 263, 264, 269
conflict of interest, 262
conflict resolution, 245
conformity, 131, 200
confrontation, 38
Confucius, 41, 43
confusion, 6, 40, 75, 260
Congress, 283, 295, 300, 304
conjecture, 3, 5, 16, 20, 36, 237
connectivity, 21, 22, 86, 95, 234
consciousness, 19, 82, 94, 102, 212, 232, 234, 254
consensus, 30, 84, 85, 91, 93, 137, 245, 260
consent, 4, 14, 19, 20, 32, 33, 35, 37, 127, 131, 133, 134, 136, 198, 199, 203, 205, 219, 228, 232, 233, 239, 240, 242, 253, 260, 271
conservation, 45, 70, 91, 139, 210, 248
consolidation, 194, 263, 300
constraints, 57, 59, 72, 78, 83, 84, 85, 98, 110, 113, 127, 133, 154, 155, 173, 176, 177, 180, 181, 183, 195, 201, 203, 205, 221, 231, 247, 248, 253, 254, 261, 264, 280, 305, 311
construction, 15, 39, 56, 63, 68, 80, 104, 127, 132, 181, 186, 234, 276, 307
consultants, 108
consulting, 95, 291
consumer goods, xi
consumers, 18, 19, 62, 63, 67, 69, 79, 80, 81, 85, 88, 111, 123, 139, 163, 210, 212, 241, 242, 244, 254

consumption, xi, 10, 17, 18, 20, 29, 35, 36, 56, 57, 59, 61, 68, 79, 83, 86, 87, 91, 94, 126, 132, 137, 138, 139, 144, 196, 222, 226, 231, 241, 247, 250, 254, 256, 257, 262, 269, 292
consumption rates, 138
contamination, 6, 7, 195, 197, 210, 216, 265
contiguity, 13, 18, 95, 233
continuity, 11, 24, 31, 51, 59, 80, 86, 108, 113, 126, 127, 129, 167, 184, 187, 193, 194, 205, 221, 248, 250, 251, 253, 263, 267, 291, 305
contracts, 77, 79, 87, 125, 187, 234
convergence, 47, 143
conversion, 37, 66, 236
cooking, 214
Copenhagen, 274, 292
Coriolis effect, 221
corn, 117
corporations, 4, 5, 53, 70, 72, 73, 82, 116, 121, 122, 131, 132, 135, 144, 232, 237, 238, 240, 245, 249, 252, 253, 256, 259, 260, 261, 263, 271, 274
correlation, 239, 246
correlations, xiii, 272
corrosion, 195
corruption, 236
cost effectiveness, 163
cost-effective, 109, 111
costs, vii, 13, 25, 26, 52, 53, 56, 59, 60, 61, 62, 69, 70, 72, 73, 89, 90, 91, 93, 100, 117, 118, 125, 131, 132, 136, 139, 140, 149, 161, 163, 165, 172, 181, 183, 187, 188, 194, 196, 200, 213, 227, 228, 229, 238, 242, 281, 307
cotton, 13, 136, 309
Council of Ministers, 274
counsel, 244
counterbalance, 62
coupling, 45, 47, 102, 106, 109, 111, 157, 158, 162, 167, 168, 169, 170, 171, 173, 174, 175, 182, 216, 218, 219, 221, 224
covering, 36, 54, 57, 61, 91, 112, 113, 117, 130, 133, 139, 149, 170, 185, 189, 197, 202, 211, 219, 239, 263
crack, 262
CRC, 273, 274, 276, 280, 282, 283, 285, 287, 290, 293, 296, 311
creativity, 53, 143, 163, 185
credibility, 61, 92, 136, 258
credit, 218, 263, 276, 306
credit card, 218
creep, 221
critical points, 176
crops, 6, 7, 22, 43, 46, 210

cross-border, 121, 132, 135, 232, 237, 238, 252, 259, 271
cross-cultural, 279
cross-linking, 28, 35, 151
cross-sectional, 26, 141
crust, 248
cultivation, 25, 133, 222, 225, 226, 227, 236, 242, 266, 303
cultural factors, x, xii
culture, 10, 11, 29, 31, 36, 38, 69, 92, 102, 117, 134, 135, 222, 225, 226, 227, 232, 234, 249, 265, 281, 284, 311
currency, 72
current balance, 254
customers, 57, 58, 64, 73, 100, 101, 111, 114, 148, 151, 164, 184, 186, 210
cybernetics, 294, 298
cycles, 13, 20, 23, 54, 55, 102, 103, 126, 147, 149, 153, 158, 161, 162, 166, 168, 171, 174, 175, 177, 178, 181, 182, 183, 191, 195, 198, 200, 202, 208, 215, 216, 217, 241, 242, 285, 311
cycling, 24
cynicism, 78

D

dairy, 219
damping, 7, 8
danger, 179, 201, 247, 264
data collection, 35, 107, 144, 182, 194
data processing, 99
data transfer, 143, 224
database, 156, 283
dating, 38, 72, 74, 77, 100, 116, 154, 155, 160, 168, 173, 174, 177, 194, 205, 206, 217, 221, 224, 225, 250, 269
death, 30, 118, 128, 245, 248, 274, 285
debt, 273
decay, 3, 5, 6, 7, 23, 29, 30, 31, 43, 44, 50, 60, 62, 87, 127, 128, 133, 138, 144, 170, 212, 224, 241, 245, 249, 250, 260
decentralisation, 163
decision making, xii
decision support tool, 71
decision task, 150
decisions, 21, 71, 77, 78, 90, 104, 115, 121, 152, 208, 289, 312
declarative knowledge, 47, 103, 104, 105
decomposition, 109
deconstruction, 237
decoupling, 169, 174, 176
deduction, 27

deep-sea, 226, 300, 302
defects, 141
deficiency, 214, 245, 246
definition, 21, 27, 48, 98, 120, 124, 139, 141, 155, 163, 173, 183, 206, 233, 261
degradation, 60, 228
delivery, x, 28, 52, 53, 54, 56, 57, 58, 60, 62, 64, 65, 66, 73, 74, 75, 76, 77, 79, 80, 81, 90, 94, 95, 96, 99, 100, 101, 102, 103, 105, 106, 107, 108, 109, 110, 112, 114, 115, 125, 126, 131, 138, 141, 144, 150, 152, 153, 154, 155, 157, 158, 159, 163, 164, 165, 184, 186, 187, 188, 189, 190, 191, 192, 207, 213, 217, 219, 228, 243, 251, 268, 279, 293, 295, 304
democracy, xiii, 4, 14, 20, 33, 34, 36, 50, 96, 97, 119, 120, 122, 127, 136, 137, 139, 140, 237, 238, 240, 242, 246, 253, 258, 260, 261, 264, 265, 266, 269, 272, 306, 308
density, 47
deposition, 70
depression, 15
deprivation, 276
deregulation, 239
derivatives, xv
desert, 256, 311
desertion, 121
designers, xvi, 105, 113
destruction, 92, 199, 228
detection, 9, 59, 60, 173, 174, 191, 192, 193, 201, 210, 211
developed countries, 259
developing countries, xi, xv, 17, 18, 46, 235, 236, 269
deviation, 59
diamond, 196, 273, 305, 307
diamonds, 117
dichotomy, 310
dictatorship, 239
differentiation, 128, 279
diffusion, 13, 121, 203
digital subscriber line, 277
directives, 61, 79, 227, 229, 230
disability, 15
disappointment, vii
disaster, 284
discipline, xii, 290
discontinuity, 7, 8, 48
Discovery, 292
discriminatory, 123
diseases, 210, 212
dislocation, 215
dispatcher, 199, 294
displacement, 174

disposables, 19, 61, 62, 63, 92, 94, 136, 166, 246, 254
disputes, 44
distortions, 236
distress, 263
distributed computing, 290
distribution, 22, 25, 28, 44, 47, 93, 117, 135, 137, 163, 165, 177, 189, 190, 223, 229, 238, 242, 249, 259, 288
divergence, 285
diversification, 128
diversity, 17, 120, 179, 229, 241, 242, 248
diving, 195, 222, 223, 226, 302
division, xiii, xv, 15, 21, 53, 54, 55, 72, 121, 122, 153, 234, 263, 269
DNA, 255
doctors, viii, 237
dogs, 23
domestic chores, 63, 209, 215
domestication, 233, 250
dominance, 7, 8, 50, 66, 236, 239, 263
dominant strategy, 13
donkey, 233
doping, 134
dosing, 171
drainage, 271, 301
dream, viii
drugs, 18, 211, 212, 219, 236, 242
drying, 24
dualism, 67, 252
duality, 288
dumping, 56, 57, 61, 62, 68, 74, 85, 87, 89, 91, 136, 227, 228, 230
duplication, 11, 32, 69, 107, 109, 149, 176, 179, 193, 236, 251, 255, 256
durability, 250
duration, 74, 110, 238
duties, xii, 4, 6, 7, 11, 13, 25, 33, 42, 59, 65, 68, 69, 74, 76, 77, 79, 80, 81, 85, 88, 89, 90, 93, 98, 105, 109, 112, 113, 114, 118, 119, 120, 124, 126, 130, 131, 132, 136, 138, 139, 140, 144, 150, 157, 165, 166, 174, 177, 179, 183, 188, 190, 195, 197, 200, 205, 208, 209, 211, 214, 215, 216, 223, 228, 229, 230, 237, 243, 253, 255, 265
dynamic control, 297
dynamical system, 180, 292, 299
dynamical systems, 180, 292, 299

E

ears, 256

earth, x, xi, 1, 2, 3, 5, 7, 8, 9, 13, 17, 18, 20, 21, 22, 23, 24, 28, 29, 30, 32, 35, 39, 43, 48, 50, 61, 78, 82, 83, 86, 89, 90, 97, 116, 118, 124, 126, 127, 128, 130, 132, 140, 201, 209, 231, 237, 238, 242, 248, 249, 250, 253, 254, 256, 257, 258, 260, 265, 267, 268, 269, 277, 279, 280, 281, 291, 307, 308, 312
East Asia, 43, 233, 234
eating, 41
ecological, ix, xiii, 18, 57, 58, 61, 83, 84, 85, 92, 126, 129, 130, 132, 133, 136, 186, 242, 247, 253, 263, 264, 280, 281, 287, 312
Ecological Economics, 282
ecologism, 17, 23, 24
ecology, 13, 51, 53, 54, 66, 81, 82, 83, 84, 85, 91, 116, 119, 121, 130, 132, 133, 136, 139, 184, 247, 254, 257, 258, 262, 264, 266, 268, 269, 272, 274, 275, 276, 277, 278, 290, 307, 308, 309
e-commerce, 282
economic change, 274, 286, 288, 309, 310
economic crisis, 306
economic cycle, 125
economic growth, 200, 203, 285, 291, 309, 310, 311
economic institutions, 289, 312
economic integration, 263
economic policy, 261
economic problem, ix
economic systems, 4, 5, 15, 18, 22, 241
economic transformation, 281, 308
economic welfare, 13
economics, xiii, 3, 15, 16, 25, 32, 33, 34, 35, 36, 91, 92, 118, 122, 125, 148, 240, 241, 249, 252, 253, 258, 262, 267, 272, 273, 274, 280, 282, 284, 285, 286, 289, 306, 308, 311
ecosystem, xi, 280
egoism, 78
Egypt, 233
elaboration, 134
elasticity, 131
elderly, viii
elders, 49
electrical power, 216, 256
electricity, 13, 22
electromagnetic, 31, 44
electromagnetic wave, 31, 44
electromagnetic waves, 31, 44
email, vii
embezzlement, 14
emission, 17, 192, 217, 218, 259
employees, 25, 214, 237, 252
employment, 4, 5, 15, 16, 89, 91, 282, 285, 305

empowered, 260
empowerment, 57, 149, 150, 231
encoding, 103, 105, 128, 193
endoscope, 206
endurance, 126, 127, 179, 195
end-users, 19, 151, 209, 286
energy, 1, 2, 3, 9, 13, 23, 26, 29, 30, 31, 32, 36, 45, 51, 60, 63, 80, 81, 87, 90, 91, 94, 97, 116, 126, 127, 128, 129, 133, 136, 138, 139, 141, 170, 195, 196, 224, 232, 244, 250, 251, 254, 255, 256, 257, 265, 266, 267, 272, 287
energy constraint, 254
energy consumption, 196
energy efficiency, 287
energy supply, 224
engagement, 56, 110, 119, 134, 136, 156, 165, 170, 171, 175, 179, 180, 205, 225, 270
England, 6, 8
entanglement, 184
enterprise, xii, 5, 52, 54, 56, 57, 63, 64, 65, 66, 68, 69, 73, 74, 75, 77, 78, 79, 80, 81, 85, 88, 89, 90, 93, 95, 99, 100, 101, 102, 103, 104, 105, 106, 108, 109, 110, 111, 112, 113, 114, 115, 116, 122, 125, 131, 133, 134, 138, 140, 141, 142, 144, 145, 150, 155, 158, 161, 163, 164, 187, 188, 227, 229, 230, 243, 244, 259, 269, 270, 273, 280, 282, 283, 284, 288, 309, 310
enterprise shares, 5
entertainment, 13, 67, 135, 212, 243, 246
entrepreneurs, 12, 67
entrepreneurship, x, 15, 16, 28, 30, 32, 42, 63, 66, 67, 68, 69, 70, 73, 75, 76, 81, 85, 90, 94, 113, 114, 115, 116, 122, 125, 129, 131, 135, 137, 145, 184, 186, 229, 236, 239, 243, 246, 251, 252, 257, 261, 263, 268, 269, 274, 277, 281, 283, 284, 285, 288, 307, 310, 311
entropy, 23, 30, 31, 36, 44, 87, 90, 127, 128, 133, 138, 241, 245, 248, 249, 250
environment, vii, ix, xii, xiii, 1, 2, 18, 19, 20, 23, 40, 45, 48, 49, 52, 54, 57, 58, 60, 62, 64, 65, 66, 73, 74, 77, 79, 80, 81, 82, 85, 86, 88, 89, 90, 91, 101, 103, 105, 106, 109, 110, 112, 113, 115, 116, 124, 128, 133, 148, 150, 155, 163, 173, 184, 192, 197, 211, 212, 217, 222, 260, 268, 270, 277, 279, 280, 281, 282, 284, 288, 290, 294, 297, 307, 308, 309, 311
environmental effects, 17
environmental impact, 58, 132, 141, 196, 254, 280
environmental issues, 280
environmental policy, 34, 61, 62, 87, 94, 96, 138
environmental protection, 92, 148, 187, 312

environmental regulations, 284
enzymes, 255
epidemic, 22, 210
equality, 33, 123
equilibrium, 13, 14, 118, 125, 140, 196, 224, 245, 260
equities, 259
equity, 14, 16, 118, 259
ergonomics, 55, 154, 183, 189
erosion, 6, 7, 24
estimating, 211
ethics, 91, 127
Eurocentric, 239
Europe, 4, 6, 8, 12, 22, 24, 34, 37, 41, 118, 233, 277, 283, 285, 287, 291, 305, 311, 312
European Commission, 34
European Union, 91
Europeans, viii, 39
evil, 46, 127, 133, 263
evolution, xiii, 1, 2, 14, 27, 31, 43, 47, 52, 56, 81, 103, 125, 133, 154, 192, 193, 196, 202, 209, 217, 225, 241, 242, 251, 252, 255, 256, 265, 272, 273, 277, 279, 290, 291, 307, 310, 311
evolutionism, 20, 37, 128, 248, 250, 251, 257
exclusion, 216, 217, 246
excuse, 78, 135, 238, 264
execution, 56, 57, 58, 60, 81, 99, 100, 101, 103, 107, 108, 117, 147, 150, 152, 154, 156, 157, 159, 162, 172, 173, 176, 177, 178, 183, 194, 201, 202, 203, 204, 205, 207, 218, 240
exercise, 39, 129, 228, 252, 259, 263
expert systems, 47, 48
expertise, 5, 26, 29, 56, 67, 73, 103, 105, 119, 155, 166, 167, 185, 192, 204, 205, 206, 208, 213, 214, 226
exploitation, x, 1, 2, 3, 7, 9, 11, 13, 14, 22, 24, 30, 32, 43, 50, 61, 65, 70, 82, 87, 94, 97, 104, 107, 111, 112, 113, 114, 128, 134, 138, 142, 157, 163, 164, 169, 195, 199, 226, 227, 236, 239, 272
explosions, 196
extended-artefact, 280
external costs, 307
externalities, 52, 54, 62, 63, 64, 65, 66, 69, 70, 75, 76, 78, 87, 88, 90, 106, 110, 112, 113, 115, 116, 141, 148, 150, 154, 162, 212, 230, 241, 252, 253
extinction, 128, 247
extremism, 263
eyes, x, 31

F

fabric, 117, 165, 166, 296
fabrication, 154
failed states, 307
failure, 7, 8, 37, 59, 60, 63, 107, 109, 152, 157, 177, 198, 199, 200, 203, 264, 275, 307
fairness, 14, 21, 35, 39, 40, 43, 44, 49, 50, 53, 68, 80, 86, 92, 118, 125, 132, 133, 141, 235, 242, 245, 254
fair-trade, 60, 63, 85
faith, 46, 48, 136, 264, 309
family, 25, 44, 117, 175, 204
family budget, 25
famine, 34, 276
Far East, 43
farmers, 25, 203, 243
farming, 6, 7, 10, 20, 21, 23, 24, 25, 26, 43, 49, 133, 200, 202, 209, 210, 212, 223, 224, 226, 227, 232, 233, 236, 241, 250, 256, 303
farms, 23, 133, 224
fatalism, 243, 245
fatigue, 98, 193
faults, 78, 197
fear, ix, 20, 23, 31, 44, 49, 78, 132, 134, 244
fee, 93, 137, 188
feedback, vii, 47, 65, 110, 120, 143, 166, 168, 169, 171, 172, 173, 174, 182, 183, 193, 205, 225
feeding, vii, 80, 102, 136, 159, 160, 161, 166, 173, 181, 190, 206, 217, 296
fees, 10, 62, 86, 90, 91, 93, 95, 135, 136, 139, 141, 187, 193
feet, 205
Fertile Crescent, 233, 234
fertilizers, 6, 7, 24, 37, 128
fidelity, vii, 70, 126
fighters, 245
film, 47
finance, 4, 6, 7, 11, 14, 17, 18, 26, 56, 82, 106, 123, 126, 141, 240, 273, 275, 285
financial capital, xv, 4, 5, 7, 8, 9, 11, 12, 13, 72, 84, 89, 117, 235, 259
financial crisis, 311
financial markets, 82, 291, 306
financial resources, 69, 93, 253
financial system, ix
firms, 70, 81, 126, 286
fish, 222, 226, 227, 236, 266
fishing, 224, 227, 236
fission, 37, 251
fitness, 91, 99, 100, 101, 111, 141, 142, 163, 164, 192, 193, 238

flame, 196
flexibility, xi, xii, 56, 57, 70, 99, 100, 101, 102, 103, 104, 105, 106, 107, 108, 109, 110, 111, 112, 116, 148, 149, 150, 152, 153, 154, 156, 157, 158, 159, 161, 162, 163, 171, 178, 183, 191, 201, 202, 203, 208, 217, 269, 270, 280, 285, 286, 293, 295, 297, 298
flexible manufacturing, xii, 158, 163, 164, 270, 292, 293, 294, 295, 297
flight, 72, 253
flow, xi, 3, 17, 25, 35, 45, 51, 53, 54, 55, 56, 57, 61, 62, 63, 70, 76, 77, 78, 79, 97, 99, 100, 101, 102, 104, 106, 107, 115, 122, 131, 136, 139, 141, 149, 150, 151, 152, 155, 157, 158, 159, 160, 161, 165, 166, 184, 186, 188, 189, 190, 191, 192, 200, 203, 214, 229, 230, 243, 251, 253, 268, 270, 289, 291, 311
fluctuations, 69, 276
fluid, 46
focusing, 79, 96, 231, 241, 254
food, 21, 117, 128, 133, 171, 214, 215, 216, 234
foodstuffs, 11, 22, 23, 25, 30, 117, 127, 128, 225, 242, 250, 256
footwear, 63
Ford, 42, 45, 53, 151, 282
foreign exchange, 285
foreign policy, 277
foreigner, 21
foreigners, 234
fortification, 116
fossil, 30, 251
fossil fuel, 251
Fourier, 45
fragility, 51, 232
fragmentation, 13, 118, 130, 132, 265, 279
framing, 157, 236
France, 6, 8, 12, 15, 24, 25, 207
free trade, 82, 264
free will, 33
freedom, 11, 18, 25, 33, 34, 49, 50, 150, 168, 172, 174, 215, 216, 218, 221, 223, 238, 261, 299, 306
freezing, 263
friction, 170, 173, 176, 205, 220, 221
fruits, 210, 302
fuel, 217, 218, 219, 228, 251, 256
fuel type, 217
full employment, 16
funding, 5, 6, 8, 39
funds, 259
fungi, 211
fusion, vii, 88, 129, 170, 173, 181, 182, 198, 205, 208, 211, 225

futures, 259, 278
fuzzy logic, 47, 49
fuzzy sets, 47

G

gait, 194, 215
garbage, 78, 139, 215, 216, 228, 311
gas, 171
gene, 256
generation, xi, 18, 21, 28, 31, 66, 82, 83, 95, 122, 128, 138, 168, 176, 232, 251, 257, 259, 273, 287, 297, 303
genes, 256
genetic code, 32, 128, 255, 256
Geneva, 195
Geneva Convention, 195
geography, 234, 250, 278
Georgia, 283
Germany, 15, 93, 206, 296, 302
gift, 37
gifts, 91, 148
global competition, 17, 83, 292, 312
global economy, 291
global markets, 186
global village, xi, 4, 5, 13, 14, 18, 33, 35, 36, 50, 67, 78, 83, 85, 95, 96, 98, 124, 126, 127, 130, 132, 133, 135, 136, 145, 231, 232, 236, 238, 239, 240, 248, 253, 257, 258, 260, 262, 265, 266, 268, 269
global warming, 308
goals, 58, 59, 61, 63, 71, 84, 90, 93, 95, 98, 100, 103, 104, 111, 112, 139, 144, 151, 156, 158, 163, 166, 171, 172, 178, 182, 183, 210, 242, 244, 250, 263
God, 38, 254
gold, 11, 16, 259
goods and services, 9, 19, 83, 124
governance, 33, 119, 235
government, 4, 5, 11, 14, 15, 16, 21, 25, 27, 33, 34, 36, 50, 68, 72, 82, 84, 87, 92, 119, 120, 121, 122, 127, 131, 132, 136, 139, 189, 193, 232, 235, 237, 258, 259, 260, 261, 262, 263, 264, 269, 274
governors, 49, 162, 173, 294
grading, 6, 7, 22, 69, 71, 114, 115, 131, 133, 142, 150, 159, 162, 169, 205, 272
grain, 256
grains, 256
grants, 13, 32, 58, 59, 101, 125, 191, 198, 206, 216, 235, 268
Great Britain, 7, 8, 12
Greece, 39, 46, 49, 302

greed, x
greenhouse, 37, 128, 133, 210, 211, 242, 271, 303
Greenhouse, 302
greening, 284
grid services, 143, 185
gross domestic product, 63
ground water, 199
grounding, 260
group activities, xii
grouping, 64
groups, viii, 18, 19, 83, 127, 134, 140, 188, 263
guerrilla, 200
guidance, 38, 303
guidelines, 34, 70, 84, 124, 131, 274
Guinea, 233

H

habitat, 23, 126, 195, 196, 209, 210, 222, 224, 225, 226, 250, 258
Haifa, 301, 304
handling, 9, 49, 85, 98, 107, 131, 147, 150, 151, 152, 153, 157, 161, 165, 166, 167, 168, 171, 181, 191, 195, 201, 203, 204, 205, 206, 207, 208, 210, 211, 216, 217, 228, 229, 271, 294, 299
hands, vii, viii, 147, 193, 196, 205, 299
handwriting, 189
hanging, 194, 196
happiness, 43, 55, 261, 289, 312
haptic, 174, 185, 193, 203, 205
harm, 40, 42, 45, 232, 235
harmony, 40, 42, 45, 232, 235
Harvard, 273, 274, 275, 276, 277, 278, 282, 283, 284, 285, 286, 287, 288, 289, 290, 307, 310
harvest, 209, 210
harvesting, 46, 209, 256, 301, 302
healing, 185, 203, 204, 213
health, 59, 82, 181, 210, 216, 217, 236, 237, 287, 289, 312
health care, 287
hearing, 185
heart, 206
Hebrew, 234
hedge funds, 259
hedging, xv
hegemony, 12, 13, 17, 18, 84, 232, 239, 263
height, 219
Henry Ford, 45
heterogeneity, 66, 210
heterogeneous, 16, 173, 185, 230, 231

heuristic, 28, 47, 103, 104, 153, 154, 156, 158, 162, 175, 176
higher quality, 126, 134
high-level, 113
high-speed, 127
hip, 119
hiring, 252
holistic, 44, 45, 46, 48, 49, 113, 235, 287
holistic approach, 48, 287
Holland, 292, 294, 295
homogeneity, 109, 141
homogenisation, 107
honesty, 118
Hong Kong, 33
horizon, 52, 69, 100, 101, 102, 104, 107, 109, 144, 158, 159, 202, 203, 234
horse, 306
horses, 233
hospital, 203
hospitals, 220
hostility, 129, 227
house, 273, 284, 286, 288, 305, 307, 310
household, 52, 61, 215
housing, 14, 117, 150, 220, 306
hub, 22, 95
human capital, xv, 4, 5, 6, 7, 8, 9, 14, 15, 16, 22, 27, 29, 66, 67, 72, 73, 82, 89, 235, 244, 252, 258
human development, 236
human resources, 120
humanitarian, 199, 200, 276, 277
humanitarian intervention, 277
humanity, 23, 34, 35, 226, 248, 255, 258, 261
humans, vii, 143, 144, 185, 197, 215, 225, 277
humidity, 210, 211
humus, 6, 7, 24
hunting, 153, 242
husbandry, 211
hybrid, 15, 119, 166, 169, 171
hydro, 266
hydrocarbon, 256
hydrocarbons, 266
hydrodynamic, 224
hydrogen, 217
hygiene, 214
hypothesis, xiii, 23, 44, 45, 97, 104, 128, 135, 155, 213, 232, 233, 237, 245, 252, 263, 271, 273

I

IBM, 102, 273, 287
ice, 126

ICT, 18, 23, 203
id, xv, 33, 131
IDEA, 275, 281, 304, 309, 310, 311
identification, 72, 189, 219, 225, 248
identity, 38, 42, 44, 49, 187, 234, 242, 263
Illinois, 310
illusions, 130
ILO, 34
images, 31, 204, 205, 258
imbalances, 17, 130, 254
IMF, 17
immersion, 206
immigration, 123
impact analysis, 57, 166
impact assessment, 18, 27, 68, 115, 253, 262
impact monitoring, 58, 66, 186, 226, 227, 250
imperialism, 16, 27, 124, 130, 259, 308
implementation, 78, 105, 113, 156, 167, 173, 176, 180, 198, 199, 207, 209, 210, 214, 216, 219, 254, 275, 293, 295
implicit knowledge, 69
in situ, 221
incentive, 16, 93, 123, 139
incentives, 15, 62, 80, 93, 139
inclusion, 55, 90, 105, 113, 115
income, 9, 15, 16, 18, 78, 82, 91, 121, 123, 132, 138
indebtedness, 15, 35
independence, 12, 37, 42, 48, 92, 170, 226, 242, 251
indexing, 173, 181
India, 12, 13, 28, 121, 130, 232, 233, 281, 305, 308
Indian, 39
Indiana, 282, 290
indication, 173
indices, 77, 93, 106, 109, 153
indigenous, 24, 312
indigenous knowledge, 312
indirect effect, 284
individual action, 48
individual differences, 277
individual rights, 36
individualism, 37, 38, 45, 48
Indochina, 233
Indonesia, 17, 233
industrial application, vii
industrial production, 280
industrial revolution, x, xi, xiii, xv, 1, 2, 3, 4, 7, 8, 12, 14, 17, 20, 22, 24, 25, 26, 28, 30, 36, 37, 39, 45, 48, 53, 86, 97, 120, 125, 141, 226, 232, 233, 235, 244, 249, 250, 255, 258, 259, 265, 267, 268, 269, 272, 277, 290, 305

industrial transformation, 16, 30, 122, 240, 268, 283
industrialisation, 121, 123, 127, 306
industry, ix, x, xi, xii, xiii, xv, xvi, 1, 2, 3, 17, 18, 25, 26, 27, 28, 36, 37, 43, 48, 50, 51, 52, 53, 54, 55, 66, 67, 72, 82, 91, 93, 97, 98, 99, 102, 107, 110, 112, 116, 118, 120, 121, 122, 125, 126, 127, 129, 147, 150, 151, 152, 156, 157, 159, 162, 164, 171, 203, 227, 229, 230, 231, 232, 235, 241, 242, 249, 250, 251, 259, 260, 262, 264, 265, 266, 267, 268, 269, 271, 272, 276, 284, 287, 291, 296
inefficiency, 50
inequality, 308
inert, 6, 7
inertia, 174, 175, 176
inferences, 39, 46
infinite, 45, 236
inflation, 16, 121
information age, 310
information and communication technologies, 11, 27, 129, 130, 239
information and communication technology, 189
information economy, 278, 311
information processing, 229
information seeking, 286
information sharing, 74, 144, 149, 242
Information System, 277, 288, 289, 290
information systems, 56, 149, 273, 274, 282, 285, 287, 289, 290, 291, 311
information technology, 72, 292
Information Technology, 295
infrared, 221
infrastructure, xii, 27, 57, 64, 66, 68, 79, 85, 99, 114, 144, 156, 208
infringement, 240
inheritance, 83, 104
inherited, 6, 10, 138, 242
inhibition, 101
injection, 194, 211
injury, iv
innovation, viii, xi, xiii, xvi, 18, 24, 32, 35, 45, 51, 60, 64, 68, 70, 72, 83, 85, 89, 93, 97, 98, 99, 101, 112, 127, 129, 132, 133, 135, 136, 142, 144, 145, 147, 148, 151, 152, 154, 157, 164, 185, 189, 196, 197, 203, 226, 231, 232, 236, 245, 246, 247, 251, 252, 255, 257, 266, 267, 269, 272, 281, 282, 284, 286, 291, 292, 311, 312
inoculation, 211
insecticides, 24
insects, 211
insertion, 175, 197, 198, 205, 206

insight, xvi, 6, 7, 9, 10, 13, 21, 28, 29, 30, 40, 46, 50, 67, 72, 75, 109, 125, 137, 176, 206, 208
inspection, xii, 153, 158, 171, 173, 183, 213, 258, 303
inspiration, 26, 256
instabilities, 134
instability, 3, 13, 16, 131, 231, 240, 249
instinct, 124
institutions, 5, 14, 21, 42, 84, 118, 122, 132, 135
instruction, 56, 147, 200, 237, 255
instruments, xv, 55, 61, 87, 126, 136, 186, 193, 199, 205, 214, 239, 259, 260, 293, 297
insulation, 137
insurance, 87
intangible, xii, xiii, xvi, 9, 11, 19, 23, 26, 27, 28, 29, 30, 43, 51, 58, 67, 69, 70, 72, 83, 86, 115, 119, 131, 137, 236, 242, 243, 244, 246, 250, 251, 254, 257, 259, 269, 271, 283
integration, xii, 13, 21, 65, 70, 71, 73, 74, 75, 77, 78, 105, 106, 108, 110, 111, 114, 115, 130, 148, 149, 150, 161, 163, 164, 166, 168, 171, 182, 184, 186, 187, 188, 189, 204, 206, 263, 264, 270, 274, 283, 296, 297, 298, 310, 312
integrity, 79, 118, 264
Intel, 282
intellectual property, 19, 70, 83, 84, 236
intelligence, 9, 20, 30, 31, 32, 37, 48, 56, 57, 85, 90, 98, 99, 110, 116, 136, 142, 143, 144, 145, 151, 167, 173, 180, 181, 183, 184, 185, 186, 191, 201, 204, 205, 208, 212, 213, 214, 219, 227, 232, 248, 249, 250, 251, 255, 257, 258, 261, 265, 266, 271, 272, 280, 291, 293, 294, 295, 296, 301, 304
intelligence gathering, 167
intentions, 42
interaction, 28, 66, 71, 72, 143, 145, 174, 191, 287
interactions, 144, 156, 185, 221, 290, 310
interactivity, 66, 143
interdisciplinary, 274
interest groups, 19, 134
interface, 76, 95, 103, 104, 143, 170, 173, 177, 183, 184, 190, 194, 198, 199, 204, 205, 213, 218, 221, 271
interference, 195
international investment, 291, 311
International Monetary Fund, 17
international trade, 92
internet, 13, 22, 64, 68, 186, 265, 282, 286
Internet, 286
interoperability, 115
interpersonal relations, 41, 119
interstate, 11, 252

intervention, 15, 16, 188, 204, 205, 206, 207, 208, 213, 244, 251, 253
interview, 149
interviews, 125
intrinsic, 3, 10, 13, 29, 40, 85, 153, 154, 164, 175, 184, 192, 194, 231, 233, 249, 255, 262, 263
intrinsic value, 164
introspection, 31
intrusions, 125, 237
invariants, 44
invasive, viii, 188, 203, 204, 300, 301
inventions, 23, 83, 142, 235, 236, 248, 251, 255, 257
inventiveness, 152, 265
inventories, 165
inventors, 29, 67
investment, 5, 10, 17, 27, 56, 58, 73, 78, 99, 100, 108, 111, 112, 125, 133, 148, 149, 154, 157, 159, 160, 161, 162, 163, 164, 165, 166, 167, 172, 183, 187, 188, 189, 191, 209, 212, 216, 229, 252, 259, 291, 309, 311
investors, 67, 69, 122
invisible hand, 13
IRP, 280, 281
irrigation, 6, 7, 24, 44
ISC, 301, 304
island, 218, 219
ISO, 238
isotropy, 176
Italy, xi, 78, 127, 138, 139, 228, 229, 270, 292, 297, 298, 299, 308
iteration, 163

J

Japan, 12, 47, 51, 163, 164, 232, 275, 276, 279, 281
Japanese, 41, 42, 233
jobs, 25, 68, 72, 74, 88, 93, 101, 102, 106, 109, 117, 125, 131, 143, 147, 150, 152, 161, 165, 166, 171, 175, 176, 177, 179, 181, 183, 184, 185, 188, 195, 197, 198, 200, 213, 215, 221, 225, 251
joints, vii, 168, 169, 174, 175, 217
judge, 200
judges, 33, 237, 310
Jun, 297
jurisdiction, 33, 237, 239, 240
justice, 33, 308

K

kernel, 49, 104, 150
Keynes, 15, 305, 306
Keynesians, 306
KILT model, 141, 142, 253, 254, 262
kinematics, 167, 168, 171, 174, 175, 176, 195, 206, 223, 298, 300
kinetics, 44
King, 282, 308
knowledge acquisition, 156, 213
knowledge capital, 11, 98, 99
knowledge economy, 98
knowledge transfer, 11
knowledge-based economy, 289
Kobe, 295
Korean, 41, 233
Kurds, 237
Kyoto protocol, 89, 91

L

labour, xiii, 4, 5, 16, 23, 26, 27, 43, 45, 53, 62, 89, 98, 99, 102, 121, 133, 141, 147, 154, 158, 171, 184, 191, 193, 203, 242, 255, 291, 297, 311
labour market, 121
land, xv, 10, 11, 21, 22, 24, 25, 43, 126, 139, 197, 200, 201, 202, 236, 243, 244, 251, 255, 277
landfill, 62, 89, 94, 136, 197, 198, 228, 301, 304
landfill management, 304
landfills, 61, 126, 197, 300, 301
language, 40, 41, 135, 198, 233, 234, 248, 284, 305, 306
laser, viii, 217
Latin America, 17, 308
law, 3, 11, 14, 16, 18, 25, 31, 32, 33, 34, 49, 61, 68, 70, 72, 74, 78, 80, 86, 89, 90, 92, 116, 118, 125, 128, 142, 168, 174, 221, 227, 237, 240, 244, 245, 248, 252, 254, 258, 259, 260, 262, 268, 290
law suits, 80
laws, 21, 23, 31, 36, 38, 42, 44, 45, 46, 48, 49, 64, 94, 131, 132, 153, 168, 169, 172, 241, 247, 249, 253, 255, 257
lawsuits, 33
layoffs, 308
leachate, 197, 300
leadership, 14, 27, 53, 77, 130, 232, 245, 259, 263
learning, viii, 9, 31, 38, 41, 49, 71, 104, 105, 119, 162, 179, 182, 282, 290

learning process, 49
legal aspects, 72
legality, 13, 50, 118, 123, 124, 126, 127, 131, 132, 139, 232, 258
leisure, 219, 311
liberal, 310
liberty, 82, 238
licensing, 290
life cycle, x, 280
life quality, xv, 3, 5, 10, 14, 135, 138, 139, 212, 265, 266
lifecycle, 27, 52, 54, 57, 58, 59, 60, 62, 65, 66, 68, 75, 79, 81, 84, 85, 86, 87, 88, 89, 90, 91, 94, 113, 114, 115, 116, 129, 131, 133, 134, 136, 138, 144, 149, 154, 155, 156, 161, 164, 165, 166, 178, 183, 184, 185, 186, 187, 188, 189, 192, 212, 241, 243, 244, 256, 268, 280, 281, 283, 285, 289, 304
life-cycle, 274
life-cycle, 311
lifespan, 185, 186, 187
lifestyle, xvi, 11, 16, 20, 51, 52, 54, 56, 57, 61, 64, 65, 66, 76, 80, 81, 84, 85, 86, 87, 90, 91, 97, 98, 105, 110, 111, 112, 135, 138, 142, 145, 148, 149, 150, 152, 153, 154, 155, 156, 158, 162, 178, 180, 184, 188, 191, 232, 251, 261, 268, 272, 277, 278, 279, 307
lifestyles, 22, 268
limitation, xvi, 12, 15, 135, 175, 217, 238, 241
limitations, 24, 57, 94, 124, 127, 138, 161, 170, 219, 240, 243, 244, 262, 275
linear, 45, 47, 48, 49, 55, 140, 141, 168, 169, 182
linear dependence, 140, 141
linear model, 141
linguistic, 38, 40, 150, 234
linkage, 22, 80, 219
links, 9, 21, 22, 26, 27, 28, 45, 59, 60, 64, 74, 80, 90, 94, 95, 104, 135, 169, 175, 190, 192, 233, 253, 259
liquid hydrogen, 217
litigation, 33
livestock, 6, 7, 25, 43
living standard, ix, 17, 148
living standards, 17, 148
loans, 89, 126, 135, 141, 235
lobbying, 252
local authorities, 85, 252
local government, 4, 5, 82, 121, 139, 236, 261
localised, 13, 73, 108, 125, 201, 247, 250, 264, 273
localization, xvi
location, 40, 47, 64, 126, 143, 185, 194, 198, 199, 205, 206, 217, 218, 219, 224, 229, 233

locomotion, 220
logical reasoning, 38, 48
logistics, xii, xvi, 19, 22, 31, 36, 51, 56, 57, 60, 61, 62, 63, 65, 66, 69, 86, 87, 89, 90, 91, 93, 95, 99, 100, 101, 110, 114, 138, 139, 144, 148, 158, 162, 186, 187, 200, 207, 209, 227, 228, 229, 241, 243, 244, 254, 268, 270, 271, 274, 280, 286, 294, 304
London, 12, 120, 274, 275, 276, 277, 278, 279, 281, 282, 283, 284, 285, 286, 287, 289, 290, 291, 292, 297, 305, 306, 307, 308, 309, 310, 311, 312
long distance, 10, 22, 95
long period, 23, 249
long-distance, 6, 8, 21, 95
losses, 63, 102, 119, 195, 211
Lovelock, 280
low-level, 219
loyalty, 33, 34, 84, 85, 118, 120, 136, 231, 240, 264

M

machinery, 10, 12, 14, 26, 95, 96, 191, 192, 212, 213
machines, 53, 158, 189, 191, 202, 256
magnesium, 211
magnetic, iv, 37, 44, 128
magnetic field, 37, 128
magnets, 173
main line, 160
maintenance, 24, 28, 57, 59, 63, 65, 68, 74, 80, 88, 89, 92, 95, 102, 104, 107, 112, 144, 148, 156, 158, 163, 165, 173, 186, 187, 188, 192, 193, 195, 196, 213, 229, 271, 293, 302
Maintenance, 186, 195
maintenance tasks, 186
maize, 233, 256
mandatory targets, 65, 74, 76, 77, 80, 84, 86, 89, 92, 94, 228, 241
manifold, 100, 101, 105, 112, 155, 158, 162, 197, 215, 278
manifolds, 101, 155, 158, 162
manipulation, 67, 133, 147, 148, 151, 153, 154, 157, 159, 160, 164, 167, 168, 169, 170, 171, 173, 174, 175, 176, 179, 180, 181, 182, 193, 203, 204, 206, 251, 258, 298, 300
manufactured goods, 22, 43, 86, 125
manufacturer, 58, 61, 73, 75, 78, 137, 174, 188, 222
manufacturing, x, xi, xii, 1, 2, 3, 20, 26, 45, 53, 55, 65, 88, 91, 99, 102, 104, 106, 107, 112, 113, 118, 123, 140, 141, 148, 149, 150, 152, 154, 157, 158, 159, 160, 161, 162, 163, 164, 165, 173, 181, 212, 251, 253, 259, 268, 270, 275, 278, 280, 281, 282, 284, 285, 290, 291, 292, 293, 294, 295, 296, 297, 298, 310, 311
manure, 210
mapping, 5, 48, 109, 167, 168, 172, 194
margin of error, 23
marginal costs, 13
marginal utility, 68
marine environment, 303
maritime, 6, 8, 12
market economy, 68, 135
market opening, 12
market prices, 62, 90
market share, 54, 69, 77, 84, 237
market value, 118
marketing, 56, 79, 142, 148, 152, 284
marketplace, 172
markets, viii, xv, xvi, 6, 8, 17, 121, 124, 136, 138, 162, 184, 268, 275, 283, 286, 309, 312
Markov, 47
Markov model, 47
Marx, 5, 14
Marxism, 4
mastery, 5, 10, 11, 26, 45, 53, 72, 119, 120, 126, 203, 208
material resources, xi, 3, 126, 244
mathematics, xv, 31
matrix, 169, 175
meanings, 26, 124
measurement, 11, 47, 71, 75, 76, 78, 109, 138, 161, 170, 171, 173, 174, 176, 244
measures, 15, 17, 22, 33, 40, 49, 58, 60, 61, 62, 68, 76, 80, 86, 91, 93, 94, 95, 109, 110, 122, 123, 125, 137, 141, 193, 200, 240, 244, 250, 253, 262, 263, 272, 289
media, xi, 185, 284, 302
median, 50
mediation, 48, 49, 148, 151
medicine, 25, 203, 204, 213, 300
Mediterranean, 6, 8, 233, 298
melody, 143
membership, 14, 41, 47, 48, 117, 242
memory, 103, 105
men, 6, 7, 8, 9, 10, 18, 20, 21, 23, 27, 28, 39, 46, 53, 82, 98, 117, 118, 127, 128, 144, 147, 152, 166, 179, 180, 195, 224, 232, 234, 244, 247, 251, 255, 257, 265, 266
mercantilism, 4, 6, 8, 11, 12, 24, 118, 119, 121, 132, 238, 239
merchandise, 123, 193
mergers, 18
message passing, 103, 106

messages, 67, 194
metals, 11
metaphor, 127
metric, 303
Mexican, 256
Mexico, 233
military, 120, 121, 199, 200, 223, 263, 288
millet, 233
mimicry, 23, 26, 30, 31, 32, 36, 43, 87, 98, 116, 128, 129, 138, 139, 142, 180, 232, 243, 250, 254, 281
mines, 25, 199, 201
mining, 6, 8
minorities, 33, 121, 262
minority, 13, 14, 15
mirror, 174
misconception, 307
misleading, x, 5, 44, 69, 239, 254
missions, 171, 182, 211, 219, 220, 223, 224, 226, 264, 303
MIT, 274, 279, 282, 283, 285, 291, 305, 306, 310, 311
mixing, 10, 17, 38, 85, 129, 168, 256, 272
mobile device, 216
mobile phone, 185
mobile robot, viii, 197, 202, 221
mobility, 63, 166, 176, 201, 202, 206, 215, 220, 222, 223, 246, 291, 298, 311
models, xi, xv, 5, 10, 15, 37, 38, 39, 44, 45, 47, 48, 49, 52, 55, 90, 103, 104, 106, 109, 110, 113, 115, 130, 140, 141, 142, 143, 154, 155, 158, 165, 170, 172, 173, 174, 176, 180, 181, 185, 186, 204, 205, 207, 217, 244, 253, 255, 257, 262, 274, 278, 282, 288, 304
modernity, 291, 312
modulation, 78, 80, 169, 171, 173, 174, 192, 221, 234
modules, 65, 103, 104, 115, 131, 143, 154, 155, 156, 158, 160, 168, 171, 173, 178, 191, 192, 195, 208, 213, 220
modus operandi, 11
moisture, 210
momentum, 80, 91
money, xv, 3, 4, 5, 6, 7, 8, 11, 16, 26, 27, 60, 117, 120, 122, 123, 138, 139, 141, 251, 259, 275, 276, 289, 305, 312
mood, 49
morals, 32, 33
morphology, 194
mortgage, 259
Moscow, 298, 302, 303, 304
motion, 40, 168, 169, 170, 174, 178, 182, 195, 204, 207, 213, 218, 221, 223, 224

motivation, viii, 11, 38, 52, 53, 54, 70, 75, 106, 121, 145, 149, 184, 200, 227, 258
motives, 17
motors, 168, 221, 223
mountains, 234, 284
movement, 304
MPA, 189
multicultural, 39
multinational companies, 238
multinational corporations, xii, 131, 239
multinational enterprises, 290, 310
multiplicity, 233, 234
muscles, vii

N

naming, 31, 97, 180
nation, x, xi, 4, 6, 7, 8, 12, 15, 17, 22, 24, 25, 32, 34, 35, 37, 44, 49, 82, 84, 85, 91, 97, 119, 120, 121, 122, 124, 125, 127, 130, 131, 132, 134, 231, 232, 237, 238, 239, 242, 245, 252, 260, 263, 269, 271, 275, 284
nation building, 120, 121
nation states, 275
national debt, 122
national income, 123
national product, 252
nationality, 135
natural capital, xi, xv, 4, 5, 7, 8, 10, 11, 13, 16, 17, 18, 19, 20, 22, 27, 29, 31, 34, 35, 61, 62, 78, 82, 83, 84, 85, 87, 88, 89, 93, 94, 120, 121, 123, 124, 126, 127, 128, 129, 130, 131, 133, 134, 136, 138, 140, 149, 186, 231, 239, 243, 244, 248, 250, 253, 254, 258, 260, 262, 264, 268, 269
natural evolution, 20, 31, 247, 248, 256
natural laws, 255
natural resources, viii, xvi, 4, 65, 82, 89, 222, 244, 254, 266
negative outcomes, 37, 245
neglect, 35, 52, 141
negligence, 193
negotiation, 70, 73, 74, 75, 80, 93, 124, 135, 161, 187
Netherlands, 93, 283, 288
network, xi, 28, 44, 69, 72, 76, 79, 95, 143, 177, 185, 193, 229, 253, 284
networking, 13, 21, 64, 70, 71, 72, 76, 87, 94, 95, 113, 143, 185, 186, 187, 188, 268, 289, 290, 310
neutral yield, 19, 31, 129, 138
New World, 233

New York, i256, 273, 274, 275, 276, 277, 278, 279, 280, 281, 282, 283, 284, 285, 286, 287, 288, 289, 290, 291, 292, 296, 305, 306, 307, 308, 309, 310, 311, 312
next generation, vii, 137, 256, 289
NGOs, 284
noble metals, 11
nodes, 21, 22, 71, 72, 74, 76, 77, 78, 79, 95, 131, 253
noise, 192, 196
non invasive, 191
non-human, 31
non-invasive, 192
nonlinearities, 141, 169, 170
non-linearity, 167, 170, 171
non-renewable, 1, 2, 6, 7, 8, 30, 43, 116, 238, 240, 244, 256
non-renewable resources, 1, 2
normal, 59, 110, 192
normal conditions, 59
norms, 211
North Carolina, 291, 311
North Sea, 195, 197
novelty, 135, 180, 232
nuclear, 37, 127, 128, 129, 134, 164, 171, 193, 251
nuclear power, 171, 193
nuclear power plant, 193
nursing, 220, 300
nursing home, 220
nuts, 41

O

oat, 233
objectivity, 80, 86, 92
obligation, 6, 50, 63, 93, 133, 135, 187, 208, 212, 228, 244, 256, 262
obligations, 43, 49, 63, 71, 85, 89, 117, 122, 136, 187, 191
observations, vii, 40, 194, 214, 225, 266
obsolete, 90
obstruction, 82
occlusion, 197
oceans, 227
off-the-shelf, 160, 191
oil, 16, 195, 309, 312
old-fashioned, 212
oligopolies, 24, 252
oligopoly, 18, 82, 84, 125, 253
olive, 233
omission, x, 268
one-to-one mapping, 168

online, vii, 74, 101, 104, 108, 152, 158, 173, 178, 191, 199, 202, 211, 270, 297
opacity, 61, 62, 227
open space, 189
operator, 11, 29, 120, 180, 183, 193, 196, 198, 202, 203, 225
opium, 46
opposition, 13, 26, 27, 31, 33, 39, 46, 47, 48, 50, 53, 72, 134, 135, 141, 192, 228, 236, 241, 246, 247, 249, 257, 258, 263
optical, 174, 181, 190, 216, 217
optimal performance, 207
optimism, 129
optimization, 288
organ, 207, 278, 311
orientation, vii, 32, 114, 118, 148, 166, 167, 181, 182
oscillation, 221
oscillations, 218, 219
otters, 263
outsourcing, 73, 121, 126, 286
overload, 186
overproduction, 107
oversight, 268
ownership, 3, 5, 6, 7, 11, 14, 19, 32, 33, 34, 35, 72, 82, 83, 84, 89, 118, 126, 136, 138, 228, 236, 243, 253, 266, 290
oxygen, 248, 256

P

packaging, 286
packets, 189, 190
palliative, 84
paradigm shift, viii, xi, 11, 12, 13, 20, 26, 27, 29, 65, 113, 208, 241
paradox, 18, 83, 309
parallel processing, 153
parallelism, 36, 173, 175, 176, 177, 178, 207
parasites, 211, 212
parents, 256
Paris, 207, 235, 275, 276, 278, 279, 289, 300, 303, 305, 307, 308, 309, 310
particles, 235
partition, 12, 21
partnership, 5, 70, 71, 74, 75, 76, 77, 91, 106, 108, 117, 120, 230, 274, 282
partnerships, 117
passenger, 193
passive, 30, 56, 63, 150, 176, 208, 245, 263, 264
pasture, 233
pastures, 201
patents, 10, 12, 18, 27, 70, 83, 236

path planning, 166, 172, 182, 215, 221, 301
path tracking, 215, 223
pathogenic, 197, 211, 212
pathogenic agents, 197, 212
pathogens, 210, 211, 212
pathology, 211
patients, viii, 204, 207
patriotism, 200
pattern recognition, 218
pedal, 304
penalties, 50
penalty, 27, 82, 184
perception, 40, 48, 50, 143, 185, 197, 243, 264
perceptions, 34
periodic, 71, 192
permit, xvi, 40, 47, 56, 64, 76, 84, 89, 91, 97, 108, 126, 133, 135, 153, 159, 165, 219, 222, 224, 234, 239, 246, 247, 248, 249, 255, 256, 266, 301
Persia, 233
personal autonomy, 42
personal benefit, 123
personal identity, 49, 234
personal relations, 151
personal responsibility, 35
personal wealth, 82, 240
perturbations, 24, 45
Peru, 233
pervasive computing, 289
pessimists, 24
pesticides, 24, 37, 128
pharmaceutical, 283, 295
philosophers, 41, 44, 117, 240
philosophical, xiii, 39
philosophy, 60, 103, 106, 199, 257, 284, 285, 288
phone, 185
phonemes, 234
photon, 44
physical world, 37, 144, 245, 248
physicists, 257
physics, 23, 30, 31, 36, 42, 44, 49, 241, 257
pig, 233
piracy, 124
pitch, 219, 223, 224
planetary, viii
plankton, 266
planned investment, 163
planning, 14, 15, 34, 35, 38, 42, 54, 55, 57, 58, 60, 71, 76, 88, 100, 101, 102, 103, 106, 107, 109, 115, 116, 136, 152, 153, 156, 157, 158, 159, 160, 163, 166, 168, 170, 172, 173, 174, 177, 182, 183, 191, 196, 198, 199, 202, 203, 204, 206, 207, 210, 215, 216, 221, 223, 230,
240, 243, 244, 246, 252, 258, 274, 282, 290, 295, 301
plants, 27, 37, 95, 100, 103, 127, 128, 129, 136, 150, 151, 159, 163, 164, 171, 195, 210, 211, 212, 214, 233, 251, 252, 256, 292, 295, 300, 303
plasma, 171
plastic, 197
plastics, 160
platforms, 185, 220
Plato, 40, 41, 44, 45
plausibility, 47, 248, 262
play, xii, xiii, xv, 7, 8, 25, 39, 41, 54, 88, 98, 106, 113, 115, 116, 128, 131, 132, 135, 180, 189, 209, 214, 226, 247, 255, 257
pleasure, 44
ploughing, 24
poisoning, 17
poisonous, 211, 217
poisons, 211, 244, 265
policy reform, 17
political aspects, 139
political instability, 238
political power, 6, 8
political stability, 253
political uncertainty, 263
politicians, xi
politics, 33, 275, 289, 290, 291
polling, 240
pollutant, 217
pollutants, 138, 254
pollution, x, xi, xiii, xv, xvi, 1, 2, 5, 7, 8, 10, 16, 17, 18, 20, 23, 24, 29, 35, 36, 42, 43, 52, 53, 61, 79, 82, 83, 85, 86, 87, 89, 91, 93, 97, 116, 118, 126, 132, 138, 139, 140, 141, 166, 187, 195, 196, 212, 222, 227, 231, 240, 241, 243, 244, 247, 249, 250, 253, 254, 257, 262, 266, 267, 268, 269, 272
pools, 197
poor, 33, 34, 79, 91, 121, 123, 134, 175, 188, 209, 237, 305
population, xi, 17, 20, 29, 37, 85, 133, 211, 212, 222, 231, 233, 241, 246, 257, 262, 266, 269, 307
populism, 16
posture, 153, 170
potato, 201, 233
potatoes, 235
poultry, 226
poverty, 17, 117, 132, 235, 238, 279, 282, 307, 308, 309
power, 6, 7, 8, 11, 14, 15, 28, 72, 81, 93, 131, 171, 193, 194, 196, 198, 201, 203, 216, 220,

221, 223, 224, 237, 240, 245, 252, 256, 259, 263, 271, 273, 277, 278, 284, 288, 289, 291, 312
power plants, 193
powers, 12, 121, 239, 264, 288
pragmatic, 250
prediction, 23
preference, 45, 78, 91
president, 102
pressure, 58, 173, 175, 195, 196, 219, 238
prevention, 61, 195, 264
preventive, 59, 60, 134, 211
prices, 13, 15, 26, 62, 73, 90, 117, 122, 125, 126, 252
primacy, 12
privacy, 15, 35, 36, 69, 79, 80, 93, 94, 131, 132, 144, 185, 230, 237, 285
private, 14, 15, 17, 19, 32, 33, 34, 35, 36, 41, 78, 120, 124, 134, 235, 236, 240, 259
private ownership, 14, 32, 33, 34, 236, 240
private property, 15
privatisation, 235, 236, 237, 238, 308
proactive, 23, 59, 60, 85, 101, 102, 111
probability, 47, 132, 257
probe, 207, 208
problem solving, 54, 55, 67, 98, 106, 266, 271
problem-solving, 39, 49, 114, 148, 152, 153, 156, 162, 208, 213, 214, 271
procedural knowledge, 47, 104, 105, 113, 155, 176, 207
process control, 55, 113, 174, 210
process innovation, 93
producers, 12, 52, 62, 65, 67, 68, 85, 90, 93, 123, 125, 126, 131, 155, 165, 166, 183, 186, 187, 209, 244, 268
product design, 52, 54, 280
product market, 288
production, x, 3, 13, 14, 15, 17, 26, 27, 45, 52, 53, 55, 56, 63, 64, 65, 69, 72, 74, 81, 88, 99, 104, 105, 106, 110, 113, 117, 120, 123, 125, 136, 141, 142, 149, 151, 152, 153, 154, 157, 158, 159, 160, 161, 162, 163, 164, 165, 181, 184, 222, 232, 241, 242, 251, 256, 274, 276, 279, 280, 288, 292
production costs, 72, 117
productivity, 1, 2, 6, 8, 13, 16, 24, 26, 30, 55, 62, 63, 68, 70, 79, 84, 94, 100, 102, 105, 107, 108, 109, 110, 115, 117, 123, 128, 129, 130, 133, 138, 139, 140, 141, 148, 149, 150, 154, 157, 159, 163, 170, 172, 174, 177, 185, 189, 191, 200, 201, 202, 203, 222, 225, 226, 235, 238, 243, 248, 251, 254, 256, 260, 273, 274, 280, 284, 286

product-process-environment, xii, 54, 65, 66, 89, 90, 106, 112, 113, 115, 148, 150, 155, 270
products-services, 69, 79, 81, 87, 88, 90, 101, 115, 250, 299, 304
professional development, 267
profit, 5, 7, 8, 9, 12, 13, 15, 16, 19, 21, 24, 25, 27, 28, 32, 33, 35, 50, 51, 56, 63, 69, 72, 79, 81, 82, 117, 118, 119, 125, 126, 130, 134, 139, 165, 173, 178, 185, 188, 227, 230, 235, 236, 237, 238, 239, 240, 245, 247, 249, 252, 256, 259, 260, 262, 271, 275, 290
profit maximisation, 240
profitability, 57, 60, 71, 79, 90, 149, 152, 162, 163, 181, 187
profits, 5, 11, 14, 19, 20, 34, 48, 58, 68, 69, 80, 90, 119, 154, 167, 170, 173, 175, 177, 197, 234, 236, 241, 243, 251
progenitors, 233
program, 289, 294
programmability, 99
programming, 47, 70, 95, 103, 104, 147, 154, 158, 162, 166, 168, 172, 173, 182, 220, 280, 292, 293, 297
progressive tax, 15, 33, 91, 138
proliferation, 70, 128, 257
promoter, 247
pronunciation, 41
propagation, 102, 114, 156, 191, 287
property, iv, 14, 19, 25, 30, 35, 46, 68, 70, 72, 81, 83, 84, 90, 128, 142, 156, 187, 217, 236, 252, 254
property rights, 25
prophylaxis, 210, 212
proposition, 70, 134, 244
propulsion, 224
prosperity, 9, 12, 17, 56, 281, 284
protection, xiii, xvi, 6, 11, 15, 17, 18, 19, 27, 28, 32, 33, 34, 36, 48, 50, 53, 54, 57, 58, 63, 64, 68, 69, 70, 72, 73, 74, 76, 77, 79, 80, 81, 82, 83, 84, 85, 91, 92, 93, 94, 96, 101, 110, 113, 115, 117, 118, 119, 122, 123, 124, 125, 126, 132, 133, 134, 138, 139, 144, 148, 149, 155, 184, 187, 190, 191, 197, 200, 210, 212, 216, 217, 219, 226, 230, 232, 236, 237, 266, 272, 312
protectionism, 16
protocol, 204, 206
protocols, 11, 77, 80, 89, 91, 95, 203, 204, 206, 208
prototype, 194, 217, 295, 301, 303
prototyping, 68, 103, 288
proxy, 149
psychology, 37, 128, 245

public, 10, 15, 16, 17, 32, 33, 35, 36, 46, 50, 78, 91, 93, 132, 189, 214, 227, 228, 235, 285
public companies, 93, 132
public corporations, 91
public debt, 16
public expenditures, 17
public service, 189, 227, 228
public welfare, 36
pumps, 215
purchasing power, 15

Q

quality assurance, 184
quality control, 165, 183, 204
quality of life, ix, xi, 1, 2, 3, 7, 9, 10, 24, 34, 51, 83, 85, 116, 126, 137, 226, 231, 241, 247, 249, 259, 269
quarantine, 243
question mark, xv, 259
questioning, 269
quotas, 14, 31, 123, 124, 138

R

radiation, 128
radio, 218, 219
radioactive waste, 251
rail, 127, 193, 211
random, 21, 22, 69, 95, 234
range, 4, 15, 34, 36, 47, 52, 54, 70, 71, 73, 74, 75, 77, 93, 100, 106, 107, 108, 112, 113, 119, 127, 137, 149, 150, 152, 154, 157, 179, 180, 183, 191, 240, 247, 248, 250, 263, 266
rating agencies, 259
rationality, 118, 134, 260
raw material, xiii, xvi, 1, 2, 3, 5, 7, 8, 13, 16, 20, 22, 23, 29, 30, 31, 43, 50, 83, 86, 89, 97, 118, 119, 123, 126, 133, 140, 150, 222, 227, 244, 272, 273
raw materials, xiii, xvi, 1, 2, 3, 5, 7, 8, 13, 16, 20, 22, 23, 29, 30, 31, 43, 50, 83, 86, 89, 97, 118, 119, 123, 133, 140, 150, 222, 227, 244, 272, 273
reactivity, 95, 151
reading, viii, xiii, xvi, 4, 70, 189, 191, 226, 228, 234, 251, 267, 273, 277, 279, 285, 305, 309
real time, 47
reality, 9, 18, 46, 76, 90, 95, 97, 103, 104, 110, 112, 120, 142, 143, 148, 154, 155, 156, 170, 185, 189, 193, 204, 215, 249, 258, 261, 263, 279, 283, 298, 300, 303, 306

reasoning, 32, 38, 39, 42, 44, 46, 48, 49, 103, 151, 152, 153, 213, 258
recall, 43, 121, 142
recalling, 112
reception, 9
recession, x, 15
recessions, 15
reciprocity, 260
reclamation, 23, 65, 68, 83, 199, 201, 202, 227, 228, 229, 241, 246, 303
Reclamation, 195
recognition, 6, 8, 31, 47, 59, 60, 143, 149, 173, 185, 190, 191, 211, 218, 243
recovery, 3, 17, 19, 23, 27, 28, 29, 31, 32, 52, 54, 57, 60, 61, 62, 65, 66, 68, 74, 80, 81, 84, 86, 87, 88, 89, 90, 91, 93, 94, 95, 105, 108, 110, 111, 112, 113, 114, 115, 126, 129, 133, 138, 139, 153, 154, 157, 159, 161, 162, 165, 173, 182, 183, 186, 187, 188, 195, 199, 201, 205, 209, 212, 213, 215, 223, 227, 228, 229, 230, 231, 244, 245, 254, 268, 281, 295, 304
recreation, 67
recycling, 63, 80, 288, 304
reductionism, xvi, 2, 3, 23, 28, 29, 37, 38, 39, 42, 43, 44, 45, 46, 48, 49, 50, 51, 53, 54, 55, 66, 76, 95, 103, 107, 113, 116, 142, 147, 148, 150, 151, 152, 155, 156, 161, 162, 164, 172, 189, 235, 262, 265, 268, 270, 271, 272, 277
redundancy, 22, 57, 59, 95, 111, 153, 157, 159, 160, 164, 166, 167, 169, 170, 171, 172, 173, 175, 176, 181, 182, 193, 205, 206, 208, 220, 221, 225, 298
reference frame, 4, 207
reflection, 264
reforms, xii
regeneration, 60, 241
regional, 93, 189, 190, 191, 221, 228, 245, 263
regression, 9, 256
regular, 90, 107, 156, 160, 182, 192, 205, 208, 224, 225, 234, 247
regulation, 11, 17, 19, 21, 25, 34, 35, 36, 52, 57, 79, 80, 82, 86, 90, 91, 92, 94, 101, 117, 123, 124, 131, 132, 137, 139, 182, 213, 217, 219, 228, 236, 237, 240, 252, 261
regulations, xiii, xvi, 4, 16, 17, 19, 33, 36, 51, 54, 61, 65, 72, 81, 82, 91, 92, 113, 124, 129, 130, 133, 149, 150, 154, 187, 214, 215, 227, 236, 253, 254, 262, 264, 272
regulators, 73
rehabilitation, 195
rejection, 18, 83, 97, 127, 135, 249
relapses, 25
relationship, 29, 67, 70, 76, 119, 310

relationships, 115, 136, 155, 257
relatives, 49
relevance, 51, 69, 99, 109, 142, 154, 166, 167, 180, 184, 213, 214, 233, 249, 259, 264
reliability, xii, 22, 24, 32, 33, 55, 59, 63, 78, 81, 90, 92, 102, 109, 113, 137, 148, 165, 166, 178, 183, 189, 191, 192, 193, 196, 197, 199, 200, 202, 204, 206, 207, 208, 215, 225, 237, 265
religious belief, 238
religious beliefs, 238
re-materialising, 126, 127
remediation, 1, 2, 23, 60, 68, 78, 87, 89, 136, 138, 196, 197, 213, 227, 229, 250, 300, 301
renewable resource, 6, 7, 24, 209, 250, 254, 258
repair, xii, 60, 63, 68, 222
reparation, 186
repeatability, vii, 171
replication, 172
reproduction, 255
reputation, 70
research and development, 192, 311
Research and Development, 292
reservation, 43, 65, 83, 89, 116, 140, 163, 176
reserves, 30, 222, 255
residuals, 193, 195, 211, 228, 230
residues, 92
resistance, 14, 50, 132
resolution, 105, 245
resource allocation, 13
resource management, 90, 132, 245
responsibilities, 48, 76, 228, 243, 251
responsiveness, 33, 38, 41, 48, 50, 54, 57, 91, 111, 118, 119, 131, 136, 188, 230, 237, 242, 244, 249, 253, 254, 257
restitution, 59, 104, 154, 171, 191, 192, 193, 198, 208
restructuring, xi, 1, 2, 68, 69, 87, 95, 96, 121, 139, 270, 292
retirement, 15
returns, 33, 55, 117, 125, 142, 165, 263
reusability, 196
revenue, 5, 14, 17, 24, 25, 26, 91, 189, 190, 258
rewards, 117, 118, 136
rhetoric, 38
ribosome, 255
rice, 9, 16, 233
rifting, 253
risk, xv, 19, 32, 37, 76, 90, 102, 119, 137, 195, 196, 197, 203, 211, 213, 218, 222, 226, 240, 245, 251, 259, 263
risk management, xv
risks, ix, xv, 61, 91, 114, 131, 132, 156, 181, 195, 210, 212, 217, 235, 238, 247, 251, 274, 311

rivers, 23, 48, 153, 284
RNA, 255
robbery, 236
robotic, vii, xii, xvi, 145, 148, 150, 151, 156, 164, 165, 166, 171, 172, 176, 178, 180, 182, 183, 184, 188, 189, 191, 197, 201, 202, 205, 206, 207, 208, 209, 211, 212, 214, 216, 217, 223, 225, 226, 227, 254, 270, 271, 292, 293, 296, 297, 298, 299, 300, 301, 302, 303
robotic arm, 298, 300
robotic surgery, 208, 301
robotics, vii, xii, 101, 112, 144, 145, 147, 148, 152, 166, 167, 169, 170, 172, 179, 180, 181, 183, 184, 188, 193, 203, 204, 207, 209, 211, 212, 215, 216, 222, 223, 226, 227, 255, 270, 271, 292, 297, 298, 299, 301, 302, 311
Robotics, vii, 98, 188, 193, 195, 197, 209, 281, 286, 292, 293, 294, 295, 297, 298, 299, 300, 301, 302, 303, 304
robust design, 288
robustness, 107
rocky, 194, 300
rods, 194, 197, 198, 299
Roman Empire, 245
Rome, viii, 117, 118, 276
rotations, 221
roughness, 179
routines, 92, 192, 198
routing, 100, 157, 158, 159, 160
royalties, 18, 70, 83
rural, 21
Russia, 14, 121, 130, 232, 233, 299
Russian, 304, 305

S

SAE, 282, 288
safeguard, 15, 18, 33, 34, 36, 38, 57, 61, 68, 80, 83, 85, 118, 121, 122, 123, 125, 133, 134, 191, 192, 194, 225, 237, 261, 266, 303
safety, 31, 37, 59, 79, 83, 87, 133, 179, 184, 195, 196, 199, 200, 202, 204, 208, 211, 216, 217, 218, 219, 226, 229, 241, 248, 253, 262, 303
salaries, 252
sales, 25, 57
salinity, 24
sample, 11, 19, 61, 145, 188, 190, 209, 214, 217
sand, 256, 266
SARA, 271, 298, 302
satellite, 185
satisfaction, 14, 21, 28, 55, 56, 57, 58, 63, 68, 73, 74, 76, 77, 80, 82, 88, 91, 114, 116, 122, 123,

125, 133, 138, 139, 144, 150, 164, 184, 187, 192, 210, 212, 213, 237, 243, 244, 289, 296
saturation, 56, 57, 121, 153, 162
savings, 159
scaffolding, 194
scaffolds, 194
scaling, 141, 205, 294
scatter, 82, 212
scattering, 36, 47, 195
scheduling, 108, 114, 162, 166, 295
schema, 23
schemas, 172
school, 10, 39, 46
scientific knowledge, 44
scientific method, 94
seabed, 195
search, 187, 267, 274, 278, 307
seaweed, 236, 266
secret, 277
Secretary of the Treasury, 25
securities, 259
security, 79, 80, 190, 204, 213, 214, 285, 291
sedentary, 21
seeds, 23, 24, 212
segmentation, 21, 35, 45, 111, 234
segregation, 35, 190, 256, 278
seizure, 300
selecting, 4, 152, 158, 167, 176, 202, 223, 257, 294
self, 237, 300
self-assessment, 71, 77, 78
self-consistency, 112
self-interest, 13, 15, 240
self-management, 240
self-regulation, 25, 132, 252
self-reproduction, 255
seller, 11, 16
Senegal, 233
sensing, vii, 143, 154, 182, 301
sensitivity, 140, 208
sensors, vii, xii, 31, 55, 59, 143, 169, 173, 174, 181, 182, 183, 185, 194, 195, 196, 198, 199, 204, 205, 211, 216, 217, 221, 225, 301
sentences, 38, 41
separation, 30, 40, 53, 72, 142, 192, 230, 252
September 11, 262
sequencing, 45, 101, 107, 108, 109, 152, 158, 159, 176, 178, 183, 207, 210, 215
series, xvi, 17, 18, 36, 38, 40, 48, 49, 52, 53, 54, 55, 56, 61, 62, 64, 70, 73, 78, 84, 88, 106, 111, 113, 129, 135, 145, 147, 153, 156, 159, 161, 164, 165, 167, 168, 172, 174, 179, 188, 190, 192, 195, 196, 198, 200, 201, 202, 203, 209,

210, 212, 216, 219, 226, 230, 235, 239, 241, 247, 253, 259, 268, 292, 299
service provider, 180, 213
service robots, viii, 180, 270, 303
services, 9, 19, 27, 28, 58, 64, 77, 81, 83, 85, 92, 96, 102, 124, 126, 137, 141, 143, 144, 157, 171, 180, 185, 189, 219, 227, 228, 237, 241, 243, 244, 259, 275, 276, 280, 281, 282, 283, 286, 288, 290, 293, 294, 310
settlements, 36, 238
sex, 248
shade, 94
shame, 228
shape, 46, 64, 72, 81, 89, 175, 195, 205, 219, 221, 263, 269
shaping, 41, 49, 54, 63, 64, 151, 169, 170, 171, 172, 173, 181, 205, 290, 310
shareholders, 122, 240, 252
shares, 5, 118, 177, 187
sharing, xii, 14, 70, 74, 109, 112, 117, 119, 123, 144, 149, 150, 161, 177, 235, 240, 242, 252, 260, 276, 288
sheep, 233
shelter, 15, 132, 237
shipping, 165, 166
shock, 170
shocks, 175
shores, 222
short period, 52
shortage, 20, 102, 116, 118, 134, 135, 209
short-term, 240, 247
shoulder, 216
shunts, 3
shuttles, 153, 157, 159, 160, 219
side effects, 137, 195, 242, 268
sign, 22, 36, 54, 124, 134, 265
signals, xv, 43, 45, 169, 170, 173, 181, 193
signs, 14, 28, 86, 87, 123, 130
silicon, 256, 266
similarity, 95, 132
simulation, 31, 52, 100, 101, 102, 103, 104, 105, 106, 107, 108, 109, 110, 111, 112, 113, 114, 115, 148, 154, 156, 158, 161, 162, 167, 170, 172, 177, 178, 181, 183, 205, 224, 225, 227, 230, 283, 288, 291, 292, 294, 295, 296, 298, 303, 304, 311
Singapore, 33, 280
singular, 31, 168, 175, 176
singularities, 219, 299
sites, 228, 229, 234, 251
skeleton, 128
skills, ix, 149
slavery, 117, 120, 124

slaves, viii, 126, 204
slums, 285
SME, 213, 304
SMEs, 188
smoke, 213
smoking, 46
smoothing, 257
snaps, 27
social benefits, 96, 126
social capital, 28, 307
social change, 274
social class, 14
social costs, 136
social environment, ix
social factors, x
social group, 44
social network, 44
social obligations, 43
social responsibility, 237
social structure, ix
sociology, 276, 308
software, 28, 30, 67, 68, 70, 72, 95, 99, 103, 104, 106, 112, 161, 164, 166, 168, 173, 176, 185, 199, 203, 270, 287, 293
soil, 195, 199, 210, 211, 221
solar, 256
solar energy, 256
solidarity, 19, 33, 36, 43, 44, 48, 49, 50, 86, 91, 96, 116, 117, 124, 135, 234, 237, 242, 247, 253, 260, 261, 264, 269
sorting, 60, 90, 92, 159, 161, 165, 172, 189, 190, 191, 211, 302, 303
sound speed, 226
sounds, 37
South America, 39
Southampton, 294
sovereignty, 10, 33
Soviet Union, 14
soy, 256
Spain, 6, 8
spatial, 176, 304
specialisation, 12, 13, 64, 72, 108, 110, 111, 128, 164, 187, 189, 195, 223, 241, 261, 270, 300
specialization, 278
speciation, 248
species, 128, 210, 247, 248, 256, 276
spectrum, 226
speculation, 10
speech, 41, 143, 205, 234, 235
speed, 73, 127, 169, 170, 175, 194, 196, 197, 202, 219, 220, 221, 223, 250, 256, 298
spheres, 18, 19, 55, 113
sponsor, 84

spreadsheets, 113
Sri Lanka, 200, 201
St. Petersburg, 298
stability, 6, 7, 8, 13, 20, 22, 27, 45, 49, 95, 99, 128, 168, 169, 174, 180, 182, 206, 211, 215, 234, 235, 241, 251, 253, 261
stages, 108, 113, 120, 125, 172, 264
stakeholder, 68
stakeholders, 19, 67, 70, 89, 96, 240, 252, 253, 260
standard model, 39
standards, 10, 11, 13, 14, 16, 31, 38, 52, 54, 55, 60, 71, 80, 81, 86, 91, 92, 93, 109, 118, 126, 130, 138, 139, 140, 142, 149, 150, 160, 161, 184, 187, 189, 192, 215, 242, 244, 249, 252, 253, 260, 263, 291
starvation, 5, 135
state intervention, 14, 15
statistics, 47, 49
steady state, 74, 88, 140
steel, 193, 307
sterile, 256
stiffness, vii, 107, 169, 174, 175, 176, 195, 206
stochastic, 28
stock, xv, 122, 138, 294, 295
stock exchange, 122
stock price, xv
stockpile, 199
storage, 9, 161, 190, 217, 229, 288
storms, 266
strategic management, 108
strategies, xi, 17, 58, 59, 104, 113, 125, 133, 154, 166, 167, 169, 170, 171, 173, 174, 201, 202, 205, 206, 215, 225, 245, 276, 278, 281, 282, 283, 286, 290, 295, 309
stratification, 194
streams, 224, 229
strength, 33, 70, 123, 125, 130, 187, 232, 245, 252, 264
stress, 40, 41, 66, 138, 180, 233
stroke, 175
structuring, 31, 69
students, 39
subjective, 40, 46, 83
sub-prime, 311
subsidiarity, 35, 261
subsidies, 15, 62, 84, 309
subsidy, 15, 17
subsistence, 43, 132, 209, 234, 244
substances, 6, 7
substitutes, 152
substitution, 5
summaries, xii

sunflower, 233
superimpose, 69
superiority, 37
superposition, 55
supervision, 203, 204, 220
supervisor, 175, 178, 225
supervisors, 59, 81, 108
suppliers, 19, 63, 64, 73, 75, 76, 80, 87, 113, 114, 131, 150, 186, 192, 214, 237, 252
supply chain, xii, 25, 27, 29, 52, 54, 56, 57, 58, 59, 60, 61, 62, 63, 64, 65, 66, 67, 68, 69, 74, 75, 76, 77, 78, 79, 80, 82, 87, 88, 89, 90, 92, 93, 99, 100, 102, 103, 104, 105, 107, 108, 109, 110, 111, 112, 113, 115, 123, 126, 131, 133, 137, 138, 139, 144, 145, 148, 154, 155, 162, 166, 183, 184, 186, 187, 188, 192, 212, 227, 229, 243, 251, 253, 254, 267, 268, 270, 272, 274, 281, 282, 286, 288, 292, 309, 312
support services, 77
suppression, 14
surgeons, 203, 204, 205, 208
surgery, viii, 188, 203, 204, 205, 206, 207, 208, 271, 300
surgical, 180, 203, 204, 206, 207, 208, 301
surgical intervention, 204, 206
surging, 260
surprise, 16, 254, 265
surveillance, 28, 66, 79, 144, 213, 221, 237
survival, xiii, 4, 18, 43, 50, 51, 70, 87, 91, 116, 124, 126, 132, 149, 179, 180, 233, 248, 264, 265, 266, 272
suspense, 152
suspensions, 221
sustainability, xvi, 1, 2, 7, 8, 15, 19, 29, 30, 34, 37, 43, 50, 54, 58, 61, 67, 68, 80, 81, 82, 84, 94, 97, 115, 121, 126, 132, 133, 134, 141, 148, 154, 166, 184, 236, 249, 254, 258, 274, 275, 280, 281, 293, 299, 303, 304
sustainable development, 226, 276, 284, 285, 290, 310
sustainable economic growth, ix
sustainable growth, xiii, 5, 13, 29, 30, 32, 33, 36, 37, 51, 52, 66, 75, 87, 89, 90, 94, 96, 112, 113, 115, 118, 121, 128, 129, 134, 137, 142, 162, 182, 226, 232, 236, 247, 249, 251, 255, 257, 265, 266, 271, 272, 274, 276, 282, 283, 288, 306, 307, 310
Sweden, 93, 309
switching, 30, 34, 52, 86, 107, 108, 119, 131, 175, 194, 195, 226, 240
symbols, 61
symmetry, 13, 140, 219
symptom, 94, 192

symptoms, 24, 59, 211
synergistic, 204
synergy, 142
synthesis, 30, 36, 94, 128, 232, 250
system analysis, 54
systems, viii, xii, 4, 9, 10, 11, 15, 17, 18, 28, 47, 56, 66, 80, 82, 83, 92, 95, 103, 110, 131, 132, 133, 138, 139, 143, 144, 149, 151, 161, 169, 184, 185, 192, 217, 228, 236, 242, 262, 268, 273, 274, 276, 280, 281, 282, 283, 285, 287, 288, 289, 290, 291, 292, 297, 304, 311, 312

T

talent, 10, 23, 45, 53, 197, 219
tangible, xii, 7, 8, 11, 23, 27, 28, 32, 35, 58, 72, 128, 131, 236, 271
tangible resources, 7, 8
Taoism, 40
targets, 55, 56, 57, 58, 61, 65, 66, 74, 76, 77, 80, 84, 86, 87, 88, 89, 90, 92, 93, 94, 111, 113, 116, 138, 153, 161, 166, 186, 187, 200, 202, 203, 215, 227, 228, 229, 230, 241, 243, 244, 245, 254, 261, 268
tariffs, 123
task performance, 183
task-orientation, 99, 169
taste, 185
tax collection, 60, 81
tax system, 138, 139, 266
taxation, 15, 16, 35, 89, 132, 139
taxes, 6, 8, 12, 15, 16, 25, 26, 36, 80, 86, 89, 117, 120, 122, 133, 138, 139, 227, 237, 241, 244
taxonomic, 192
taxonomy, 192
teachers, 237
teaching, 39, 40, 99
technical change, 278
technicians, 165, 186
technological change, 273, 277
technological progress, 127
technological revolution, 287
technology, viii, x, xiii, 10, 18, 26, 29, 37, 38, 52, 54, 56, 57, 63, 66, 69, 79, 81, 82, 83, 94, 97, 98, 99, 100, 102, 107, 112, 116, 121, 126, 127, 129, 134, 137, 141, 142, 144, 145, 147, 148, 151, 152, 153, 162, 164, 165, 171, 172, 179, 181, 182, 183, 184, 188, 189, 193, 197, 202, 203, 204, 206, 208, 219, 223, 224, 226, 231, 232, 236, 246, 251, 255, 258, 259, 265, 266, 267, 268, 271, 273, 274, 276, 282, 283, 286, 287, 288, 290, 304, 310
telecommunication, 127

telephone, 13, 22
Tellus Institute, 283
temperature, 210, 211, 213
temporal, 70, 213
Tennessee, 233
territorial, 6, 8, 21, 22
territory, 21, 24, 189, 219, 229, 233, 247
terrorism, 13, 35, 200, 239, 245, 263
terrorist, 200, 202
textile, 6, 8, 12, 13
textiles, 63, 165
thermodynamics, 44
thinking, ix, 85, 157, 179, 189, 260, 278, 281
third party, 79, 81, 85, 93, 94, 133, 186
thorns, 118
threat, 23, 33, 48, 200, 245, 264, 305
threatened, 128
threatening, xi, 226
threats, ix, 77, 200, 264
threshold, 22, 60, 102, 224
thresholds, 13, 52, 59, 94, 102, 140, 141, 170, 173, 175, 177, 181, 192, 202, 221
timber, 192, 234
timing, xvi, 177, 221, 270
title, 14, 35
tobacco, 46
Tokyo, 276, 279, 280, 281
tolerance, 119, 120, 153, 183, 184
tomato, 235
top-down, 14, 22, 41, 78, 95, 151, 238
topology, 79, 80
torque, vii, 170, 198, 207, 221
torus, 219
total product, 54
toxic, 7, 8, 136, 210, 211, 212
Toyota, 42, 45, 151
toys, 63
tracking, 101, 106, 114, 115, 164, 173, 199, 215, 217, 223, 224, 234
traction, 221
trade, xi, 1, 2, 6, 8, 11, 12, 17, 19, 21, 22, 24, 25, 27, 28, 29, 34, 35, 57, 60, 63, 67, 69, 70, 79, 80, 82, 83, 84, 85, 86, 90, 91, 92, 100, 107, 118, 119, 120, 122, 124, 125, 126, 131, 139, 141, 187, 191, 199, 212, 228, 229, 232, 234, 235, 238, 240, 241, 252, 254, 259, 263, 264, 266, 269, 285, 305, 306, 311
trade agreement, 107, 131, 264
trade union, 125, 126, 191, 259
trademarks, 236
trade-off, 254

trading, xi, 6, 7, 8, 10, 11, 12, 22, 29, 36, 49, 63, 70, 72, 117, 118, 124, 136, 163, 184, 188, 212, 227, 234, 235, 238, 243, 251, 259, 261, 264
tradition, 10, 24, 192, 308
traffic, 193, 219
training, 38, 43, 61, 70, 71, 72, 73, 110, 120, 125, 135, 147, 182, 200, 201, 279, 286
traits, 99, 181, 206, 234, 267, 270
trajectory, 168, 169, 170, 172, 175, 220, 221
trans, 16, 28, 35, 64, 119, 121, 131, 132, 135, 136, 237, 238, 239, 252, 257, 260, 262, 263
transaction costs, 187
transactions, 11, 12, 62, 72, 113, 186, 187, 251, 259
transducer, 175
transfer, 11, 16, 19, 25, 28, 91, 95, 96, 121, 122, 131, 132, 135, 143, 159, 160, 171, 187, 197, 201, 204, 211, 224, 232, 240, 242, 243
transformation, xi, xv, xvi, 1, 2, 3, 4, 16, 21, 25, 28, 30, 43, 44, 46, 48, 52, 72, 97, 116, 119, 122, 126, 133, 141, 149, 168, 218, 226, 238, 240, 251, 254, 256, 258, 267, 268, 269, 270, 278, 281, 283, 306, 308
transformation processes, 3, 30
transformations, 1, 2, 20, 22, 23, 25, 27, 29, 42, 45, 111, 162, 168, 174, 241, 248, 250, 258, 265, 266
transition, 31, 47, 54, 72, 107, 110, 135, 253, 278
transition period, 135
translation, 26, 110
transmission, 9, 10, 13, 67, 93, 130, 169, 184, 189, 206, 207, 225, 253, 299
transparency, x, xiii, 19, 32, 36, 57, 67, 68, 69, 80, 84, 86, 95, 96, 104, 113, 115, 116, 120, 128, 129, 131, 135, 136, 139, 149, 174, 183, 184, 186, 188, 192, 214, 216, 227, 228, 230, 237, 248, 250, 251, 253, 254, 265, 267
transparent, 31, 48, 77, 80, 85, 86, 92, 93, 119, 129, 132, 136, 242, 250, 271
transport, 21, 160, 168, 212, 220, 224, 228, 237, 240, 245, 246, 307
transportation, 13, 119, 123, 130, 157, 158, 159, 160, 161, 229, 242
traps, 183
travel, 22, 152, 159, 190, 193
Treasury, 25
trees, 209, 256, 266
trial, 38, 190
tribes, 21
trickle down, 231, 232, 235
troubleshooting, 173, 212, 213
trucks, 217, 218
true/false, 47, 49

trust, x, 5, 23, 24, 33, 36, 37, 41, 42, 51, 54, 136, 142, 213, 214, 240, 244, 262
trusts, 13, 49
trustworthiness, 92, 208
tuition, 135
tungsten, 171
turbulence, 284
typology, 189
TYPUS metric, 141, 142, 236, 242, 253, 260, 262

U

uncertainty, 46, 47, 54, 59, 107, 110, 167, 204, 263, 264, 299
underwater robotics, 195
underwater vehicles, 223, 304
unemployment, 15, 125
UNEP, 280
unification, 219, 312
uniform, 21, 22, 61, 102, 140, 234, 236, 253
unions, 238, 264
United Kingdom, 24
United States, 91
universal law, 48
universe, 23, 38, 40, 42, 44, 45, 48, 49, 128, 132, 238, 247, 256, 262, 266, 309
universities, 102
updating, xi, 47, 65, 108, 114, 151, 154, 156, 161, 190
urbanisation, 234
use-and-dump, 61, 94, 166, 246

V

vacuum, 215
validation, 47, 103, 113, 203, 205
validity, 47, 62, 81, 85, 86, 251
values, xii, 47, 62, 105, 126, 193, 259
variability, 111, 150, 153, 156, 163, 234
variables, 102, 169, 181
variation, 175, 207, 239, 242
vegetables, 24, 211, 247, 250
vegetation, 248, 265
vehicles, 62, 92, 153, 159, 201, 217, 219, 222, 228, 271, 303, 304
velocity, 170, 176, 196, 221
venture capital, 70, 259
versatility, 99, 101, 104, 105, 106, 107, 108, 109, 111, 112, 114, 149, 150, 153, 158, 160, 163, 166, 167, 169, 173, 180, 182, 202, 208, 224, 226
vertical integration, 274

vibration, 192, 196
Victoria, 12
video surveillance, 213
village, 14, 18, 28, 36, 44, 49, 133, 136, 149, 237, 262, 264, 265, 266
violence, 53, 132
violent, 242
virtual enterprise, 64, 65, 66, 70, 71, 73, 75, 76, 77, 78, 87, 88, 106, 131, 187, 212, 229, 230, 275, 287, 289, 311
virtual reality, 110, 112, 143, 154, 155, 156, 170, 185, 189, 193, 204, 298, 300
viruses, 211
visible, 62, 72, 93, 100, 139, 214
vision, vii, viii, xiii, 19, 74, 76, 77, 123, 132, 133, 134, 182, 183, 185, 191, 193, 206, 217, 265, 272, 274, 299
visual perception, 199
voice, 137, 185
voids, 219
voters, 16, 33, 34, 35, 43, 50, 85, 91, 121, 122, 124, 131, 238, 261, 262
voting, 50, 85, 120, 122

W

wages, 15, 16, 26, 73, 84, 117, 122, 123, 150, 165, 166, 183, 200, 237, 252, 259
walking, 37, 125
war, 12, 13, 130, 131, 134, 137, 245, 260, 261, 264, 277, 281, 307
warehousing, 296
warfare, 12, 200
Washington Consensus, 17, 82
waste treatment, 227, 228, 229
wastes, xi, 80, 86, 242, 256
water, 109, 129, 134, 196, 197, 215, 222, 224, 226, 227, 256, 303, 311
waterways, 7, 8
wavelet, 192, 193, 301
weakness, 12, 18, 91, 134, 136, 263
wealth, ix, x, xi, xiii, xv, xvi, 1, 2, 3, 7, 8, 9, 11, 13, 14, 16, 17, 19, 20, 21, 22, 24, 25, 26, 28, 29, 31, 32, 34, 35, 37, 38, 43, 45, 50, 66, 67, 72, 82, 83, 84, 85, 94, 117, 121, 122, 123, 125, 126, 131, 132, 134, 136, 137, 138, 140, 148, 209, 231, 232, 238, 239, 240, 243, 244, 245, 247, 249, 252, 253, 264, 269, 272, 273, 275, 276, 278, 281, 282, 287, 289, 308, 312
wealth distribution, 13, 137, 238, 247, 249, 264
weapons, 134, 199, 245
wear, 29, 63, 193

web, 12, 22, 28, 57, 64, 68, 95, 96, 136, 143, 185, 186, 189, 199, 227, 230, 242, 252, 266, 282, 286
web-based, 227, 286
welding, 171, 173
welfare, 16, 124, 125, 134, 240, 246, 261, 306, 312
wellbeing, xv, 5, 9, 10, 13, 14, 21, 22, 25, 121, 130, 134, 244, 245, 248
western countries, 163, 246
western culture, xi
wheat, 233
wild animals, 21
wind, 24, 129
wine, 117
winning, 3, 14, 24, 25, 42, 45, 50, 51, 52, 53, 57, 63, 64, 70, 73, 84, 98, 125, 127, 147, 151, 181, 184, 200, 203, 234, 236, 245, 270, 271, 281, 292, 312
wireless, 143
wisdom, 40, 46, 49
withdrawal, 7, 8, 29, 133, 228, 236

workers, 14, 15, 16, 26, 42, 45, 56, 67, 69, 72, 98, 120, 121, 122, 123, 125, 147, 149, 150, 151, 152, 165, 166, 191, 194, 200, 252
workflow, 200, 201
workforce, xii, xv, 16, 53, 54, 56, 99, 119, 121, 125, 126, 258, 273
working conditions, 186
workplace, 123
World Bank, 276, 279, 306
World Wide Web, 287
worry, xi, 39, 45, 46, 197, 249
writing, 234
wrongdoing, 193
WTO, 34, 238

Y

yield, 19, 31, 44, 83, 128, 129, 136, 138, 139, 140, 159, 163, 185, 221, 222, 226, 258, 259